# Lecture Notes in Computer Scier

Commenced Publication in 1973
Founding and Former Series Editors:
Gerhard Goos, Juris Hartmanis, and Jan van Leeuwen

Lecture Notes in Computer Science

Barbara Catania   Giovanna Guerrini
Jaroslav Pokorný (Eds.)

# Advances in Databases and Information Systems

17th East European Conference, ADBIS 2013
Genoa, Italy, September 1-4, 2013
Proceedings

Springer

Volume Editors

Barbara Catania
Giovanna Guerrini

University of Genoa
DIBRIS - Dipartimento di Informatica, Bioingegneria,
Robotica e Ingegneria dei Sistemi
Via Dodecaneso 35, 16146 Genoa, Italy
Email: {barbara.catania; giovanna.guerrini}@unige.it

Jaroslav Pokorný
Charles University
Department of Software Engineering
Malostranské nám. 25
11800 Praha, Czech Republic
E-mail: pokorny@ksi.mff.cuni.cz

ISSN 0302-9743                          e-ISSN 1611-3349
ISBN 978-3-642-40682-9                  e-ISBN 978-3-642-40683-6
DOI 10.1007/978-3-642-40683-6
Springer Heidelberg New York Dordrecht London

Library of Congress Control Number: 2013946716

CR Subject Classification (1998): H.2, H.4, H.3, C.2, J.1, H.2.8

LNCS Sublibrary: SL 3 – Information Systems and Application, incl. Internet/Web
and HCI

*Typesetting:* Camera-ready by author, data conversion by Scientific Publishing Services, Chennai, India

Printed on acid-free paper

Springer is part of Springer Science+Business Media (www.springer.com)

# Preface

This volume contains a selection of the papers presented at the 17th East-European Conference on Advances in Databases and Information Systems (ADBIS 2013), held during September 1–4, 2013, in Genoa, Italy.

The ADBIS series of conferences aims to provide a forum for the dissemination of research accomplishments and to promote interaction and collaboration between the database and information systems research communities from Central and East European countries and the rest of the world. The ADBIS conferences provide an international platform for the presentation of research on database theory, development of advanced DBMS technologies, and their advanced applications. ADBIS 2013 continued the ADBIS series held in St. Petersburg (1997), Poznań (1998), Maribor (1999), Prague (2000), Vilnius (2001), Bratislava (2002), Dresden (2003), Budapest (2004), Tallinn (2005), Thessaloniki (2006), Varna (2007), Pori (2008), Riga (2009), Novi Sad (2010), Vienna (2011), and Poznań (2012).

The program of ADBIS 2013 included keynotes, research papers, a Big Data special session, a Doctoral Symposium, and thematic workshops. Short papers, Big Data special session papers, and workshop papers are published in a companion volume.

The main conference attracted 92 paper submissions from 43 different countries representing all the continents. In a rigorous reviewing process, 26 papers were selected for inclusion in these proceedings as full contributions. Moreover, we selected 14 papers as short contributions. All papers were evaluated by at least three reviewers. The selected papers span a wide spectrum of topics in the database field and related technologies, ranging from semantic data management and similarity search, to spatio-temporal and social network data, data mining and data warehousing, and data management on novel architectures (GPU, parallel DBMS, cloud and MapReduce).

The volume also includes three invited papers for the conference keynote talks from distinguished researchers, namely, Martin Theobald (University of Antwerp, Belgium), Anastasia Ailamaki (Ecole Polytechnique Federale de Lausanne, Switzerland), and Bela Stantic (Griffith University, Australia).

We would like to express our thanks to everyone who contributed to the success of ADBIS 2013. We thank the authors, who submitted papers to the conference. We were also dependent on many members of the community offering their time in organizational and reviewing roles - we are very grateful for the energy and professionalism they exhibited. Special thanks go to the Program Committee members as well as to the external reviewers, for their support in evaluating the papers submitted to ADBIS 2013, ensuring the quality of the scientific program. A "super" thanks goes to our "super" reviewers, who accepted to complete reviews in a shorter period and to provide additional reviews on papers with conflicting reviews. Thanks also to all the colleagues involved in the conference organization, as well as the workshop organizers. A special thank you is deserved by the Steering Committee and, in particular, its Chair, Leonid Kalinichenko, for their help and guidance.

Finally, we thank Springer for publishing the proceedings containing invited and research papers in the LNCS series. The Program Committee work relied on EasyChair, and we thank its development team for creating and maintaining it, it offers great support for the reviewing process.

The conference would not have been possible without our sponsors and supporters: DIBRIS - Dipartimento di Informatica, Bioingegneria, Robotica e Ingegneria dei Sistemi, Università di Genova, Camera di Commercio di Genova, COOP Liguria, Comune di Genova, and CTI Liguria.

Last, but not least, we thank the participants of ADBIS 2013 for having made our work useful.

September 2013                                                    Barbara Catania
                                                                Giovanna Guerrini
                                                                 Jaroslav Pokorný

# Organization

## General Chair

Barbara Catania                    University of Genoa, Italy

## Program Committee Co-chairs

Jaroslav Pokorný                   Charles University in Prague, Czech Republic
Giovanna Guerrini                  University of Genoa, Italy

## Workshop Co-chairs

Themis Palpanas                    University of Trento, Italy
Athena Vakali                      Aristotle University of Thessaloniki, Greece

## PhD Consortium Co-chairs

Alfons Kemper                      Technical University of Munich, Germany
Boris Novikov                      St. Petersburg University, Russia

## ADBIS Steering Committee Chair

Leonid Kalinichenko                Russian Academy of Science, Russia

## ADBIS Steering Committee

Paolo Atzeni, Italy                          Joris Mihaeli, Israel
Andras Benczur, Hungary                      Tadeusz Morzy, Poland
Albertas Caplinskas, Lithuania               Pavol Návrat, Slovakia
Barbara Catania, Italy                       Boris Novikov, Russia
Johann Eder, Austria                         Mykola Nikitchenko, Ukraine
Hele-Mai Haav, Estonia                       Jaroslav Pokorný, Czech Republic
Theo Härder, Germany                         Boris Rachev, Bulgaria
Mirjana Ivanovic, Serbia                     Bernhard Thalheim, Germany
Hannu Jaakkola, Finland                      Gottfried Vossen, Germany
Marite Kirikova, Latvia                      Tatjana Welzer, Slovenia
Mikhail Kogalovsky, Russia                   Viacheslav Wolfengagen, Russia
Yannis Manolopoulos, Greece                  Robert Wrembel, Poland
Rainer Manthey, Germany                      Ester Zumpano, Italy
Manuk Manukyan, Armenia

## Organizing Committee

### Publicity Chair

Marco Mesiti                    University of Milan, Italy

### Web Chair

Federico Cavalieri              University of Genoa, Italy

### Local Arrangements Chair

Paola Podestà                   IMATI-CNR, Genoa, Italy

### Local Organizing Committee

Alessandro Solimando            University of Genoa, Italy
Beyza Yaman                     University of Genoa, Italy

### Supporting Companies and Institutions

Dipartimento di Informatica, Bioingegneria, Robotica e Ingegneria dei Sistemi
Università di Genova
Camera di Commercio di Genova
Coop Liguria
Comune di Genova
CTI Liguria

## Program Committee

Suad Alagic                     University of Southern Maine, USA
Manish Kumar Anand              Salesforce, USA
Andreas Behrend                 University of Bonn, Germany
Ladjel Bellatreche              LIAS/ENSMA, France
Michela Bertolotto              University College Dublin, Ireland
Nicole Bidoit                   Université Paris Sud 11, France
Maria Bielikova                 Slovak University of Technology in Bratislava,
                                   Slovakia
Iovka Boneva                    University of Lille 1, France
Omar Boucelma                   LSIS- CNRS, France
Stephane Bressan                National University of Singapore
Davide Buscaldi                 Universitè Paris Nord 13, France
Albertas Caplinskas             Vilnius University, Lithuania
Boris Chidlovskii               XRCE, France
Ricardo Ciferri                 Federal University of São Carlos, Brazil
Alfredo Cuzzocrea               ICAR-CNR and University of Calabria, Italy
Todd Eavis                      Concordia University, Canada
Johann Eder                     University of Klagenfurt, Austria

Alvaro A.A. Fernandes      The University of Manchester, UK
Pedro Furtado             University of Coimbra, Portugal
Johann Gamper             Free University of Bozen-Bolzano, Italy
Matjaz Gams               Jozef Stefan Institute, Slovenia
Anastasios Gounaris       Aristotle University of Thessalonik, Greece
Goetz Graefe              HP Labs, USA
Adam Grzech               Wroclaw University of Technology, Poland
Hele-Mai Haav             Tallinn University of Technology, Estonia
Melanie Herschel          Université Paris Sud 11, France
Theo Härder               TU Kaiserslautern, Germany
Mirjana Ivanovic          University of Novi Sad, Serbia
Hannu Jaakkola            Tampere University of Technology, Finland
Leonid Kalinichenko       Russian Academy of Science, Russia
Alfons Kemper             TU München, Germany
Maurice van Keulen        University of Twente, The Netherlands
Marite Kirikova           Riga Technical University, Latvia
Margita Kon-Popovska      Ss Cyril and Methodius University, Macedonia
Georgia Koutrika          HP Labs, USA
Stanislaw Kozielski       Silesian University of Technology, Poland
Jan Lindström             IBM Helsinki, Finland
Yannis Manolopoulos       Aristotle University of Thessaloniki, Greece
Rainer Manthey            University of Bonn, Germany
Giansalvatore Mecca       University of Basilicata, Italy
Marco Mesiti              University of Milan, Italy
Paolo Missier             Newcastle University, UK
Bernhard Mitschang        University of Stuttgart, Germany
Irena Mlynkova            Charles University in Prague, Czech Republic
Martin Nečaský            Charles University in Prague, Czech Republic
Anisoara Nica             SAP, Canada
Nikolaj Nikitchenko       Kiev State University, Ukraine
Boris Novikov             University of St. Petersburg, Russia
Kjetil Nørvåg             Norwegian University of Science and
                            Technology, Norway
Torben Bach Pedersen      Aalborg University, Denmark
Dana Petcu                West University of Timisoara, Romania
Evaggelia Pitoura         University of Ioannina, Greece
Elisa Quintarelli         Politecnico di Milano, Italy
Peter Revesz              University of Nebraska, USA
Stefano Rizzi             University of Bologna, Italy
Henryk Rybiński           Warsaw University of Technology, Poland
Ismael Sanz               Universitat Jaume I, Spain
Kai-Uwe Sattler           TU Ilmenau, Germany
Klaus-Dieter Schewe       Software Competence Center, Austria
Marc H. Scholl            University of Konstanz, Germany
Holger Schwarz            Universität Stuttgart, Germany
Bela Stantic              Griffith University, Australia

Yannis Stavrakas                    IMIS, Greece
Janis Stirna                        Stockholm University, Sweden
Ernest Teniente                     Universitat Politècnica de Catalunya, Spain
Goce Trajcevski                     Northwestern University, USA
Olegas Vasilecas                    Vilnius Gediminas Technical University,
                                     Lithuania
Krishnamurthy Vidyasankar           Memorial University, Canada
Gottfried Vossen                    Universität Münster, Germany
Fan Wang                            Microsoft, USA
Tatjana Welzer                      University of Maribor, Slovenia
Robert Wrembel                      Poznan University of Technology, Poland
Esteban Zimányi                     Université Libre de Bruxelles, Belgium

## Additional Reviewers

Fekete, David                       Miu, Tudor
Flores, Enrique                     Móro, Róbert
Jean, Stéphane                      Pissis, Solon
Koncilia, Christian                 Rabosio, Emanuele
Kramár, Tomáš                       Reggio, Gianna
Lechtenbörger, Jens                 Solimando, Alessandro
Mansmann, Svetlana                  Trinkunas, Justas
Mazuran, Mirjana                    Zeleník, Dušan

# Table of Contents

## Data Mining

## OLAP

## XML Data Processing

## Querying

## Similarity Search

## GPU

## Querying in Parallel Architectures

## Performance Evaluation

# Distributed Architectures

# 10 Years of Probabilistic Querying – What Next?

Martin Theobald[1], Luc De Raedt[2], Maximilian Dylla[3],
Angelika Kimmig[2], and Iris Miliaraki[3]

[1] Universiteit Antwerpen, Middelheimlaan 1, 2020 Antwerp, Belgium
martin.theobald@ua.ac.be
[2] Katholieke Universiteit Leuven, Celestijnenlaan 200A, 3001 Heverlee, Belgium
{luc.deraedt,angelika.kimmig}@cs.kuleuven.be
[3] Max Planck Institut Informatik, Campus E1.4, 66123 Saarbrücken, Germany
{dylla,miliaraki}@mpi-inf.mpg.de

**Abstract.** Over the past decade, the two research areas of *probabilistic databases* and *probabilistic programming* have intensively studied the problem of making structured probabilistic inference scalable, but—so far—both areas developed almost independently of one another. While probabilistic databases have focused on describing tractable query classes based on the structure of query plans and data lineage, probabilistic programming has contributed sophisticated inference techniques based on knowledge compilation and lifted (first-order) inference. Both fields have developed their own variants of—both exact and approximate—top-$k$ algorithms for query evaluation, and both investigate query optimization techniques known from SQL, Datalog, and Prolog, which all calls for a more intensive study of the commonalities and integration of the two fields. Moreover, we believe that *natural-language processing* and *information extraction* will remain a driving factor and in fact a longstanding challenge for developing expressive representation models which can be combined with structured probabilistic inference—also for the next decades to come.

**Keywords:** Probabilistic databases, probabilistic programming, natural-language processing, information extraction.

## 1 Introduction

Over the past decade, the two fields of *probabilistic databases* and *probabilistic programming* have studied similar problems but developed almost independently of one another. Although this can be explained by a different focus of the two communities (probabilistic databases focus on querying large databases, whereas probabilistic programming has focused largely on learning statistical models from data), there are ample opportunities for a fruitful cross-fertilization between these two domains. Datalog, for example, the logical query language used in databases, lies at the very intersection of relational algebra (which forms also the semantic basis of SQL) and pure logic-programming languages such as Prolog. Probabilistic programming is closely related to statistical relational learning [27,56] and today encompasses a rich variety of formalisms including functional languages, such as Church [28] and IBAL [52], and logical ones, such as Prism [62], ICL [53] and ProbLog [12,22,39]. Throughout the

B. Catania, G. Guerrini, and J. Pokorný (Eds.): ADBIS 2013, LNCS 8133, pp. 1–13, 2013.
© Springer-Verlag Berlin Heidelberg 2013

present paper we shall—within these families—largely focus on probabilistic extensions of logic programming languages such as ProbLog, which is motivated by the close relationship to database query languages such as Datalog.

With this research statement, we specifically motivate for studying the commonalities between the two worlds of probabilistic databases and probabilistic programming. While probabilistic databases benefit from a mature database infrastructure for *data storage* and *query answering* (including various forms of index structures, support for updates, multi-user concurrency control and transaction management), the field of probabilistic programming has intensively studied advanced techniques for probabilistic inference such as *knowledge compilation* and *lifted inference*. In particular state-of-the-art techniques for lifted inference (i.e., the task of inferring probabilities of queries and shared query components from their first-order representation) have so far not been applied in the context of probabilistic databases. Our aim is to focus on integrating their techniques for query processing and probabilistic inference, both at the propositional level (by exploiting recent developments for knowledge compilation) and at the first-order representation (thus investigating new techniques for top-$k$ query processing and lifted inference). Moreover, we believe that *natural-language processing* and *information extraction* will remain a major driving factor for structured probabilistic inference also for the next decades to come. Albeit natural language is inherently structured and—at least in most cases—follows well-defined grammatical rules, an abundance of ambiguities arises in resolving this structure and in detecting the actual meaning of statements expressed in natural language. While these ambiguities are only partly due to shortcomings of current tools, many statements expressed in natural language remain hard to resolve even for humans and often allow for a variety of interpretations.

## 2   Probabilistic Databases

Managing uncertain data via probabilistic databases (PDBs) has evolved as an established field of research in the database community in recent years. The field meanwhile encompasses a plethora of applications, which are ranging from *scientific data management*, *sensor networks*, *data integration*, to *information extraction* and *knowledge management* systems [65,71]. While classical database approaches benefit from a mature and scalable infrastructure for the management of relational data, probabilistic databases aim to further combine these well-studied data management strategies with efficient inference techniques that exploit given independence assumptions among database tuples whenever possible. PDBs adopt powerful query languages from relational databases, including relational algebra, the Structured Query Language (SQL), and logical query languages such as Datalog. Besides the two prevalent types of tuple-independent and block-independent probabilistic database models investigated in the PDB literature [65], PDBs may in fact capture arbitrary correlations among database tuples, and in particular recent PDB developments clearly lie at the intersection of databases and probabilistic graphical models [36,37,63]. The Trio probabilistic database system [46,72], which was developed at Stanford University back in 2006, was the first such system that explicitly addressed the integration of data management (using SQL as query language), lineage (aka. "provenance") management via Boolean formulas, and probabilistic inference based on the lineage of query answers. In particular, the so-called form of *How-lineage* (see [7] for an overview) captures lineage as propositional formulas that trace which relational operations were involved in deriving each individual query answer—in the sense of "How was this query answer derived?" The underlying data model of Trio, coined Uncertainty and Lineage Database (ULDB) [3], was the first PDB approach that was shown to provide a *closed and complete* probabilistic extension to the relational data model which supports all relational (i.e., SQL-based) operations. These principles directly carry over to Datalog (and logic programming), as it is shown in the following example.

*Example 1.* Figure 1 (from [18]) depicts a tuple-independent probabilistic database which consists of the extensional relations *Directed*, *ActedIn*, *Category*, and *WonAward*, as well as the database views $\nu_1$–$\nu_4$ (depicted in Datalog notation, explicitly showing also the quantifiers for all variables) which define the intensional relations *KnownFor*, *BestDirector* and *ActedOnly*. View $\nu_1$, for example, expresses that directors are known for a movie category if they occur in the relation *BestDirector* together with a movie of that category. Likewise, view $\nu_2$ expresses that actors are known for movies that won a best picture award, but only if they appear together in the *ActedOnly* relation together with that movie. Evaluating the query *KnownFor*($X$, *Crime*), thus asking for directors or actors $X$ who are known for *Crime* movies, over the database tuples and views shown in Figure 1 involves an iterative rewriting of the intensional query predicate *KnownFor* via the views and the extensional relations. By applying a form of top-down SLD resolution [18], we first observe that the head literals of both $\nu_1$ and $\nu_2$ unify with the query literal and thus rewrite *KnownFor*($X$, *Crime*) as a Boolean combination of the body literals of the two views, as it is shown in Figure 2. In the following two SLD steps, we resolve the remaining intensional literals *BestDirector*($X, Z$) and *ActedOnly*($X, Z$) via views $\nu_3$ and $\nu_4$ in a similar way. Finally, by binding the distinguished query variable $X$ to the corresponding constants of those tuples that unify with the remaining extensional literals, we obtain the two probabilistic query answers *KnownFor*(*Coppola*, *Crime*) and *KnownFor*(*Tarantino*, *Crime*) with their lineages $((t_2 \wedge t_8) \wedge t_{12})$ and $(t_{10} \wedge (t_6 \wedge \neg t_3) \wedge t_{13})$, respectively.

A key in making probabilistic inference scalable to the extent that is needed for modern data management applications lies in the identification of tractable query classes [9,10] and in the adaptation of known inference techniques from probabilistic graphical models to a relational-data setting [63]. While for specific classes of queries, probabilistic inference can directly be coupled with the relational operations [10,61], the performance may very quickly degenerate when the underlying independence assumptions do not apply or are too difficult to resolve. Despite the polynomial complexity for the data computation step that is involved in finding answer candidates to probabilistic queries, the confidence computation step, i.e., the actual probabilistic inference, for these answers is known to be of exponential cost already for fairly simple select-project-join (SPJ) queries in SQL. Consequently, much of the recent research in PDBs has focused on establishing a *dichotomy of query plans* [9] for which confidence computations are either of polynomial runtime or are $\#\mathcal{P}$-hard [60,65]. Generally, $\#\mathcal{P}$ denotes an exponential complexity class of counting problems which in the case of a probabilistic

| Directed | | |
|---|---|---|
| Director | Movie | p |
| $t_1$ Coppola | ApocalypseNow | 0.8 |
| $t_2$ Coppola | Godfather | 0.9 |
| $t_3$ Tarantino | PulpFiction | 0.7 |

| ActedIn | | |
|---|---|---|
| Actor | Movie | p |
| $t_4$ Brando | ApocalypseNow | 0.6 |
| $t_5$ Pacino | Godfather | 0.3 |
| $t_6$ Tarantino | PulpFiction | 0.4 |

| WonAward | | |
|---|---|---|
| Movie | Award | p |
| $t_7$ ApocalypseNow | BestScript | 0.3 |
| $t_8$ Godfather | BestDirector | 0.8 |
| $t_9$ Godfather | BestPicture | 0.4 |
| $t_{10}$ PulpFiction | BestPicture | 0.9 |

| Category | | |
|---|---|---|
| Movie | Category | p |
| $t_{11}$ ApocalypseNow | War | 0.9 |
| $t_{12}$ Godfather | Crime | 0.5 |
| $t_{13}$ PulpFiction | Crime | 0.9 |
| $t_{14}$ Inception | Drama | 0.6 |

$\nu_1 : \forall X, Y\ KnownFor(X, Y) \quad :- \exists Z\ BestDirector(X, Z), Category(Z, Y)$

$\nu_2 : \forall X, Y\ KnownFor(X, Y) \quad :- \exists Z\ WonAward(Z, BestPicture), ActedOnly(X, Z), Category(Z, Y)$

$\nu_3 : \forall X, Y\ BestDirector(X, Y) :- Directed(X, Y), WonAward(Y, BestDirector)$

$\nu_4 : \forall X, Y\ ActedOnly(X, Y) \quad :- ActedIn(X, Y), \neg Directed(X, Y)$

**Fig. 1.** A tuple-independent probabilistic database about movies with a number of extensional and intensional relations

databases resolves to $\#\mathcal{SAT}$, i.e., the problem of counting the number of variable assignments that satisfy a Boolean formula.

Thus, efficient strategies for probabilistic ranking and early pruning of low-confidence query answers (see [42] for an overview of various ranking semantics in PDBs) remain a key challenge also for the scalable management of uncertain data via PDBs. Recent work on efficient confidence computations in PDBs has addressed the $\#\mathcal{P}$-hardness of general SQL queries mainly from two ends, namely by restricting the class of queries that are allowed, i.e., by focusing on so-called *safe query plans* [10], or by considering a specific class of tuple-dependencies, commonly referred to as *read-once functions* [64][1]. Intuitively, safe query plans denote a class of queries for which confidence computations can directly be coupled with the relational operators and thus be performed by an extensional query plan. Read-once formulas, on the other hand, denote a class of Boolean lineage formulas which can be factorized (in polynomial time) into a form where every variable in the formula (each representing a database tuple) appears at most once, thus again permitting efficient confidence computations.

While safe plans focus on the characteristics of the query structure, and read-once formulas focus on the logical dependencies among individual data objects, top-$k$ style pruning approaches have recently been proposed as an alternative way to address confidence computations in PDBs [5,18,51,57]. These approaches aim to efficiently identify the top-$k$ most probable answers, using lower and upper bounds for their marginal probabilities, without the need to actually compute the exact probabilities of these answers. Earlier works by Suciu, Dalvi and Ré [57,5] addressed this by approximating the probabilities of the top-$k$ answers using Monte-Carlo-style sampling techniques. Olteanu and Wen [51], on the other hand, recently extended their approach of decomposing propositional formulas in order to derive a similar (but exact) bounding approach based

---

[1] These have been shown to be equivalent in [50] for unions of conjunctive queries without self-joins but, in general, read-once formulas may allow for efficient inference for a wider class of queries by considering individual tuple-dependencies.

on finding shared subqueries and by employing partially expanded ordered-binary-decision-diagrams (OBDDs) [24]. In particular the latter top-$k$ algorithm [51] can effectively circumvent the need for exact confidence computations and can still—in many cases—return the top-ranked query answers in an exact way. However, as opposed to top-$k$ approaches in deterministic databases [21,32,66], none of the former top-$k$ approaches in PDBs saves upon the data computation step that is needed to find potential answer candidates. Thus, extensive materialization is required for queries with multiple nested subqueries or over multiple levels of potentially convoluted views.

In [18], Dylla et al.—for the first time in the context of PDBs—develop an integrated approach that combines both the data and confidence computation steps needed for intensional query evaluations, i.e., for queries where confidences cannot be computed via an efficient extensional query plan. By introducing a new notion of *first-order lineage formulas*, this approach iteratively computes lower and upper bounds for the marginal probabilities of all query answers that are represented by such a first-order formula. Lineage formulas are derived from a top-down grounding algorithm of queries in Datalog, which incrementally expands the query literals against the views and the extensional relations until the top-$k$ query answers with the highest marginal probabilities are determined. The algorithm supports the full expressiveness of Datalog, including recursion and stratified negation, which (without recursion) forms also the core of relational algebra and SPJ queries in SQL (however without grouping and aggregations, see [23,25] for extensions required to support full relational algebra in a PDB setting). Figure 2 depicts the construction of such a first-order lineage formula for the example PDB shown in Figure 1, as it is triggered by the query $KnownFor(X, Crime)$. The figure also explicitly depicts the existential quantifiers that are introduced for local variables occurring only in the bodies of the view definitions.

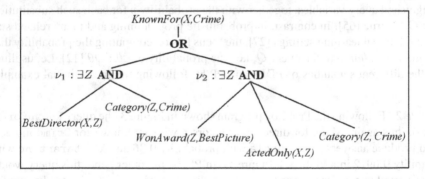

**Fig. 2.** A partially grounded lineage formula for the query $KnownFor(X, Crime)$

## 3  Probabilistic Programming

Probabilistic programming (PP) extends programming languages with probabilistic primitives, thus resulting in rich languages for probabilistic models. Probabilistic logic programming languages based on Prolog, such as Prism [62], ICL [53] and ProbLog

```
0.8::directed(coppola,apocalypseNow).        0.3::wonAward(apocalypseNow,bestScript).
0.9::directed(coppola,godfather).            0.8::wonAward(godfather,bestDirector).
0.7::directed(tarantino,pulpFiction).        0.4::wonAward(godfather,bestPicture).
                                             0.9::wonAward(pulpFiction,bestPicture).

0.6::actedIn(brando,apocalypseNow).          0.9::category(apocalypseNow,war).
0.3::actedIn(pacino,godfather).              0.5::category(godfather,crime).
0.4::actedIn(tarantino,pulpFiction).         0.9::category(pulpFiction,crime).
                                             0.6::category(inception,drama).

knownFor(X,Y)   :- bestDirector(X,Z), category(Z,Y).
knownFor(X,Y)   :- wonAward(Z,bestPicture), actedOnly(X,Z), category(Z,Y).
bestDirector(X,Y) :- directed(X,Y), wonAward(Y,bestDirector).
actedOnly(X,Y)  :- actedIn(X,Y), \+directed(X,Y).
```

**Fig. 3.** A ProbLog program equivalent to the PDB in Figure 1

[12,22,39], have been among the first such extensions studied. These are originally based on Sato's distribution semantics [62] which in turn exhibits surprising similarities to the possible-worlds semantics [1], as it is used in PDBs today. PP approaches combine a logic program with probabilistic facts, and are thus closely related to PDBs that associate probabilities to tuples. In fact, the afore described PDB maps directly to PP and ProbLog, as illustrated in Figure 3, where we have one probabilistic fact for each relational tuple in the tables of the PDB and the same clauses (this time however in Prolog notation) as before.

Despite their closely related semantics, both PDBs and PPs have so far focused on different types of probabilistic queries. In a (probabilistic) database setting, one is typically interested in finding *all solutions* to a query $Q$, i.e., on generating all query answers, where the aim is to find tuples of constants that serve as substitutions for the distinguished query variables, together with the probability $p$ for each such substitution $\theta$ that $Q\theta$ is true [65]. In contrast, in probabilistic programming and many related statistical relational learning settings [27], the focus lies on computing the probability that there *exists a solution* to the query $Q$, i.e., the probability of $\exists \theta : Q\theta$ [12]. Let us illustrate the different semantics of PDBs and PP by following up on our initial example.

*Example 2.* Following the ProbLog program shown in Figure 3, the query knownFor(X, crime), thus again asking for directors or actors X who are known for Crime movies, would produce answers X = coppola with probability 0.36 and X = tarantino with probability 0.0972 in a PDB (see Figure 1). In PP, on the other hand, this query would be interpreted to ask for the existence of any directors or actors X who are known for crime movies, and would thus be answered by the single probability 0.422.

Moreover, in probabilistic programming and statistical relational learning, it is common to compute *conditional probabilities*, thus answering queries of the form $P(Q|E)$, where $E$ denotes the evidence, i.e., particular observations that have been made. A second important inference setting in PP, as in probabilistic graphical models, is maximum-a-posteriori (MAP)—often also referred to as most-probable-explanation (MPE)—inference, where the task is to find the most likely joint truth-value assignment to a set of ground query atoms under a given evidence. PDBs, on the other hand, consider top-$k$ queries, which ask for the $k$ most likely answer substitutions for a given query, so far however without considering evidence or any form of conditioning [18,51]. Both the fields of PDBs and PP have developed their own semantics and algorithms for top-$k$ queries over probabilistic data [18,39,51,58], and thus we believe that their combination constitutes a very natural and intriguing subject for further research. Finally, while PP conditions on evidence given at query time, the use of hard logical constraints is common in databases; one then has to condition the probabilities of queries in order to account also for the "impossible worlds" (i.e., those that violate the constraints) [40]. A detailed semantic categorization of these queries (together with their answering mechanisms) thus will likely reveal a unifying framework for query answering and probabilistic inference in both PDBs and PP. Given the closeness in semantics, a key difference between PDBs and PPs, however is that PP languages have the expressive power of a programming language and are, unlike PDBs, Turing equivalent.

In contrast to PDBs, PP so far has mostly focused on general-purpose probabilistic inference techniques, including approaches based on *knowledge compilation*, but did not yet investigate the complexity of evaluating different query structures such as safe plans or read-once formulas. This focus on inference techniques, both exact and approximate, is in part due to the close connection of PP to statistical relational learning and graphical models. In PP, parameters (or probabilities of tuples) are often learned from data rather than provided as part of the model, which typically includes solving large numbers of inference tasks during learning and thus requires efficient inference. Knowledge compilation (see [11] for an overview of techniques) is a standard technique in graphical models and PP [12,22,39,68], but has only fairly recently been considered in PDBs [23,34,35,50]. These approaches aim to compile a Boolean formula into a more succinct representation formalism, over which probabilistic inference can then be performed in polynomial time with respect to the size of the compiled data structure. The work on ProbLog has been the first to use knowledge compilation for inference in a PP setting. It compiles given information about (in)dependencies among individual data objects that contribute to the answer of a probabilistic query into a compact data structure that allows for efficient probability computation. Early work in PP has used OBDDs [12,39], while the state of the art employs even more succinct structures such as $d$-DNNFs [22]. While ProbLog uses a single type of target structure that is independent of the query structure, work in PDBs has more explicitly considered different classes of query structures, based on query plans or data lineage, for which probabilistic inference is either tractable or $\#P$-hard [65]. Thus, there is indeed a close connection between knowledge compilation in PP and the management of lineage formulas in PDBs that has so far not been investigated explicitly.

Probabilistic programs and statistical relational models represent knowledge at an abstract level that makes abstraction of specific (types of) objects and the contexts in which they appear. In this way, it is possible to identify groups of "similar" data objects that can be handled interchangeably. This closely corresponds to finding symmetries in the underlying logical representation. Recent techniques for *lifted inference* take advantage of such symmetries to avoid redundant computations and thus mitigate the cost of inference [68]. The problem of lifted probabilistic inference was first introduced by Poole [54] and has attracted a lot of attention since then (see [13,45] for some seminal approaches and [38] for an overview). In PP and statistical relational learning, lifted inference has been well-studied, but in PDBs this is still largely absent (but see [33]). The advantage of lifted inference is that, for certain classes of PP, inference is guaranteed to be polynomial in the size of the domains (i.e., it is "domain-liftable"), whereas non-lifted inference would be exponential. Van den Broeck [14,67] has contributed the first positive and negative results concerning (domain-)liftability. A state-of-the-art technique for lifted inference is weighted first-order model counting (WFOMC), which is based on the concept of first-order $d$-DNNF circuits [15]. Such a circuit represents the possible worlds or answers to a given query. A simple example of a first-order formula and the corresponding circuit is shown in Figure 4 (from [67]). The theory states that smokers are only friends with other smokers. The circuit introduces a new domain variable $D$ which denotes a subset of the domain $People$. It states that there exists such a $D$ for which (i) all people in $D$ are smokers, (ii) no other people are smokers, and (iii) smokers are not friends with non smokers. The circuit makes abstraction of the possible subsets $D$. Only the size of the subset matters for performing probabilistic inference (and hence the circuit captures symmetries between all sets of the same size).

$$\forall X, Y \in People : smokes(X) \wedge friends(X, Y) \Rightarrow smokes(Y).$$

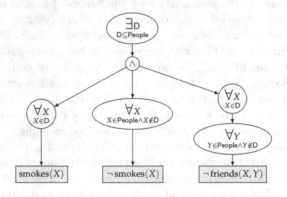

**Fig. 4.** A first-order formula and its compact representation as a first-order $d$-DNNF

By comparing the first-order lineage structure in Figure 2 and the $d$-DNNF structure in Figure 4, it is clear that the methods developed in both fields are heading in the same direction. Therefore, we expect that the more mature lifted techniques known from PP can help to also accelerate probabilistic inference for first-order lineage formulas in PDBs.

## 4   Information Extraction

Natural language inherently is highly structured. This simple matter stands in a sometimes surprising contrast to common database jargon, in which natural-language text generally is coined by the term "unstructured data". Any natural-language statement,

such as a piece of text, a sentence, or even just a short phrase—at least when formulated in a meaningful way—follows a well-defined syntactic structure which can be
resolved—at least in most cases—by current tagging and parsing tools. Nearly every
text snippet, be it an entire Wikipedia article or just a short tweet, contains mentions
of named entities, which "just" need to be disambiguated and be put into proper relationships to other entities. Today, a majority of the syntactic variants of statements
expressible in natural language are already resolved correctly by current dependency
parsers and link grammars, while the remaining cases are often difficult to resolve even
for humans. Ambiguities in the parsing structures can be captured—at least to some
extent—by probabilistic models such as conditional random fields (CRFs) and probabilistic context-free grammars (PCFGs), which are usually based on hidden Markov
models (HMMs) (see [41,55] for an introduction). State-of-the-art parsers are able to
learn the transition probabilities among states and partly even allow for learning the
production rules of the underlying grammar from training sentences.

Information extraction (IE) aims to bridge the gap between natural language and
structured (typically relational) representation formalisms. Recent, so-called *domain-
oriented*, IE approaches [4,30,47] extract triplets of facts captured in the popular Resource Description Framework (RDF). Each such RDF fact consists of a pair of known
entities (or a pair of an entity and a string literal, respectively) and a canonical relation
that connects this pair. Domain-oriented IE approaches operate over a well-defined set
of canonical entities and relationships (such as entities represented by Wikipedia articles
and relations occurring as infobox attributes) and often require just a handful of fairly
simple regular expressions for the extraction step. Conversely, approaches in the field of
*open* IE [2,6] aim to relax the requirement of producing a canonical set of target entities
and relations, which generally improves extraction recall but also makes it much harder
to integrate their output with structured inference. These approaches employ series of
natural-language processing tools, such as part-of-speech (POS) tagging, named-entity
recognition (NER), and partly more sophisticated techniques such as semantic role labeling (SRL) [20]. Only few approaches exist that aim to bridge the gap between the
two worlds of domain-oriented and open IE by mining for hierarchies of verbal phrases
that capture the semantic relationships among entities [48]. Despite its simplicity, RDF
will likely not prevail as a representation formalism for natural language. RDF inherently reaches its limitations (or at least quickly loses its conciseness) when capturing
different modalities of facts, such as temporal, spatial or other modifiers, or when capturing more than just binary relationships among entities. The ClausIE [8] system, for
example, is an interesting approach that directly turns the dependency-based parse trees
of natural-language sentences into higher-arity, but sparse, relations. From a more relational perspective, learning logical deduction rules via inductive logic programming
(ILP) [26], which again intersects with the field of logic programming, provides a
very promising application of structured inference in large and incomplete knowledge
bases. Information extraction and scalable probabilistic inference have been addressed
by a plethora of approaches in machine learning [17,43,44,49,59], but only relatively
few works so far address the exact interplay of information extraction tools (based on
HMMs and CRFs) and inference in probabilistic databases [29,69,70]. In summary, we
advocate that turning natural language into a machine-readable and processable format

(a process coined "machine reading" in [19] and "language learning" in [6]) constitutes one of the best showcases for managing uncertain data we have seen so far. This application domain will certainly remain a major challenge for both structured and probabilistic inference also for next decades to come. The integration of domain-oriented and open IE techniques, on the other hand, is a key for applying these inference techniques at Web scale.

## 5 Conclusions

The methods and technologies developed for probabilistic inference both in databases and in logic programming are becoming mature and scalable and—already today—in many cases allow for exact probabilistic inference over query answers derived from many thousands of variables. Probabilistic databases currently support a wide range of queries formulated in expressive declarative query languages such as SQL, SPARQL, Datalog and function-free Prolog, as well as XPath and restricted subsets of XQuery. Probabilistic programming provides sophisticated knowledge compilation techniques and initial approaches for lifted (first-order) inference, with judiciously tuned approximation algorithms for the cases when exact inference remains intractable. Information extraction will be the driver to foster indexing and querying natural-language contents with temporal, spatial, and other contextual annotations that go beyond just RDF facts. Structured machine learning with scalable probabilistic inference and natural language modeling are two longstanding problems in artificial intelligence and machine learning [16]. Scaling-out probabilistic inference via distributed main-memory platforms for data storage and querying (see, e.g., [43]) will provide major challenges also for future research. Parallel query processing techniques, using both vertical and horizontal partitioning schemes (see, e.g., [31,73]), have a great potential to serve as a solid database backend for these inference techniques as well. The integration of all of the above will require major contributions from the fields of databases, machine learning, and natural-language processing. It's about time to join our forces!

## References

1. Abiteboul, S., Kanellakis, P., Grahne, G.: On the representation and querying of sets of possible worlds. Theor. Comput. Sci. 78(1), 159–187 (1991)
2. Banko, M., Cafarella, M.J., Soderland, S., Broadhead, M., Etzioni, O.: Open information extraction from the Web. In: IJCAI, pp. 2670–2676 (2007)
3. Benjelloun, O., Sarma, A.D., Halevy, A.Y., Theobald, M., Widom, J.: Databases with uncertainty and lineage. VLDB J. 17(2), 243–264 (2008)
4. Bizer, C., Lehmann, J., Kobilarov, G., Auer, S., Becker, C., Cyganiak, R., Hellmann, S.: DBpedia - a crystallization point for the Web of Data. J. Web Sem. 7(3), 154–165 (2009)
5. Boulos, J., Dalvi, N.N., Mandhani, B., Mathur, S., Ré, C., Suciu, D.: MYSTIQ: a system for finding more answers by using probabilities. In: SIGMOD, pp. 891–893 (2005)
6. Carlson, A., Betteridge, J., Kisiel, B., Settles, B., Hruschka Jr., E.R., Mitchell, T.M.: Toward an architecture for never-ending language learning. In: AAAI (2010)
7. Cheney, J., Chiticariu, L., Tan, W.-C.: Provenance in databases: Why, how, and where. Found. Trends Databases 1, 379–474 (2009)

8. Corro, L.D., Gemulla, R.: ClausIE: clause-based open information extraction. In: WWW, pp. 355–366 (2013)
9. Dalvi, N.N., Suciu, D.: The dichotomy of conjunctive queries on probabilistic structures. In: PODS, pp. 293–302 (2007)
10. Dalvi, N.N., Suciu, D.: Efficient query evaluation on probabilistic databases. VLDB J. 16(4), 523–544 (2007)
11. Darwiche, A., Marquis, P.: A knowledge compilation map. Journal of Artificial Intelligence Research 17(1), 229–264 (2002)
12. Raedt, L.D., Kimmig, A., Toivonen, H.: ProbLog: A probabilistic Prolog and its application in link discovery. In: IJCAI, pp. 2462–2467 (2007)
13. de Braz, R.S., Amir, E., Roth, D.: Lifted first-order probabilistic inference. In: Getoor, L., Taskar, B. (eds.) Introduction to Statistical Relational Learning. MIT Press (2007)
14. Van den Broeck, G.: On the completeness of first-order knowledge compilation for lifted probabilistic inference. In: NIPS, pp. 1386–1394 (2011)
15. Van den Broeck, G., Taghipour, N., Meert, W., Davis, J., De Raedt, L.: Lifted probabilistic inference by first-order knowledge compilation. In: IJCAI, pp. 2178–2185 (2011)
16. Dietterich, T.G., Domingos, P., Getoor, L., Muggleton, S., Tadepalli, P.: Structured machine learning: the next ten years. Machine Learning 73(1), 3–23 (2008)
17. Domingos, P., Lowd, D.: Markov Logic: An Interface Layer for Artificial Intelligence. Synthesis Lectures on Artificial Intelligence and Machine Learning. Morgan & Claypool Publishers (2009)
18. Dylla, M., Miliaraki, I., Theobald, M.: Top-k query processing in probabilistic databases with non-materialized views. In: ICDE, pp. 122–133 (2013)
19. Etzioni, O., Banko, M., Cafarella, M.J.: Machine reading. In: AAAI Spring Symposium: Machine Reading, pp. 1–5 (2007)
20. Etzioni, O., Fader, A., Christensen, J., Soderland, S.: Mausam: Open information extraction: The second generation. In: IJCAI, pp. 3–10 (2011)
21. Fagin, R., Lotem, A., Naor, M.: Optimal aggregation algorithms for middleware. J. Comput. Syst. Sci. 66(4), 614–656 (2003)
22. Fierens, D., Van den Broeck, G., Thon, I., Gutmann, B., De Raedt, L.: Inference in probabilistic logic programs using weighted CNF's. In: UAI, pp. 211–220 (2011)
23. Fink, R., Han, L., Olteanu, D.: Aggregation in probabilistic databases via knowledge compilation. PVLDB 5(5), 490–501 (2012)
24. Fink, R., Olteanu, D.: On the optimal approximation of queries using tractable propositional languages. In: ICDT, pp. 174–185 (2011)
25. Fink, R., Olteanu, D., Rath, S.: Providing support for full relational algebra in probabilistic databases. In: ICDE, pp. 315–326 (2011)
26. Galárraga, L.A., Teflioudi, C., Hose, K., Suchanek, F.M.: AMIE: association rule mining under incomplete evidence in ontological knowledge bases. In: WWW, pp. 413–422 (2013)
27. Getoor, L., Taskar, B.: An Introduction to Statistical Relational Learning. MIT Press (2007)
28. Goodman, N.D., Mansinghka, V.K., Roy, D.M., Bonawitz, K., Tenenbaum, J.B.: Church: A language for generative models. In: UAI, pp. 220–229 (2008)
29. Guptaand, R., Sarawagi, S.: Creating probabilistic databases from information extraction models. In: VLDB, pp. 965–976 (2006)
30. Hoffart, J., Suchanek, F.M., Berberich, K., Weikum, G.: YAGO2: A spatially and temporally enhanced knowledge base from Wikipedia. Artif. Intell. 194, 28–61 (2013)
31. Huang, J., Abadi, D.J., Ren, K.: Scalable SPARQL querying of large RDF graphs. PVLDB 4(11), 1123–1134 (2011)
32. Ilyas, I.F., Beskales, G., Soliman, M.A.: A survey of top-$k$ query processing techniques in relational database systems. ACM Comput. Surv. 40, 11:1–11:58 (2008)

33. Jha, A.K., Gogate, V., Meliou, A., Suciu, D.: Lifted inference seen from the other side: The tractable features. In: NIPS, pp. 973–981 (2010)
34. Jha, A.K., Suciu, D.: Knowledge compilation meets database theory: compiling queries to decision diagrams. In: ICDT, pp. 162–173 (2011)
35. Jha, A.K., Suciu, D.: On the tractability of query compilation and bounded treewidth. In: ICDT, pp. 249–261 (2012)
36. Jha, A.K., Suciu, D.: Probabilistic databases with MarkoViews. PVLDB 5(11), 1160–1171 (2012)
37. Kanagal, B., Deshpande, A.: Lineage processing over correlated probabilistic databases. In: SIGMOD, pp. 675–686 (2010)
38. Kersting, K.: Lifted probabilistic inference. In: ECAI, pp. 33–38 (2012)
39. Kimmig, A., Demoen, B., De Raedt, L., Costa, V.S., Rocha, R.: On the implementation of the probabilistic logic programming language ProbLog. Theory and Practice of Logic Programming 11, 235–262 (2011)
40. Koch, C., Olteanu, D.: Conditioning probabilistic databases. PVLDB 1(1), 313–325 (2008)
41. Lafferty, J.D., McCallum, A., Pereira, F.C.N.: Conditional Random Fields: Probabilistic models for segmenting and labeling sequence data. In: ICML, pp. 282–289 (2001)
42. Li, J., Saha, B., Deshpande, A.: A unified approach to ranking in probabilistic databases. PVLDB 2(1), 502–513 (2009)
43. Low, Y., Gonzalez, J., Kyrola, A., Bickson, D., Guestrin, C., Hellerstein, J.M.: Distributed GraphLab: A framework for machine learning in the cloud. PVLDB 5(8), 716–727 (2012)
44. McCallum, A., Schultz, K., Singh, S.: FactorIE: Probabilistic programming via imperatively defined factor graphs. In: NIPS, pp. 1249–1257 (2009)
45. Milch, B., Zettlemoyer, L.S., Kersting, K., Haimes, M., Kaelbling, L.P.: Lifted probabilistic inference with counting formulas. In: AAAI, pp. 1062–1068 (2008)
46. Mutsuzaki, M., Theobald, M., de Keijzer, A., Widom, J., Agrawal, P., Benjelloun, O., Sarma, A.D., Murthy, R., Sugihara, T.: Trio-One: Layering uncertainty and lineage on a conventional DBMS. In: CIDR, pp. 269–274 (2007)
47. Nakashole, N., Theobald, M., Weikum, G.: Scalable knowledge harvesting with high precision and high recall. In: WSDM, pp. 227–236 (2011)
48. Nakashole, N., Weikum, G., Suchanek, F.M.: Discovering and exploring relations on the Web. PVLDB 5(12), 1982–1985 (2012)
49. Niu, F., Ré, C., Doan, A., Shavlik, J.W.: Tuffy: Scaling up statistical inference in Markov Logic Networks using an RDBMS. PVLDB 4(6), 373–384 (2011)
50. Olteanu, D., Huang, J.: Using OBDDs for efficient query evaluation on probabilistic databases. In: Greco, S., Lukasiewicz, T. (eds.) SUM 2008. LNCS (LNAI), vol. 5291, pp. 326–340. Springer, Heidelberg (2008)
51. Olteanu, D., Wen, H.: Ranking query answers in probabilistic databases: Complexity and efficient algorithms. In: ICDE, pp. 282–293 (2012)
52. Pfeffer, A.: IBAL: A probabilistic rational programming language. In: IJCAI, pp. 733–740 (2001)
53. Poole, D.: The independent choice logic for modelling multiple agents under uncertainty. Artificial Intelligence 94(1-2), 7–56 (1997)
54. Poole, D.: First-order probabilistic inference. In: IJCAI, pp. 985–991 (2003)
55. Rabiner, L.R.: A tutorial on hidden Markov models and selected applications in speech recognition. Proceedings of the IEEE, 257–286 (1989)
56. De Raedt, L., Frasconi, P., Kersting, K., Muggleton, S.H. (eds.): Probabilistic Inductive Logic Programming. LNCS (LNAI), vol. 4911. Springer, Heidelberg (2008)
57. Ré, C., Dalvi, N.N., Suciu, D.: Efficient top-k query evaluation on probabilistic data. In: ICDE, pp. 886–895 (2007)

58. Renkens, J., Van den Broeck, G., Nijssen, S.: k-optimal: A novel approximate inference algorithm for ProbLog. Machine Learning 89(3), 215–231 (2012)
59. Riedel, S.: Improving the accuracy and efficiency of MAP inference for Markov Logic. In: UAI, pp. 468–475 (2008)
60. Roth, D.: On the hardness of approximate reasoning. Artif. Intell. 82, 273–302 (1996)
61. Sarma, A.D., Theobald, M., Widom, J.: Exploiting lineage for confidence computation in uncertain and probabilistic databases. In: ICDE, pp. 1023–1032 (2008)
62. Sato, T.: A statistical learning method for logic programs with distribution semantics. In: ICLP, pp. 715–729 (1995)
63. Sen, P., Deshpande, A., Getoor, L.: PrDB: managing and exploiting rich correlations in probabilistic databases. VLDB J. 18(5), 1065–1090 (2009)
64. Sen, P., Deshpande, A., Getoor, L.: Read-once functions and query evaluation in probabilistic databases. PVLDB 3(1), 1068–1079 (2010)
65. Suciu, D., Olteanu, D., Ré, C., Koch, C.: Probabilistic Databases. Synthesis Lectures on Data Management. Morgan & Claypool Publishers (2011)
66. Theobald, M., Weikum, G., Schenkel, R.: Top-k query evaluation with probabilistic guarantees. In: VLDB, pp. 648–659 (2004)
67. Van den Broeck, G.: Lifted Inference and Learning in Statistical Relational Models. PhD thesis, Informatics Section, Department of Computer Science, Faculty of Engineering Science, Katholieke Universiteit Leuven (January 2013)
68. Van den Broeck, G., Taghipour, N., Meert, W., Davis, J., De Raedt, L.: Lifted probabilistic inference by first-order knowledge compilation. In: IJCAI, pp. 2178–2185 (2011)
69. Wang, D.Z., Franklin, M.J., Garofalakis, M.N., Hellerstein, J.M.: Querying probabilistic information extraction. PVLDB 3(1), 1057–1067 (2010)
70. Wang, D.Z., Michelakis, E., Franklin, M.J., Garofalakis, M.N., Hellerstein, J.M.: Probabilistic declarative information extraction. In: ICDE, pp. 173–176 (2010)
71. Weikum, G., Theobald, M.: From information to knowledge: harvesting entities and relationships from Web sources. In: PODS, pp. 65–76 (2010)
72. Widom, J.: Trio: A system for data, uncertainty, and lineage. In: Managing and Mining Uncertain Data. Springer (2008)
73. Zeng, K., Yang, J., Wang, H., Shao, B., Wang, Z.: A distributed graph engine for Web scale RDF data. In: SIGMOD (to appear, 2013)

# Computational Neuroscience Breakthroughs through Innovative Data Management

Farhan Tauheed[1,2], Sadegh Nobari[3], Laurynas Biveinis[4],
Thomas Heinis[2], and Anastasia Ailamaki[2]

[1] Data-Intensive Applications and Systems Lab, EPFL, Switzerland
[2] Brain Mind Institute, EPFL, Switzerland
[3] National University of Singapore, Singapore
[4] Department of Computer Science, Aalborg University, Denmark

**Abstract.** Simulations have become key in many scientific disciplines to better understand natural phenomena. Neuroscientists, for example, build and simulate increasingly fine-grained models (including subcellular details, e.g., neurotransmitter) of the neocortex to understand the mechanisms causing brain diseases and to test new treatments in-silico.

The sheer size and, more importantly, the level of detail of their models challenges today's spatial data management techniques. In collaboration with the Blue Brain project (BBP) we develop new approaches that efficiently enable analysis, navigation and discovery in spatial models of the brain. More precisely, we develop an index for the scalable and efficient execution of spatial range queries supporting model building and analysis. Furthermore, we enable navigational access to the brain models, i.e., the execution of of series of range queries where he location of each query depends on the previous ones. To efficiently support navigational access, we develop a method that uses previous query results to prefetch spatial data with high accuracy and therefore speeds up navigation. Finally, to enable discovery based on the range queries, we conceive a novel in-memory spatial join.

The methods we develop considerably outperform the state of the art, but more importantly, they enable the neuroscientists to scale to building, simulating and analyzing massively bigger and more detailed brain models.

## 1  Introduction

Scientists across many different fields have started to complement their traditional methods for understanding a phenomena in nature with the simulation of spatial models of it. Simulating spatial models has become standard practice in many disciplines and applications. Examples include the simulation of peptide folding [1], star formation in astronomy [2], earthquakes in geology [3], fluid dynamics as well as the brain in neuroscience [4]. To develop a better understanding the scientists continuously increase the size and complexity of the simulations as much as their hardware allows.

B. Catania, G. Guerrini, and J. Pokorný (Eds.): ADBIS 2013, LNCS 8133, pp. 14–27, 2013.

Today's tools and algorithms required to cope with the data at the core of simulations, however, cannot cope with the data deluge resulting from bigger and more detailed simulations. While the many spatial indexes [5] developed in the past are of great help to scientists in analyzing and building models, many cannot deal with the complexity and size of today's spatial models.

A particular example of scientists who simulate a detailed model on a massive scale are the neuroscientists of the Blue Brain Project (BBP [4]). In their attempt to understand the brain, i.e., what gives rise to cognition and what mechanisms lead to brain disease, they model and simulate the rat brain (and later the human brain) in great detail. To this end, the neuroscientists build models of the neocortex in unprecedented detail and simulate electrical activity on a supercomputer (BlueGene/P with 16K cores).

To build detailed models, the neuroscientists in the BBP have analyzed the rat brain tissue in the wet lab over several years and have identified the exact electrophysiological properties and the precise morphological structure of neurons. The neuron morphology defines the branches that extend into large parts of the tissue in order to receive and send out information to other neurons. To obtain a biorealistic model (an example is illustrated in Figure 1, left) the neuroscientists put together thousands or millions of neuron morphologies, each of which is represented by thousands of small cylinders (a morphology is shown in Figure 1, right).

The models built in the BBP have quickly grown in recent years and feature several million neurons today. Although the current model size is still far from their ultimate goal of building and simulating models as big as the human brain ($\sim 10^{11}$ neurons), today's size already seriously challenges state-of-the-art spatial data management tools.

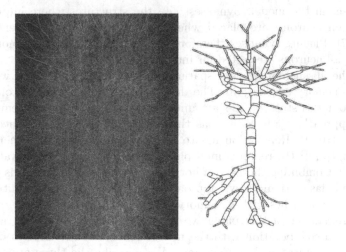

**Fig. 1.** A visualization of a model microcircuit comprised of thousands of neurons (left) and a schema of a neuron morphology modeled with cylinders (right)

More important than only the size (number of cylinders) of the models, however, is their level of detail: to build more detailed and biorealistic models the neuroscientists pack more and smaller elements into the same volume. Analyzing the detailed models with current methods is becoming a challenge because the state of the art like the R-Tree [6] do not scale to the increasingly fine-grained/dense models. To address the challenge of increasingly detailed models, we develop a new query execution strategy for dense datasets. At its core is a novel two-phased range query execution strategy where each phase is independent of data density.

The ability to efficiently execute range queries on dense data greatly speeds up building and analyzing of models. Crucial for the further analysis, however, is navigational access to the brain models, i.e., the interactive execution of series of range queries where the location of the next query depends on the result of the previous query. The neuroscientists frequently follow a branch of a neuron and execute spatial range queries for detailed analysis to validate the model. At each location, a query is executed, the data is retrieved as well as visualized before the scientist decides on the next location where a query is executed. Because spatial range query execution is disk bound, executing range query series is a very time-consuming process.

To speed up the execution of interactive range query series, data can be prefetched. State-of-the-art approaches, however, do not prefetch spatial data with high accuracy because they rely on limited information, e.g., previous query positions. We therefore develop a novel approach that prefetches spatial data with a considerably higher accuracy by using the content of previous queries (instead of only their position) and thereby achieve substantially higher prefetch accuracy.

A particular computation that needs to be run based on the query results of each spatial range query and that enables neuroscientific discovery, is placing the synapses in the model. Synapses, i.e., the structures where impulses leap over between neurons, are placed wherever two cylinder of different neurons intersect [7]. Placing synapses therefore is equivalent to an in-memory spatial join where all neurons are tested for intersection.

Given the absence of efficient spatial join methods for memory, we develop a novel in-memory spatial join. The design of our new approach considerably departs from the state of the art and avoids the overlap problem of data-oriented approaches [8,9] as well as the replication problem of space-oriented approaches [10,11]. Replication has to be avoided because it (a) increases the memory footprint, (b) requires multiple comparisons and (c) removal of duplicate results. Combining the best of both worlds, our new approach is one order of magnitude faster than known approaches and two orders of magnitude faster than known approaches with a memory footprint of the same size.

In the remainder of this paper we discuss the spatial indexing algorithms we develop in collaboration with the neuroscientists of the Blue brain project to advance computational neuroscience. We describe the three approaches we develop, FLAT [12] for efficient range query execution, SCOUT [13] for the accurate prefetching of spatial data and TOUCH [14] for efficient and scalable in-memory joins, discuss how neuroscientists use them and we demonstrate the considerable impact they have on the process of building the models.

## 2    Retrieving Dense Neuroscience Data

A crucial type of query in the model building process in the BBP is the spatial range query. Range queries are repeatedly used to visualize parts of the models or to ensure that the models built are biorealistic (testing the tissue density, synapse density or other statistics).

Because today's models of the brain are already very detailed and dense, state-of-the-art indexes to execute range queries [5] are not efficient. The efficient execution of range queries to build and validate models, however, is pivotal today and will become even more important in the future where the models will be increasingly biorealistic and thus dense as the neuroscientists will model phenomena on the subcellular level.

### 2.1    Motivation

Several spatial access methods supporting the execution of spatial range queries [5] have been developed in the past. While these approaches execute range queries efficiently on many datasets, they unfortunately do not do so on dense or detailed neuroscience models. To make matters worse, they will only scale poorly to more dense and detailed models built in the future.

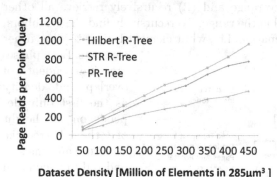

**Fig. 2.** Point query performance on R-Tree variants

Current methods are based on the hierarchical organization of the multidimensional spatial data (R-Trees and variants: STR, TGS, PR-Tree, R+-Tree, R*-Tree and others [5]) and therefore suffer from overlap [15] and dead space. Spatially close elements are stored together in the same node (on the same disk page) and a tree is recursively built such that each node has a minimum bounding rectangle (MBR) that encloses all MBRs of its children. Nodes with overlapping MBRs lead to ambiguity in the tree, i.e., to execute a range query several paths in the tree need to be followed, leading to the retrieval of an excessive number of nodes. To make matters worse, overlap becomes substantially worse with increasing density of the dataset: as the number of spatial elements in the same unit of space increases, so does the overlap of tree-based indexes.

The experiment shown in Figure 2 where we increase the density of the dataset and measure point query performance of different R-tree based approaches [16,17,18] clearly demonstrates the problem of overlap. Point queries are a good indicator of overlap in R-Trees: the number of nodes retrieved from disk should be in the order of the height of the tree (five in this case) in the absence of overlap. As the results in Figure 2 show, however, the overlap grows rapidly with increasing dataset density, translating into a higher execution time and thus degraded performance.

Despite numerous proposed improvements, e.g., reducing overlap through splitting and replicating elements [15] (thereby increasing the number of nodes in the tree and also its size on disk considerably), the fundamental problem remains the same and needs to be addressed to enable the neuroscientists to build, analyze and validate more detailed and biorealistic models.

## 2.2   FLAT Query Execution

To enable the neuroscientists in the BBP to build and analyze models of the brain on an unprecedented detailed level, we develop FLAT [12] with a two phased query execution at its core. The key insight we use is that while finding all elements in a particular range query in an R-Tree-like index suffers from overlap, finding an arbitrary element in a range query on the other hand is independent of overlap and therefore is a comparatively cheap operation. Because only one path in the tree has to be followed, the cost of finding an arbitrary element is in the order of the height of the tree.

With this insight, we develop a two phased approach where we (1) find an arbitrary element $e$ in the query range and (2) recursively retrieve all other elements (the neighbors of $e$) within the range. To recursively find the neighbors, FLAT stores neighborhood information, i.e., what element is next to what other elements. Both phases are independent of overlap and density as the first only depends on the height of the R-tree-like structure and the second only depends on the number of elements in the range query. With both phases avoiding the problem of overlap and only depending on the result size of the query, FLAT will scale to much denser/detailed models.

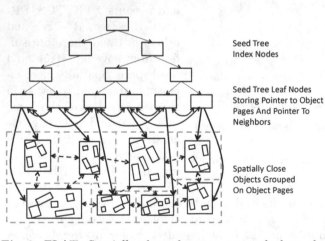

Seed Tree
Index Nodes

Seed Tree Leaf Nodes
Storing Pointer to Object
Pages And Pointer To
Neighbors

Spatially Close
Objects Grouped
On Object Pages

**Fig. 3.** FLAT: Spatially close elements are packed on the same disk page (rectangle) and pointers (arrows) are added between neighboring pages

To limit the amount of information stored, FLAT stores the neighborhood information on the level of groups of elements instead of on the level of single elements, i.e., what spatial element neighbors what other spatial elements. FLAT groups spatially close elements together (and stores them on the same disk page), indexes the groups with an R-Tree index, called the seed tree, and finally computes the neighborhood information between the groups. The neighborhood information itself is stored in the leaf nodes of the R-Tree (seed tree).

Figure 3 illustrates how queries are executed using the groups of elements and the neighborhood information: first FLAT retrieves an arbitrary group in the query range and then recursively retrieves all neighboring elements in the query range.

While some datasets may contain inherent neighborhood information (e.g., in meshes the neighborhood is stored in the edges), many (and in particular the neuroscience datasets) do not have any or only limited neighborhood information. As a consequence, to make FLAT work on arbitrary datasets, we add neighborhood information to the index in a preprocessing phase. Computing and storing neighborhood information is not an undue burden as the increase in index building time is below 10% and the space increases by only 12% (both compared to the STR [16] bulkloading approach).

## 2.3   Impact of FLAT

The impact FLAT on the work of the neuroscientists is considerable. Until recently they have not been able to build, analyze and validate models exceeding one million neurons. With FLAT they now have the ability to scale to bigger and, more importantly, to much more detailed models. An experiment with a neuroscience dataset shows the promise of FLAT. In this experiment we execute 200 small sized queries (each with a volume of $5 \times 10^{-7}\%$) used for structural analysis on a dataset where the density increases from 50 to 450 million elements in a constant volume of $285\mu m^3$. We increase the number of elements in the

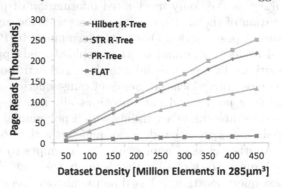

**Fig. 4.** Execution time for executing 200 the small structural analysis queries of size $5 \times 10^{-7}\%$

same volume to emulate increasingly detailed models.

As Figure 4 impressively demonstrates, FLAT already today considerably outperforms R-Tree based approaches by a factor of 8 for the densest model (with 450 million elements). More importantly, however, the trend clearly shows that FLAT will scale better to more detailed and dense models in the future. This is pivotal as it finally enables the neuroscientists to build more bio-realistic models where subcellular elements (e.g., neurotransmitters) can be precisely replicated for a more accurate simulation of brain activity.

# 3    Prefetching for Structure Following Spatial Queries

FLAT enables the neuroscientists to efficiently execute spatial range queries on todays models and also on future, even more dense & detailed spatial models. More important for many of their analysis, however, is it to execute a series of range queries: following a structure in the model, e.g., a neuron branch, they need to execute several range queries to assess the quality or validity of the model. On the result of each range query they compute different types of statistics (tissue density, synapse placement, synapse count, etc). Series of range queries are not only crucial for the neuroscientists, but also for other scientists who analyze road networks, arterial trees and others.

Executing a series of range queries is an interactive process where the user follows a structure, executes a query, computes one or several statistics, analyzes the statistics and then decides on the location of the next query and executes it. Because the series is interactive, the disk is idle during the computation of statistics (between two range queries) and data can be prefetched to speed up the series. State-of-the-art approaches, however, rely on limited information to predict the next query location and thus prefetch with low accuracy.

## 3.1    Motivation

Known approaches used to prefetch spatial data do not have good enough accuracy as they only use limited information of previous queries to predict the location of the next query. Several state-of-the-art approaches rely on the positions of past queries. One particular approach [19] uses the last query position and prefetches around it. More sophisticated approaches [20] use the last few positions, fit a polynomial into them and extrapolate the polynomial to predict the next query location. Series of range queries on neuron structures, however, are very jagged and not smooth at all. The irregular structure makes it very hard to interpolate accurately with a polynomial and consequently this class of prediction approaches does not prefetch with good accuracy.

Another class of approaches [21] attempts to learn from past user behavior by keeping track of all paths visited in the past. Prefetching, i.e., predicting the next query location, is based on the history. Because the models in our scenario are so massive, it is unlikely that any path will be visited twice, therefore making prefetching strategies based on past paths visited inaccurate.

## 3.2    Content-Aware Prefetching

To prefetch more accurately and to considerably speed up the execution of series of range queries and therefore analysis, we develop SCOUT [13]. SCOUT departs form previous approaches as it does not only consider previous query positions, but also takes into account the previous query content, knowing that the scientist follows one of the structures in the previous queries. As a consequence, SCOUT prefetches with a considerably higher accuracy, speeding up query series by a factor of up to 15×.

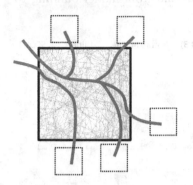

SCOUT summarizes the content of the most recent query $q$, i.e., it identifies the topological skeleton in $q$ and approximates it with a graph. The graph of $q$ represents all the structures the neuroscientist is potentially following and SCOUT therefore prefetches data at all locations where the graph leaves $q$ (exit locations). Range queries are executed to prefetch data at these locations until the user executes a new query in the series. Figure 5 shows how small range queries are executed at the exit locations of the last query.

**Fig. 5.** Prefetching of spatial data at the exit locations of the structure

As the example in Figure 5 shows, in some cases SCOUT has to prefetch in multiple locations. This is the case at the beginning of a series of queries where SCOUT cannot yet identify the one structure the neuroscientist follows. By using iterative candidate pruning, however, SCOUT can reliably identify the neuron branch followed after a few queries already.

Iterative candidate pruning exploits that all previous queries must contain the branch the scientist follows. To prefetch for the $n^{th}$ query, SCOUT thus only needs to consider the set of branches leaving the $(n-2)^{th}$ query and the set of branches entering the $n-1^{th}$ (most recent) query. The branch followed is in the intersection of both sets. As the number of queries in a series increases, the number of branches in the intersection between two consecutive queries decreases continuously and the branch the user follows can be identified reliably. Figure 6 illustrates how through iteratively reducing the set of candidates, SCOUT can reliably identify the structure the scientist follows after only a few queries.

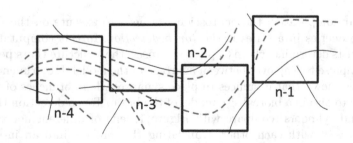

**Fig. 6.** Pruning the irrelevant structures (solid lines) from the candidate set (dashed lines) in subsequent queries (solid squares) of the series

### 3.3   Impact of SCOUT

SCOUT helps the neuroscientists to make building and analyzing models substantially faster. The speedup for different types of analysis ranges between 4× and 15× and enables a significantly faster turnaround from model building to analysis.

The experiment in Figure 7 shows this impressively by comparing the cache hit rate and the speedup with state-of-the-art approaches. In this experiment we have executed different series of range queries from different applications (visualization, model building) which differ in size of each query (30'000 to $80'000\mu m^3$) as well as the length of the series (ranging from 25 to 65).

The experiment shows that SCOUT speeds up the analysis considerably: the cache hit rate of SCOUT is 4× higher for the visualization query series resulting in a 7× higher speedup compared to EWMA [22], the fastest state-of-the-art approach (based on polynomial extrapolation). Like FLAT, SCOUT also speeds up the model building process and allows the neuroscientists to build bigger models considerably faster.

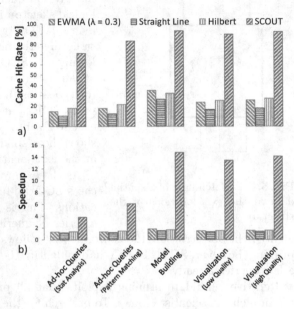

**Fig. 7.** Accuracy of the approaches for all microbenchmarks (a) and speedup of the approaches for all microbenchmarks (b)

## 4   In-Memory Spatial Join for Model Building

A particular computation the neuroscientists need to execute on the result of each of the queries in a series is the *touch detection*. In this computation the neuroscientists determine where to place synapses, the structure thats permit an electrical impulse to leap over between neurons, in the model. Experiments in the wet lab have shown that it suffices to place synapses where branches of neurons intersect [7] to obtain a biorealisitc model of the brain. Touch detection therefore needs to find cylinders (compare with Figure 1, left) of different neurons that overlap/intersect with each other, translating this process into an in-memory spatial join. While many spatial join methods have been developed for disk in recent years, no scalable in-memory approach exists.

### 4.1 Challenge

For the touch detection, the neuroscientists need to execute a spatial join on every result of the query series. Because the result of each query fits into the memory of even desktop machines, the spatial join needs to be performed in-memory.

Despite decades of research into spatial joins, only two algorithms have been developed to join two datasets in memory: the nested loop join [23] and the sweep line approach [24]. Neither of the two scales well: the nested loop join has a complexity of $O(n^2)$ whereas the sweep line approach becomes inefficient when too many elements are on the sweep line (very likely in case of dense data/detailed neuroscience models).

Approaches primarily developed for disk [25] can of course also be used in memory. Existing work can be categorized into space- or disk-oriented partitioning approaches. Besides advantages, both classes also have clear

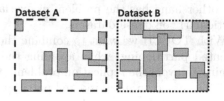

(a) The datasets $A$ and $B$

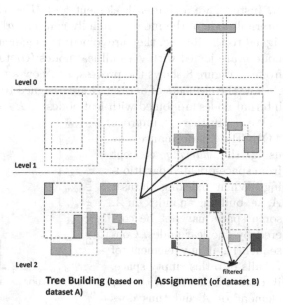

(b) Tree building, assignment and joining phases

**Fig. 8.** The three phases of : building the tree, assignment and joining

disadvantages: space-oriented approaches [10,11] generally need to replicate elements (elements that intersect with two partitions are copied to both) leading to considerable overhead and multiple detection of the same intersections whereas data-oriented approaches [8] suffer from the overlap problem of R-Trees which degrades performance considerably, particularly when used with dense datasets.

### 4.2 TOUCH: Efficient In-Memory Spatial Join

Given the lack of in-memory spatial join approaches as well as the challenges of either data- and space oriented approaches, we have developed TOUCH [14], a novel in-memory spatial join algorithm. With TOUCH we want to avoid space-oriented partitioning because it typically leads to replication of elements. Replication has to be avoided because it (a) increases the memory footprint and

(b) requires multiple comparisons between copies of elements (as well as making the removal of duplicate results necessary). Data-oriented partitioning on the other hand has the problem of overlap resulting in degraded performance, particularly on dense datasets.

With TOUCH we want to combine the best of both, space- as well as data-oriented partitioning, while avoiding the pitfalls. We use data-oriented partitioning to avoid the replication problem of space-oriented partitioning and build an index based on data-oriented partitioning (similar to an R-Tree) on the first dataset $A$ (all elements of $A$ are in the leaf nodes). To avoid the issue of overlap, we do not probe the data-oriented index for every element of the second dataset $B$. Instead, we assign each element $b$ of $B$ to the lowest (closest to the leafs) internal node of the index that fully contains $b$. Once all elements of $B$ are assigned to the R-Tree, they are joined: the elements of $B$ in a particular internal node $n$ are joined with with all leaf nodes (containing elements of $A$) reachable from $n$. Figure 8 shows the process, i.e., how an index is built based on dataset $A$, how the elements of dataset $B$ are assigned to internal nodes and finally, how internal nodes are joined with leaf nodes.

We further improve TOUCH's performance by using the *filtering* concept from space-oriented partitioning. When indexing dataset $A$, i.e, building an index on $A$, some space may not be covered by the leaf nodes. Consequently, if any element of $B$ falls into this empty space, it cannot intersect with any element of $A$ and thus does not have to be considered. Depending on the distribution of the dataset, filtering can considerably reduce the number intersection tests between elements.

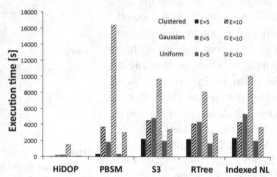

**Fig. 9.** Comparing the approaches for two different distance predicates $\epsilon$ on all datasets

## 4.3   Impact of TOUCH

TOUCH is one order of magnitude faster than known approaches and two orders of magnitude faster than known approaches with an equally small memory footprint. An experiment with neuroscience data shows this in Figure 9: in this experiment we measure how fast the spatial join can be computed on datasets of size 50 millions with disk-based join methods used in memory for two different distance predicates (we exclude the two in-memory join methods because they are too slow). We use three different datasets where the locations of the objects are distributed uniform, gaussian and clustered (100 clusters, each with a gaussian distribution). Clearly TOUCH is the fastest for either distance predicate, followed by PBSM [11]. Although PBSM is the fastest competitor, it is still one order of magnitude slower and uses considerable more memory (a factor of 8 × more).

TOUCH makes a considerable difference in the work of the neuroscientists. Without TOUCH, the biggest model they touch detected was 1 million neurons big. Today with TOUCH, they are able to touch detect a model of 10 millions and they are working on touch detecting a 33 million neuron model. Although TOUCH is very conservative with respect to memory, running touch detection on bigger models is not possible because a 33 million neuron model entirely fills the memory of their current infrastructure. We are currently investigating out of core methods where disk capacity is the bottleneck and no longer memory.

## 5    Conclusions

To gain a better understanding of how the brain works, what gives rise to cognition and what mechanisms govern dementia, the researchers of the Blue Brain Project build increasingly big, complex and detailed models of the brain. Because their models are so big and detailed, state-of-the-art methods no longer can be used to efficiently build, analyze and validate them. We have therefore developed FLAT, a method for the scalable execution of spatial range queries on dense spatial models, TOUCH, a prefetching method for spatial data used to speed up series of spatial range queries and finally TOUCH, an efficient in-memory spatial join approach.

The impact of the methods we have developed on the Blue Brain project are substantial: the neuroscientists can build bigger and more detailed models faster. The limit of models they can build has grown considerably to 33 million neurons today. Dealing with increasingly detailed and complex spatial models is not just a problem of neuroscientists, but is shared in many scientific disciplines that simulate natural phenomena and the algorithms we developed thus have impact beyond neuroscience.

Our results also demonstrate that despite decades of research in spatial data management, many problems still remain open. Increasing main memory as well as novel storage technology (in the memory hierarchy), for example, means that several spatial indexes need to be redesigned as we have shown with TOUCH. The execution of spatial range queries or nearest neighbor queries, for example, needs to be supported with radically different indexes optimized for memory.

New types of datasets (e.g., dense, complex spatial datasets) make new indexes necessary (FLAT) and new types of queries (e.g., series of range queries) also call for the development of new indexes. Nearest neighbor queries with constraints (find the nearest neuron with a given voltage), for example, are not yet supported efficiently.

Finally, also the massive scale of the brain models as well as the size of the simulation results mandates new methods for data management. Analyzing the models with spatial queries as well as the simulation output with spatio-temporal queries to find interesting phenomena needs to be supported with scalable methods. Large-scale parallel approaches to analyze massive spatial or spatio-temporal therefore will have to be developed.

# References

1. Gnanakaran, S., Nymeyer, H., Portman, J., Sanbonmatsu, K.Y., Garcia, A.E.: Peptide folding simulations. Current Opinion in Structural Biology 13(2), 168–174 (2003)
2. Gray, J., Szalay, A., Thakar, A., Kunszt, P., Stoughton, C., Slutz, D., Vandenberg, J.: Data Mining the SDSS SkyServer Database. Technical Report, MSR-TR-2002-01, Microsoft Research (2002)
3. Komatitsch, D., Tsuboi, S., Ji, C., Tromp, J.: A 14.6 Billion Degrees of Freedom, 5 Teraflops, 2.5 Terabyte Earthquake Simulation on the Earth Simulator. In: International Conference on Supercomputing, SC 2003 (2003)
4. Markram, H.: The Blue Brain Project. Nature Reviews Neuroscience 7(2), 153–160 (2006)
5. Gaede, V., Guenther, O.: Multidimensional Access Methods. ACM Computing Surveys 30(2) (1998)
6. Guttman, A.: R-trees: a Dynamic Index Structure for Spatial Searching. In: SIGMOD 1984 (1984)
7. Kozloski, J., Sfyrakis, K., Hill, S., Schürmann, F., Peck, C., Markram, H.: Identifying, Tabulating, and Analyzing Contacts Between Branched Neuron Morphologies. IBM Journal of Research and Development 52(1/2), 43–55 (2008)
8. Brinkhoff, T., Kriegel, H., Seeger, B.: Efficient Processing of Spatial Joins R-Trees. In: SIGMOD 1993 (1993)
9. Lo, M.L., Ravishankar, C.V.: Spatial Joins Using Seeded Trees. In: SIGMOD 1994 (1994)
10. Koudas, N., Sevcik, K.C.: Size Separation Spatial Join. In: SIGMOD 1997 (1997)
11. Patel, J.M., DeWitt, D.J.: Partition Based Spatial-Merge Join. In: SIGMOD 1996 (1996)
12. Tauheed, F., Biveinis, L., Heinis, T., Schürmann, F., Markram, H., Ailamaki, A.: Accelerating range queries for brain simulations. In: ICDE 2012 (2012)
13. Tauheed, F., Heinis, T., Schürmann, F., Markram, H., Ailamaki, A.: Scout: Prefetching for latent structure following queries. In: VLDB 2012 (2012)
14. Nobari, S., Tauheed, F., Heinis, T., Karras, P., Bressan, S., Ailamaki, A.: TOUCH: In-Memory Spatial Join by Hierarchical Data-Oriented Partitioning. In: SIGMOD 2013 (2013)
15. Sellis, T.K., Roussopoulos, N., Faloutsos, C.: The R+-Tree: A Dynamic Index for Multi-Dimensional Objects. In: VLDB 1987 (1987)
16. Leutenegger, S.T., Lopez, M.A., Edgington, J.: STR: a Simple and Efficient Algorithm for R-tree Packing. In: ICDE 1997 (1997)
17. Arge, L., Berg, M.D., Haverkort, H., Yi, K.: The Priority R-tree: A Practically Efficient and Worst-case Optimal R-Tree. ACM Transactions on Algorithms 4(1), 1–30 (2008)
18. Kamel, I., Faloutsos, C.: Hilbert R-Tree: An Improved R-Tree using Fractals. In: VLDB 1994 (1994)
19. Park, D.J., Kim, H.J.: Prefetch policies for large objects in a web-enabled gis application. Data & Knowledge Engineering 37(1), 65–84 (2001)
20. Chan, A., Lau, R.W.H., Si, A.: A motion prediction method for mouse-based navigation. In: International Conference on Computer Graphics, pp. 139–146.

21. Lee, D., Kim, J., Kim, S., Kim, K., Yoo-Sung, K., Park, J.: Adaptation of a Neighbor Selection Markov Chain for Prefetching Tiled Web GIS Data. In: Yakhno, T. (ed.) ADVIS 2002. LNCS, vol. 2457, pp. 213–222. Springer, Heidelberg (2002)
22. Chim, J., et al.: On caching and prefetching of virtual objects in distributed virtual environments. In: MULTIMEDIA 1998 (1998)
23. Mishra, P., Eich, M.H.: Join Processing in Relational Databases. ACM Computing Surveys 24(1), 63–113 (1992)
24. Edelsbrunner, H.: Algorithms in combinatorial geometry. Springer (1987)
25. Jacox, E.H., Samet, H.: Spatial join techniques. ACM TODS 32(1), Article 7, 44 (2007)

# Periodic Data, Burden or Convenience

Bela Stantic

Institute for Integrated and Intelligent Systems
Griffith University
B.Stantic@griffith.edu.au

**Abstract.** Periodic events seem to be an intrinsic part of our life, and a way of perceiving reality. There are many application domains where periodic data play a major role. In many of such domains, the huge number of repetitions make the goal of *explicitly* storing and accessing such data very challenging to the extent of even not being possible, in cases of open ended intervals. In this work, we present a concept to represent periodic data in an *implicit* way. The representation model we propose captures the notion of periodic granularity provided by the temporal database glossary. We define the algebraic operators, and introduce access algorithms to cope with them and also with temporal *range queries*, proving that they are correct and complete with respect to the traditional explicit approach. In an experimental evaluation we show the advantages of our approach with respect to traditional explicit approach, in terms of space usage, physical disk I/O's and query response time.

## 1 Introduction

Periodic data play a major role in many application domains, such as manufacturing, office automation, and scheduling. Day and nights repeat at regular periodic patterns, as well as seasons, and years. Accordingly, many human activities are scheduled at periodic time (e.g., office activities, scheduling of train, airplanes, lessons, etc). Due also to such a wide range of different contexts of application, it is widely agreed that adopting a 'standard' and fixed menu of granularities (e.g., minutes, hours, days, weeks, years and so on in the Gregorian calendric system) is not enough in order to provide the required expressiveness and flexibility. For instance, a while ago Soo and Snodgrass [1] emphasized that the use of a calendar depends on the cultural, legal and even business orientation of the users, and listed many examples of different calendric systems and user-defined periodic granularities, such as academic, legal, and financial year. They also stressed that different user-defined periodic granularities are usually used even in the same area, for example the definition of holidays in different companies. The number of repetitions of periodic data may be very large and in some cases repetitions may also be 'open-ended', meaning that we do not know the ending time when the repetition will stop, which is for example the case in therapies for chronical patients as it may be repeated all the life long. Therefore, in the Computer Science literature (in particular, in the areas of Databases, Logics, and Artificial Intelligence), there is a common agreement that *formalisms* are needed in order to cope with *user-defined* periodic data in an *implicit* (also termed *intensional*) way, without an explicit storing of all the repetitions. In this work we consider periodic data which have value-equivalent repetitions

B. Catania, G. Guerrini, and J. Pokorný (Eds.): ADBIS 2013, LNCS 8133, pp. 28–41, 2013.

at periodic time (e.g., the periodic schedule of trains); data that are acquired at periodic time, but may assume different values (e.g., periodic monitoring of blood pressure) are not taken into account in this work. Periodic data play an important role in Databases and a specific entry has been devoted to such a topic in the Encyclopedia of Database Systems by Springer. In the Encyclopedia [2], three main classes of Database *implicit* approaches to user-defined periodicities have been identified:

- *Deductive rule-based* approaches, using deductive rules, and approaches in classical temporal logics [3].
- *Constraint-based* approaches, using mathematical formulae and constraints (e.g., [4]), and
- *Algebraic* also termed *Symbolic*) approaches, providing a set of 'high-level' and 'user-friendly' operators (e.g., [5], [6], [7], [8], [9], [10]).

A comparison among such classes approaches is out of the scope of this work (the interested reader is referred to [2] and also to [11]). However, it worth stressing that in most approaches in the literature (in particular in all algebraic approaches) the focus is on the design of *high-level formalisms* to model (in an implicit way) user-defined periodicities in a 'commonsense' or at least 'user-friendly' way. Most of such approaches do not take into account issues such as the definition of *relational temporal algebraic operators*, extending Codd's operators to query periodic data, *range queries* and efficient *indexing* and *access* of temporal data.

In summary, although there seems to be a general agreement within the Database (and also Artificial Intelligence) literature that general-purpose implicit approaches are needed in order to cope with *user-defined* periodic data, and despite the fact that a lot of such approaches have been devised in the last two decades, none of such approaches focus specifically on the definition of a comprehensive relational approach coping with user-defined periodic data efficiently, considering a relational representation of periodic data, efficient access of such data, algebraic operators, and additional temporal operators such as range queries. However, all such issues are fundamental for the practical applicability of any Database approach considering periodic data. The goal of our work is to devise such a comprehensive approach. Specifically, our approach has been designed in such a way that:

- The data model has the expressiveness to capture all *periodic granularities*, as defined in the Database literature [12], [7],
- Algebraic and temporal query operators are correct and complete with respect to the conventional explicit approaches, in which all the repetitions of periodic data are explicitly stored.
- Extended algebraic operators operate in polynomial time, and are a consistent extension of standard nontemporal relational algebraic ones.

First property grants that the expressiveness of our data model is the one requested by the temporal Database literature. Second property grants that, although periodic data are only implicitly stored, we get the same (correct) results as one obtained with traditional (fully explicit) models. Moreover, in this paper we also provide testing, to show the advantages of our approach with respect to conventional explicit approaches, especially in terms of disk I/O's and query response time.

On the other hand, in this work:

- We do not address the treatment of the transaction time of events (i.e., the time when events are *inserted in/deleted from* the database [13]) since no periodicity issue is usually involved by it. As a matter of fact, transaction time is *orthogonal* to valid time (the time when the fact described by the tuples takes place). As a consequence, the proposed approach dealing with the periodic valid time can be extended to deal also with transaction time.
- Although we cope with *user-defined* periodic granularities we assume that each periodic granularity is directly expressed in terms of a 'bottom' granularity. Therefore, we are not interested to cope with issues concerning conversions between periodic granularities, or properties of relations between them, which is, on the other hand, a main focus of other approaches dealing with multiple granularities, [14], [8], [15], [16].

The rest of the paper is organized as follows. The next Section is a preliminary one, in which we briefly mention the implicit vs explicit approaches. In Section 3 at first we define the quasi-periodic granularities and propose an abstract implicit representation of quasi-periodic granularities, and then we propose an extended relational temporal data model coping with it. Temporal *range queries* are particularly important in the temporal Database context [17]. In Section 4 we identify different types of temporal range queries in the context of periodic data, and we devise algorithms to cope with them. In section 5, we briefly overview the current approaches to index temporal data, proposing the adoption of TD-tree for indexing our representation. In section 6, we present an experimental evaluation of our approach, showing its advantages with respect to the 'traditional' explicit approach. In section 7 we provide proof that proposed algorithm is correct and complete. Finally, section 8 addresses conclusions.

## 2  Implicit vs. Explicit Approaches to Periodic Data

There is an obvious and trivial way to cope with user-defined periodic data, namely by explicitly storing all of them. Such an approach, usually called 'explicit' (or 'extensional') approach, basically reduces periodic data to standard temporal non-periodic ones. For instance, in order to deal with an activity $X$ scheduled each day from 9 to 11am in a year, an explicit approach can simply represent the *valid time* of the activity through list all of the 365 *periods* in which the activity takes place. The obvious advantage of such an approach is its simplicity as periodic temporal data are simply coped with as standard temporal data, so that any temporal Database approach in the literature can suffice. Such approach makes all of the database implementation simpler, from indexing to query processing. However, there are at least three main disadvantages of the 'explicit' approach:

- It is not 'commonsense' and 'human-oriented': humans usually tend to abstract, so they usually prefer to manage periodic data in an implicit way. For instance, every Monday this semester from 9 to 11am.
- It is not feasible in the case of 'open-ended' periodic data, when the duration of repetitions is unknown, so that no explicit elicitation of all the data is possible.

– It can be very expensive in regard to space complexity. In many practical application areas the number of repetitions of activities is very high, so making explicit all the repetitions would be very space demanding. For example, activities on an automatized schedule in a production chain may be performed at a very high (periodic) frequency for very long duration of repetitions. Making all such data explicit might rapidly reach a critical data size even for the most efficient commercial DBMS, with dramatic consequences especially in term of physical disk I/O's.

## 3   Quasi-Periodic Granularities

We rely on the definition of granularity taken from the temporal database glossary [12] and adapted in [18], [19]. Such definitions are the basis for our treatment of periodic data.

**Definition 1.** *A granularity G quasi-periodically groups into a granularity H if:*

*(i)  G groups into H, and*
*(ii)  there exists a finite set of finite intervals $E_1, \ldots, E_z$ and positive integers n and m, where n is less than the minimum of the number of granules of H, such that for all $i \in Z$, if $H(i) = \bigcup_{r=0}^{k} G(j_r)$ and $H(i + n) \neq \emptyset$, and $i+n < min(E)$, where E is the closest existing exception after $H(i)$ (if such exception exists; otherwise $E = max(k|H(k) \neq \emptyset)$, then $H(i + n) = \bigcup_{r=0}^{k} G(j_r + m)$.*

The definition of periodic and quasi-periodic granularities in terms of a bottom granularity is:

**Definition 2.** *A periodic granularity is a granularity periodic with respect to the bottom granularity.*

In Figure 1, we graphically show the basic notions underlying our representation. The 'Explicit Representation' part of the figure shows what a periodic data actually is a sequence of time periods, which repeat regularly on the time line. Usually repetitions are bounded and only a part of the whole timeline, which we term *Frame Time* is considered. Given some periodic activity, a *range* query may ask whether such an activity has to be performed within a specific time period $Q$ or not. If we have an explicit representation of all the time periods for the periodic activity, the range query can be easily answered by looking whether there is an intersection between explicitly stored data and query window $Q$.

Considering the implicit representation, one can isolate the pattern of periods that repeat regularly in time. Only one such pattern must be explicitly stored, with the intended meaning that such a pattern repeats regularly every $P$ during the *Frame Time*. It is obvious that the the implicit representation is space-effective, but answering the above query on the basis of the implicit representation is quite complex, since standard checks for intersections between the data and query window cannot be applied. Such a challenging task will be addressed in this work. Analogous challenging problems have to be faced when coping with the other kind of queries (e.g., queries based on algebraic relational operators) on an implicit representation of periodic data.

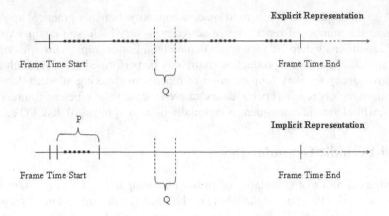

**Fig. 1.** Representations of Periodic data

## 3.1 An Implicit Representation of Quasi-Periodic Granularities

As a working example, let us consider the following user-defined granularity. Let us suppose that, in the year 2012, starting from Monday January 9th and ending on Sunday December 23rd, the 'working shift' for one employee in company is from 08:00 to 12:00, and from 14:00 to 18:00 each day from Monday to Friday, and from 08:00 to 12:00 on Saturday, let us call such a granularity 'WS'. In addition, let us suppose that person also works on Saturday evening from 14:00 to 18:00 in two specific days, say on January $14^{th}$ and $21^{st}$, let us call 'WS+' the granularity WS with such an addition.

We have that, by definition, each quasi-periodic granularity groups quasi-periodically with respect to a bottom granularity, we will assume that hours (henceforth 'HR') is the bottom granularity (but our approach is independent of such a choice). Let us consider the user-defined granularity WS. First of all, notice that each granule in WS is composed by a set of granules of HR. For instance, the first granule WS(0) of WS is the union of the granules HR(176), HR(177), HR(178), and HR(179), which represent working shift first Monday from 8:00 to 12:00, and HR(182), HR(183), HR(184), and HR(185), which represents working shift on first Monday from 14:00 - 18:00 (assuming to denote with HR(0) the first hour of January $1^{st}$, 2012). An instance of the 'grouping patterns' is:

$$WS(0) = \{HR(176) \cup HR(177) \cup HR(178) \cup HR(179) \cup$$
$$HR(182) \cup HR(183) \cup HR(184) \cup HR(185)\}$$
$$\dots$$
$$WS(5) = \{HR(296) \cup HR(297) \cup HR(298) \cup HR(299)\}$$

This pattern repeats every 168 granules of HR (each week), and 'frame time' the period of time spanning from the first and last non-empty granules in $H$ (8567 HR). However, such an initial representation can be simplified:

(a) If all granularities are expressed in terms of the bottom granularity, the bottom granularity may be left implicit. In our example, the representation of the first item may be simplified, stating, e.g., $\mathrm{WS}(0) = \{176, 177, 178, 179, 182, 183, 184, 185\}$ and so on.

(b) Second, contiguous granules of the bottom granularity can be more compactly represented as (convex) periods. For instance, in our example, the representation of the first item may be simplified, stating, e.g., $\mathrm{WS}(0) = \{[176, 179], [182, 185]\}$ and so on.

(c) Third, given a periodic granularity any 'periodic pattern' can be chosen in order to represent it. However, if we adopt the convention that the chosen 'periodic pattern' is the first one starting at the granule '0' of the bottom granularity ($\mathrm{HR}(0)$ in our example), then also the indexes of the granules of the quasi-periodic granularity may be left implicit.

Notice that simplifications (a) and (c) together are very important, since they allow us to keep all the indexes implicit in the representation, making it much more compact and easy. Finally, in the case of quasi-periodic granularities, a further component must be considered into the representation, namely, the list of the non-periodic repetitions. In other words, besides time periods which repeats periodically in time, we also optionally add a set of periods that do not follow such periodic pattern.

We propose the following implicit representation of a quasi-periodic granularity $G$. A quasi-periodic granularity $G$ is represented by a quadruple:

$$G = \langle\, P,\ I_P,\ I_A,\ FT\, \rangle$$

where $P$ is an integer representing the duration of the periodic pattern; $I_P$ is the set of the convex periods in the first 'periodic pattern' of the bottom granularity; $I_A$ is the set of the convex periods constituting the aperiodic part; $FT$ is a period constituting the frame time. In turn, a period having as first granule the bottom granule $B(i)$ and as last granule the bottom granule $B(j)$ is represented by '$[i, j]$'.

Working shift example, WS and WS+ are represented in our formalism as follows:

$$\mathrm{WS} = \langle 168, \{[176, 179], [182, 185], [200, 203], \ldots, [296, 299]\},$$
$$\{\}, [168, 8567]\rangle$$
$$\mathrm{WS}+ = \langle 168, \{[176, 179], [182, 185], [200, 203], \ldots, [296, 299]\},$$
$$\{[2029, 2033], [2197, 2201]\}, [168, 8567]\rangle$$

### 3.2 Data Model for Implicit Periodic Data

The abstract representation of quasi-periodic granularities presented in previous subsection is the basis to define our extended model, coping with quasi-periodic data in a relational environment. However, several aspects need to be investigated, and choices done. For instance, we could associate an unique identifier to each user-defined quasi-periodic granularity, and extend the data model with just an additional attribute, used in order to pair each tuple with the identifier of the granularity. One (or more) dedicated tables could then be used in order to associate with each identifier the 'implicit' description of the granularity they denote. Our model is also based on the two considerations:

- Given a periodic tuple, its 'frame time' can be interpreted, roughly speaking, as an approximation of its 'valid time', in the sense that it contains all the time periods in which the tuple holds.
- Given a quasi-periodic tuple, the 'non-periodic' part of its granularity can be simply represented by a set of time periods, i.e., of 'standard' valid times in the 'consensus' approach.

We can now define our new data model. Given any schema $R = (A_1, \ldots, A_n)$ (where $A_1, \ldots, A_n$ are standard non-temporal attributes), a periodic relation $r$ (in our example we termed Activity) is a relation defined over the schema

$$R^P = (A_1, \ldots, A_n \mid VT_S, VT_E, Per, PatID)$$

where:

- $VT_S$ is a timestamp representing the starting point of the 'frame time'
- $VT_E$ is a timestamp representing the ending point of the 'frame time'
- $Per$ is an interval, representing the duration of the repetition pattern
- $PatID$ is an identifier, denoting a periodic pattern

In addition, in order to code periodic patterns, an additional dedicated relation (a valid-time relation, in the sense of TSQL2) is needed (called Pattern relation). In Table 1 'NULL' in column $Per$ indicates that the tuples represent aperiodic component as there is no periodic repetitions and also it does not refer to periodic pattern.

The Pattern is a relation over the schema $(PatID, Start, End)$, where $PatID$ is an attribute containing identifiers denoting periodic patterns, while $Start$ and $End$ are temporal attributes denoting the starting and the ending points of the periods in the periodic pattern.

**Table 1.** Activity periodic relation – Implicit model

| ActID | Act | ActorID | VT_S | VT_E | Per | PatID |
|-------|-----|---------|------|------|------|-------|
| 1 | A | John | 168 | 8567 | 168 | P1 |
| 2 | A | John | 2029 | 2033 | Null | Null |
| 3 | A | John | 2197 | 2201 | Null | Null |

**Table 2.** Pattern relation – Implicit model

| PatID | Start | End |
|-------|-------|-----|
| P1 | 176 | 179 |
| ... | ... | ... |
| P1 | 296 | 299 |

## 4  Range Queries about Implicit Periodic Data

In the context of periodic data represented in implicit way, we identified three different types or range queries, depending on whether:

– One is interested in the non-temporal part of the tuples only. For example what activities have to be performed from May 1st to July 31, 2012, and by which employees? To this type of queries we will refer as 'atemporal range queries' henceforth.
– One is interested in the tuples and in their explicit time. Such as what activities have to be performed from May 1st to July 31, 2012, by which employees; for each of them, list all the periods they have to be performed - within the query period; We will refer as 'explicit temporal range queries'.
– One is interested in the tuples and in their implicit time. What activities have to be performed from May 1st to July 31, 2012, by which employees, and when –implicit time? This is 'implicit temporal range queries'.

**Input**: ($r$: periodic relation; $P_Q$:period)
**Output**: $r'$: periodic relation
$r' \leftarrow \oslash$;
$r_{aper} \leftarrow$ Select * From $r$ Where $Per$=NULL;
$r_{per} \leftarrow$ Select * From $r$ Where $Per \neq$ NULL;
**foreach** *tuple* $t \in r_{aper}$ **do**
    **if** ( *NOT* ($t[Atemp] \in r'$)) **then**
        Let $P_t$ be the period $[t[VT_S], t[VT_E]]$;
        **if** *(Intersects($P_t$, $P_Q$))* **then**
            Let $I_S$ and $I_E$ be the starting and the ending points of the intersection $P_t \cap P_Q$;
            $r' \leftarrow r' \cup \{(t[Atemp]|I_S, I_E, NULL, NULL)\}$;
        **end**
    **end**
**end**
**foreach** *tuple* $t \in r_{per}$ **do**
    **if** *Check_Periodic_Intersection(t, $P_Q$)* **then**
        Let $I_S$ and $I_E$ the starting and the ending points of the intersection $[t[VT_S], t[VT_E]] \cap P_Q$;
        $r' \leftarrow r' \cup \{t[Atemp]|I_S, I_E, t[Per], t[PatID]\}$;
    **end**
**end**
return ($r'$);

**Algorithm 1.** Implicit temporal range queries

Algorithms for 'atemporal range queries' can be found in [18] and 'explicit temporal range queries' on implicit model is straight forward by making explicit answer from 'implicit temporal range queries'. In following subsection we provide algorithm for implicit temporal range queries.

## 4.1   Implicit Temporal Range Queries

We describe at an abstract level the algorithm for implicit temporal range queries. The Algorithm 1 takes as input a implicit periodic relation $r$ and a query period $P_Q$ and gives as output a periodic relation containing all the tuples occurring during $P_Q$, answer is provided in implicit representation. Only parts of the valid times that intersect with the query period $P_Q$ are reported in output. To perform implicit temporal range queries there is need to check for intersection and Algorithm 2 *Check_Periodic_Intersection* has as input a periodic tuple (*PerID* $\neq$ *NULL*), and the query period $P_Q$, and checks whether there is an intersection.

**Input**: (*t*: periodic tuple; $P_Q$:period)
**Output**: boolean
let $P_t$ be the period $[t[VT_S], t[VT_E]]$;
**if** *Intersects*($P_t$, $P_Q$) **then**
    Let $P' \leftarrow P_t \cap P_Q$;
    **if** *duration*($P'$) $\geq t[Per]$ **then**
        return (TRUE);
    **else**
        $Pset \leftarrow get\_periods(t[PatID], \texttt{Periodicity})$;
        **if**
        $G\_Intersects(circular\_module(P', t[Per]), circular\_module(Pset, t[Per]))$
        **then**
            return (TRUE);
        **else**
            return (FALSE);
        **end**
    **end**
**end**

**Algorithm 2.** Check Periodic Intersection

There is also required to perform circular module, provided in Algorithm 3, to identify which patterns might intersect with query interval.

**Input**: *Set* : *set_of_periods*, *n* : *int*
**Output**: *res*: *set_of_periods*
$res \leftarrow \{\}$;
**foreach** $[s, e] \in Set$ **do**
    $s' \leftarrow s \ mod \ n$;
    $e' \leftarrow e \ mod \ n$;
    **if** $(s' \leq e')$ **then**
        $res \leftarrow res \cup \{[s', e']\}$;
    **else**
        $res \leftarrow res \cup \{[0, e'], [s', n-1]\}$;
    **end**
**end**
return (*res*)

**Algorithm 3.** Circular module

If $n = 7$ and $Set\_of\_periods = \{[11,13]\}$ then $circular\_module(Set\_of\_periods, n) = \{[4,6]\}$, while if $Set\_of\_period = [11,16]$ we have that $s' = 11 \bmod 7 = 4$ and $e' = 16 \bmod 7 = 2$ so that $circular\_module(Set\_of\_periods, n) = \{[0,2],[4,6]\}$.

## 5   Indexing Periodic Data

Different access methods have been presented in literature and some of them have been recommended for handling temporal data. Because we intend to index temporal data within commercial relational DBMS, we can only considered ones which can exploit existing structures such as the $B^+$tree and do not require any modification to the database kernel. Several such structures have been mentioned in literature. Such as methods which can map one dimensional ranges to one dimensional points, MAP21 [20], or manage the intervals by two relational indexes RI-tree method [21]), or partition the space and utilize the virtual structure TD-tree [22], [23].

The Triangular Decomposition Tree (TD-tree) is efficient access method for temporal data. In contrast to previously proposed access methods for temporal data this method can efficiently answer a wide range of query types, including point queries, intersection queries, and all nontrivial interval relationships queries, using a single algorithm without dedicated query transformations. It also can be built within commercial relational database system as it uses only builtin functionalities within the SQL:1999 standard and therefore no modification to the database kernel is required. The TD-tree is a space partitioning access method. The basic idea is to manage the temporal intervals by a virtual index structure [22].

Since we intend to index temporal periodic data (both implicit and explicit) and periodic data are temporal in nature, we will utilize methods for indexing temporal data. It has been shown in the literature that the TD-tree [22] has the best performance considering the Physical disk I/O and the query response time and at the same time can be employed within commercial RDBMS, we decided to employ the TD-tree in our implementation. Specifically, we index the $VT_S$ and $VT_E$ temporal attributes of periodic relations.

However it is worth stressing that any method, previously proposed in literature for indexing intervals, can deal only with explicit times. As a consequence, they can not be applied directly to any 'implicit' approach to periodic data, since in such approaches periods are only implicitly described. As a matter of facts, one of the main challenges of our approach was that of providing a suitable querying approach, which in conjunction with indexing methodology for indexing intervals can efficiently manage implicit periodic data.

## 6   Empirical Testing

In order to show the practical relevance of our implicit approach to efficiently manage periodic data, we have performed an extensive experimental evaluation. In particular, we have compared the performance of our approach with respect to the standard explicit one. We used Oracle built-in methods for statistics collection, analytic SQL functions, and the PL/SQL procedural runtime environment. We compared our results considering the space usage, physical I/O, CPU usage, and query response time. In order to carry on the experiments, the same periodic activities concerning sample hospital patients have been represented both in the *implicit* and *explicit* model and we performed atemporal range queries as well as detail experiments to investigate the explicit/implicit ratio when implicit approach starts to perform better. The experiments have shown that already at explicit/implicit ratio of 5 our implicit method starts to gain significant advantages as regards to both disk I/O and query response time. It is important to mention that the ratio in example where we simulated small hospital was 157. More details of other experiments can be found in [18].

## 7    Correctness and Completeness of the Method

In order to prove that the range query answering algorithms, operating on the extended temporal model, are correct and complete, we show that, given any range query and a database of relations expressed in our *implicit* data model, our algorithms provide *all and only* the results that are provided by asking the same queries on an *explicit* representation of the same data.

In the proof we will use the notation above, plus the following conventions: $Q_{Imp}$ is the set of tuples returned by algorithm 1 given an atemporal range query $Q$; similarly $Q_{Exp}$ is the set of tuples from $Exp$ that satisfy the temporal range query $Q$. To avoid confusion with interval we will use the dot notation for temporal attributes of tuples apart for the non temporal part, for which we continue to use $t[Atemp]$. In addition to improve readability we will use $c\_module$ instead of *circular_module*.

Let us at first define the relationships between the implicit and the explicit models. Let $Imp$ be the join of a relation $r$ (`Activity`) and the table `Pattern` in the implicit model, and $Exp$ be the corresponding relation in the explicit model. Then, given $Imp$, $Exp$ is defined as follows:

$$Exp = \{(t[Atemp], t.VT_s, t.VT_e) : t \in r \wedge t.per = NULL\} \cup$$
$$\{(t[Atemp], i^-, i^+) : \exists n \in \mathbb{N},$$
$$t.VT_s \leq n * t.per + (t.start - t.off) \leq t.VT_e\}$$

where $\forall n \in \mathbb{N} : \lfloor t.VT_s/t.per \rfloor \leq n \leq \lfloor t.VT_e/t.per \rfloor$

$$i^- = \max\{t.VT_s, n * t.per + (t.start - t.off)\}$$
$$i^+ = \min\{t.VT_e, n * t.per + (t.end - t.off)\}$$

and $t.off = \lfloor t.start/t.per \rfloor * t.per$.

*Soundness.* To prove that algorithm 1 is sound we have to prove that if $t \in Q_{Imp}$, then $\exists s \in Exp$, such that $t[Atemp] = s[Atemp]$ and $s \in Q_{Exp}$. Given algorithm 1, we have to consider two mutually exclusive cases: (1) $t \in r_{aper}$ and (2) $t \in r_{per}$.

Case 1: $t.per = NULL$, and by construction the tuple $s = (t[Atemp], t.VT_s, t.VT_e)$ is in $Exp$, and $[t.VT_s, t.VT_e] \cap P_Q \neq \emptyset$, thus $s \in Q_{Exp}$ such that $t[Atemp] = s[Atemp]$.

Case 2: We have two mutually exclusive subcases: (i) $duration(P') > t.per$, and (ii) $duration(P') \leq t.per$.

For (i): Let $n = \lfloor P'^- /t.per \rfloor$, $n+m = \lfloor P'^+ /t.per \rfloor$, $m \geq 1$ and $o = \lfloor t.VT_s/t.per \rfloor$. Let $pat_j = (start_j, end_j)$ be any pattern in the table Pattern. Let

$$j^- = (n-o)*t.per + start_j - t.off,$$
$$j^+ = \min\{(n-o)*t.per + (end_j - t.off), t.VT_e\}.$$

If $j^- \geq P'^-$, then we consider the tuple $s = (t[Atemp], j^-, j^+)$. Now $s \in Exp$ since $t.VT_s \leq P'^- \leq j^-$ and $[j^-, j^+] < t.per$, so $t.VT_e \geq P'^+ \geq j^+$, and for essentially the same reasons, $s \in Q_{Exp}$. If $j^- < P'^-$, then we can repeat the same argument as the previous case, but this time we set

$$j^- = (n+1-o)*t.per + (start_j - t.off),$$
$$j^+ = \min\{(n+1-o)*t.per + (end_j - t.off), t.VT_e\}.$$

For (ii): Let $pat_j = [start_j, end_j]$ be a pattern in the table Pattern such that $c\_module$ $(P', t.per) \cap c\_module(pat_j, t.per) \neq \emptyset$. Let $n^{\pm} = \lfloor P'^{\pm}/t.per \rfloor$, $r^- = \lfloor start_j/t.per \rfloor$ and $r^+ = \lfloor end_j/t.per \rfloor$. Then

$$c\_module(P', t.per) = \begin{cases} [q^-, q^+] & \text{if } n^+ = n^- \\ [0, q^+], [q^-, t.per - 1] & \text{otherwise} \end{cases}$$

$$c\_module(pat_j, t.per) = \begin{cases} [o^-, o^+] & \text{if } r^+ = r^- \\ [0, o^+], [o^-, t.per - 1] & \text{otherwise} \end{cases}$$

Since the intersection of $c\_module(P', t.per)$ and $c\_module(pat_j, t.per)$ is not empty, we have the following cases (where $E = t.per - 1$):

1. $[q^-, q^+] \cap [o^-, o^+] \neq \emptyset$ or
2. $[q^-, q^+] \cap [0, o^+] \neq \emptyset$ or
3. $[q^-, q^+] \cap [o^-, E] \neq \emptyset$ or
4. $[0, q^+] \cap [o^-, o^+] \neq \emptyset$ or
5. $[0, q^+] \cap [0, o^+] \neq \emptyset$ or
6. $[0, q^+] \cap [o^-, E] \neq \emptyset$ or
7. $[q^-, E] \cap [o^-, o^+] \neq \emptyset$ or
8. $[q^-, E] \cap [0, o^+] \neq \emptyset$ or
9. $[q^-, E] \cap [o^-, E] \neq \emptyset$.

Let us consider the tuple $e = (t[Aper], j^-, j^+)$ where, $n$ and $r$ can be set to either $n^+$ or $n^-$ and $r^+$ or $r^-$ depending on the cases above:

$$j^- = \max\{(n-r)*t.per + t.start_j - t.off, t.VT_s\}$$
$$j^+ = \min\{(n-r)*t.per + t.end_j - t.off, t.VT_e\}$$

It is immediate to verify that $e \in Exp$ (since the tuple in Imp satisfies the query $n - r \geq 0$ for the appropriate assignment of $n$ and $r$). Cases 2–9 are similar to the case above. What we have to do is to consider the appropriate values of $n$, where we have to consider $n$ or $n + 1$ depending whether the circular module returns one or two intervals.

*Completeness.* To prove completeness we have to prove that if $t \in Q_{Exp}$, then $\exists s \in Imp$, such that $t[Atemp] = s[Atemp]$ and $s \in Q_{Imp}$. We prove the property by contradiction. Suppose that the property does not hold. This means that for every tuple in $t \in Q_{Exp}$ there is not tuple in $Imp$ such that the tuple in $Imp$ generated the tuple $t$, but the tuple in $Imp$ does not satisfy the query.

Clearly the tuple $s$ in $Imp$ cannot be an aperiodic tuple, otherwise we obtain immediately a contradiction, since the $s.VT_s = t.VT_s$ and $s.VT_e = t.VT_e$, and the conditions for the explicit query and the condition in the algorithm for the implicit query are the same. Similarly we have that $s.per < P'$; otherwise $s$ satisfies the condition to be included in $Q_{Imp}$, contrary to the assumption.

## 8  Conclusions

In this work, we presented a new approach to cope efficiently with periodic data in relational databases. Specifically:

- we have presented an 'implicit' relational data model for user-defined periodic data, which is based on the 'consensus' definition of granularity in the temporal database glossary and its extension to cover periodic granularities, and it is a consistent extension of TSQL2's.
- we have taken into account range queries, providing sound and complete query answering algorithms for them;
- we have extended Codd's algebraic operators of *Cartesian product*, *Union*, *Projection*, *nontemporal selection*, and *Difference*, in order to provide a complete query language coping with implicit periodic data; Such operators are sound and complete, and are a consistent extension of BCDM (and TSQL2) algebra. Details can be found [18].
- in an extensive experimentation of our model and methodology, we showed that our 'implicit' approach overcomes the performance of traditional 'explicit' approaches both in terms of space and disk I/O's, and in terms of query response time. Moreover, we have also analysed to what extent our implicit approach is advantageous, depending on the 'explicit/implicit' ratio.

**Acknowledgements.**  The author is very grateful to Richard T. Snodgrass for many in-dept and inspiring comments on a preliminary version of this work. Also, author would like to acknowledge Paolo Terenziani and Guido Governatori for contributions on algorithms and proofs.

## References

1. Soo, M., Snodgrass, R.: Multiple calendar support for conventional database management systems. In: Proc. Int'l Workshop on an Infrastructure for Temporal Databases (1993)
2. Terenziani, P.: Temporal periodicity. In: Liu, L., Özsu, M.T. (eds.) Encyclopedia of Database Systems. Springer (2009)

3. Chomicki, J., Imielinski, T.: Finite representation of infinite query answers. ACM ToDS 18(2), 181–223 (1993)
4. Kabanza, F., Stevenne, J.M., Wolper, P.: Handling infinite temporal data. Journal of Computer and System Sciences 51, 3–17 (1995)
5. Leban, B., McDonald, D., Forster, D.: A representation for collections of temporal intervals. In: Procs. of AAAI 1986, pp. 367–371 (1986)
6. Niezette, M., Stevenne, J.M.: An efficient symbolic representation of periodic time. In: Procs. of CIKM (1992)
7. Bettini, C., De Sibi, R.: Symbolic representation of user-defined time granularities. Annals of Mathematics and Artificial Intelligence 30(1-4), 53–92 (2000)
8. Ning, P., Wang, X., Jajodia, S.: An algebraic representation of calendars. Annals of Mathematics and Artificial Intelligence 36(1-2), 5–38 (2002)
9. Terenziani, P.: Symbolic user-defined periodicity in temporal relational databases. IEEE TKDE 15(2), 489–509 (2003)
10. Egidi, L., Terenziani, P.: A flexible approach to user-defined symbolic granularities in temporal databases. In: Procs. of ACM SAC 2005, pp. 592–597 (2005)
11. Egidi, L., Terenziani, P.: A mathematical framework for the semantics of symbolic languages representing periodic time. Annals of Mathematics and Artificial Intelligence 46(3), 317–347 (2006)
12. Bettini, C., Dyreson, C.E., Evans, W.S., Snodgrass, R.T., Wang, X.S.: A glossary of time granularity concepts. In: Etzion, O., Jajodia, S., Sripada, S. (eds.) Temporal Databases: Research and Practice. LNCS, vol. 1399, pp. 406–413. Springer, Heidelberg (1998)
13. Snodgrass, R.T., Ahn, I.: A Taxonomy of Time in Databases. In: Navathe, S.B. (ed.) Proceedings of the 1985 ACM SIGMOD International Conference on Management of Data, pp. 236–246. ACM Press (1985)
14. Snodgrass, R.T.: The TSQL2 Temporal Query Language. Kluwer Academic (1995)
15. Dyreson, C.E., Snodgrass, R.T., Freiman, M.: Efficiently Supporting Temporal Granularities in a DBMS. Technical Report TR 95/07 (1995)
16. Bettini, C., Jajodia, S., Wang, S.: Time Granularities in Databases, Data Mining, and Temporal Reasoning. Springer (2000)
17. Tsotras, V.J., Jensen, C.S., Snodgrass, R.T.: An Extensible Notation for Spatiotemporal Index Queries. ACM SIGMOD Record 27(1), 47–53 (1998)
18. Terenziani, P., Stantic, B., Bottrighi, A., Sattar, A.: An intensional approach for periodic data in relational databases. Journal of Intelligent Information Systems (2013), doi:10.1007/s10844-013-0245-8
19. Stantic, B., Terenziani, P., Governatori, G., Bottrighi, A., Sattar, A.: An implicit approach to deal with periodically-repeated medical data. Journal in Artificial Intelligence in Medicine 55(3), 149–162 (2012)
20. Nascimento, M.A., Dunham, M.H.: Indexing Valid Time Databases via $B^+$-Tree. IEEE Transactions on Knowledge and Data Engineering 11(6), 929–947 (1999)
21. Kriegel, H.P., Pötke, M., Seidl, T.: Managing intervals efficiently in object-relational databases. In: Proceedings of the 26th International Conference on Very Large Databases, pp. 407–418 (2000)
22. Stantic, B., Terry, J., Topor, R., Sattar, A.: Advanced indexing technique for temporal data. In: Computer Science and Information Systems - ComSIS, vol. 7, pp. 679–703 (2010)
23. Stantic, B., Terry, J., Topor, R., Sattar, A.: Indexing Temporal Data with Virtual Structure. In: Catania, B., Ivanović, M., Thalheim, B. (eds.) ADBIS 2010. LNCS, vol. 6295, pp. 591–594. Springer, Heidelberg (2010)

# Semantic Enrichment of Ontology Mappings: A Linguistic-Based Approach

Patrick Arnold and Erhard Rahm

University of Leipzig, Germany

**Abstract.** There are numerous approaches to match or align ontologies resulting into mappings specifying semantically corresponding ontology concepts. Most approaches focus on finding equality correspondences between concepts, although many concepts may not have a strict equality match in other ontologies. We present a new approach to determine more expressive ontology mappings supporting different kinds of correspondences such as equality, is-a and part-of relationships between ontologies. In contrast to previous approaches, we follow a so-called enrichment strategy that semantically refines the mappings determined with a state-of-the art match tool. The enrichment strategy employs several linguistic approaches to identify the additional kinds of correspondences. An initial evaluation shows promising results and confirms the viability of the proposed enrichment strategy.

## 1 Introduction

There are numerous approaches and tools for ontology matching or alignment, i.e., the automatic or semi-automatic identification of semantically corresponding or matching concepts in related ontologies [14], [2]. These approaches typically utilize different techniques exploiting the linguistic and structural similarity of concepts and their neighborhood or the similarity of concept instances. All determined correspondences between two ontologies build a so-called ontology mapping. Ontology mappings are useful for many tasks, e.g., to merge related ontologies or to support ontology evolution.

A restriction of most previous match approaches is that they focus on finding truly matching pairs of concepts so that each correspondence expresses an equality relationship between two concepts. This is a significant limitation, since a more expressive mapping should also include further kinds of correspondences, such as is-a or part-of relationships between concepts. Such more expressive or semantic mappings are generally beneficial and have been shown to substantially improve ontology merging [11]. The existing approaches have even problems with finding truly equivalent concepts, since similarity-based match approaches are inherently approximative, e.g., if one assumes a match and the concept names have a string similarity above some threshold. Hence, the correspondences often express only some "relatedness" between concepts that can reflect equality or some weaker (e.g. is-a) relationship.

B. Catania, G. Guerrini, and J. Pokorný (Eds.): ADBIS 2013, LNCS 8133, pp. 42–55, 2013.

For illustration, we have shown in Figure 1 (left) the result for matching two simple ontologies with the state-of-the-art match tool COMA 3.0 (the successor of COMA++) [1], [9]. Each line represents a correspondence between two concepts. The example shows that not all such correspondences represent equality relationships, e.g., *Action_Games - Games*.

We present a new approach to determine more expressive ontology mappings supporting different kinds of correspondences such as equality, is-a and part-of relationships between ontologies. There are already a few previous approaches to identify such mappings (see Section 2), but they are still far from perfection. They have in common that they try to directly identify the different kinds of relationships, typically with the help of dictionaries such as WordNet. By contrast, we propose a so-called enrichment strategy implementing a two-step approach leveraging the capabilities of state-of-the art match tools. In a first step we apply a common match tool to determine an initial ontology mapping with approximate equality correspondences. We then apply different linguistic approaches (including the use of dictionaries) to determine for each correspondence its most likely kind of relationship. In Figure 1 (right) we illustrate how the enrichment approach can improve the mapping by identifying several is-a and inverse is-a relationships. The two-step approach has the advantage that it can work in combination with different match tools for step 1, and that it has to process relatively compact mappings instead of evaluating a large search space as for 1-step semantic match approaches. As we will see in the evaluation, we can still achieve a high match effectiveness.

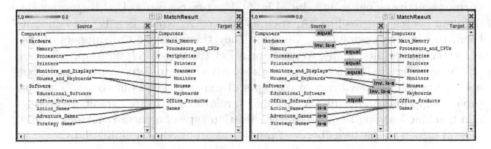

**Fig. 1.** Input (left) and output (right) of the Enrichment Engine

Our contributions are as follows:

- We propose a new two-step, enrichment approach to semantic ontology matching that refines the correspondences of an equality-based ontology mapping by different kinds of relationships between concepts (Section 3).
- We propose the combined use of four linguistic-based approaches for determining the relationship type of correspondences, including the use of background knowledge such as dictionaries (Section 4).
- We evaluate the new approach for different real-life test cases and demonstrate its high effectiveness (Section 5).

## 2   Related Work

Only a few tools and studies already try to determine different kinds of correspondences or relationships for ontology matching. S-Match [4][5] is one of the first such tools for "semantic ontology matching". They distinguish between equivalence, subset (is-a), overlap and mismatch correspondences and try to provide a relationship for any pair of concepts of two ontologies by utilizing standard match techniques and background knowledge from WordNet. Unfortunately, the result mappings tend to become very voluminous with many correspondences per concept while users are normally interested only in the most relevant ones. We tried to apply S-Match to our evaluation scenarios and report on the results in Section 5.

Taxomap [7] is an alignment tool developed for the geographic domain. It regards the correspondence types equivalence, less/more-general (is-a / inverse is-a) and is-close ("related"). It uses linguistic techniques and background sources such as WordNet. The linguistic strategies seem rather simple; if a term appears as a part in another term, a more-general relation is assumed which is not always the case. For example, in Figure 1 the mentioned rule holds for the correspondence between *Games* and *Action_Games* but not between *Monitors* and *Monitors_and_Displays*. In [12], the authors evaluated Taxomap for a mapping scenario with 162 correspondences and achieved only a low recall of 23 % and a good precision of 89 %.

Several further studies deal with the identification of semantic correspondence types without providing a complete tool or framework. An approach utilizing current search engines is introduced in [6]. For two concepts $A$, $B$ they generate different search queries like "A, such as B" or "A, which is a B" and submit them to a search engine (e.g., Google). They then analyze the snippets of the search engine results, if any, to verify or reject the tested relationship.

The approach in [13] uses the Swoogle search engine to detect correspondences and relationship types between concepts of many crawled ontologies. The approach supports equal, subset or mismatch relationships. [15] exploits reasoning and machine learning to determine the relation type of a correspondence, where several structural patterns between ontologies are used as training data.

## 3   Overview and Workflow

An ontology $O$ consists of a set of concepts $C$ and relationships $R$, where each $r \in R$ links two concepts $c_1, c_2 \in C$. In this paper, we assume that each relation in $O$ is either of type "is-a" or "part-of". We call a concept *root* if there is no other concept linking to it. A path from a root to a concept is called a *concept path*. We denote concept paths as follows: $root.concept_1.concept_2.(...).concept_n$. Each concept is referenced by its *label*.

A correspondence $C$ between two ontologies $O_1$ and $O_2$ consists of a source concept $C_S \in O_1$, a target concept $C_T \in O_2$, a relationship or correspondence type, and an optional confidence value between 0 and 1 expressing the computed

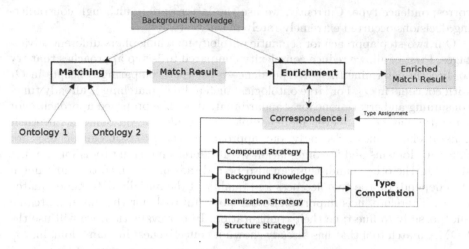

**Fig. 2.** Basic Workflow for Mapping Enrichment

likelihood of the correspondence. In this study, we consider six correspondence types: equal, is-a (or subset), inverse is-a, part-of (or composition), has-a (inverse of part-of) and related.

The basic workflow of our enrichment approach is shown in Figure 2. It consists of two steps: (initial) matching and enrichment. The matching step is performed using a common tool for ontology matching. It takes two ontologies and possibly additional background knowledge sources as input and computes an initial match result (a set of correspondences). This match result together with background knowledge sources is the input for the enrichment step. In the enrichment step, we currently apply four strategies (Compound, Background Knowledge, Itemization, Structure) for each input correspondence to determine the correspondence type. The four strategies will be described in the next section. Each strategy returns one type per correspondence or "undecided" if no type can be confirmed. From the individual results gained by the four strategies, we determine the final type and assign it to the correspondence (we apply the type which was most frequently returned by the strategies). Our final match result consists of the semantically enriched correspondences with an assigned relationship type.

The four strategies may return different correspondence types or only "undecided". In case that all strategies return "undecided", we will apply the equal type, because it is the default correspondence type from the initial ontology matching step. If there are different outcomes from the strategies (e.g., one strategy decided on equal, one on is-a and the other two returned undecided), we have different possibilities to decide on one type. We could either prioritize the relationship types or the strategies. For the latter we can use the experienced degree of effectiveness of the different strategies based on evaluation results. We could also interact with the user to request a manual decision about the correspondence type. Currently, we use the latter option, although contradicting decisions occurred extremely rarely in our tests.

Our two-step approach for semantic ontology matching offers different advantages. First of all, we reduce complexity compared to 1-step approaches that try to directly determine the correspondence type when comparing concepts in $O_1$ with concepts in $O_2$. For large ontologies, such a direct matching is already time-consuming and error-prone for standard matching. The proposed approaches for semantic matching are even more complex and could not yet demonstrate their general effectiveness. Secondly, our approach is generic as it can be used for different domains and in combination with different matching tools for the first step. On the other hand, this can also be a disadvantage since the enrichment step depends on the completeness and quality of the initially determined match result. Therefore, it is important to use powerful tools for the initial matching and possibly to fine-tune their configuration. In our evaluation, we will use the COMA match tool that has already shown its effectiveness in many domains [9].

## 4   Implemented Strategies

In the following we will introduce the four implemented strategies to determine the correspondence type. Table 1 gives an overview of the strategies and the relationship types they are able to detect. It can be seen that the Background Knowledge approach is especially valuable as it can help to detect all relationship types. All strategies are able to identify is-a correspondences.

**Table 1.** Supported correspondence types per strategy

| Strategy | equal | is-a | part-of | related |
|---|---|---|---|---|
| Compounding | | X | | |
| Background K. | X | X | X | X |
| Itemization | X | X | | |
| Structure | | X | X | |

### 4.1   Compound Strategy

In linguistics, a compound is a special word $W$ that consists of a head $W_H$ carrying the basic meaning of $W$, and a modifier $W_M$ that specifies $W_H$ [3]. In many cases, a compound thus expresses something more specific than its head, and is therefore a perfect candidate to discover an is-a relationship. For instance, a blackboard is a board or an apple tree is a tree. Such compounds are called *endocentric compounds*. There are also *exocentric compounds* that are not related with their head, such as buttercup, which is not a cup, or saw tooth, which is not a tooth. These compounds are of literal meaning (metaphors) or changed their spelling as the language evolved, and thus do not hold the is-a relation, or only to a very limited extent (e.g., airport, which is a port only in a broad sense). There is a third form of compounds, called *appositional* or *copulative* compounds, where the two words are at the same level, and the relation is rather more-general (inverse is-a) than more-specific, as in Bosnia-Herzegowina, which means both Bosnia and Herzegowina, or bitter-sweet, which means both bitter and sweet (not necessarily a "specific bitter" or a "specific sweet"). However, this type is quite rare.

In the following, let $A$, $B$ be the literals of two concepts of a correspondence. The Compound Strategy analyzes whether $B$ ends with $A$. If so, it seems likely that $B$ is a compound with head $A$, so that the relationship $B$ is-a $A$ (or $A$ inv. is-a $B$) is likely to hold. The Compound approach allows us to identify the three is-a correspondences shown in Figure 1 (right).

We added an additional rule to this simple approach: $B$ is only considered a compound to $A$ if $length(B) - length(A) \geq 3$, where $length(X)$ is the length of a string $X$. Thus, we expect the supposed compound to be at least 3 characters longer than the head it matches. This way, we are able to eliminate obviously wrong compound conclusions, like *stable* is a *table*, which we call *pseudo compounds*. The value of 3 is motivated by the observation that typical nouns or adjectives consist of at least 3 letters.

We also tested a variation of the approach where we extracted the modifier of a supposed compound and checked whether it appears in a word list or dictionary. This is expected to prevent pseudo compounds like *"nausea* is a *sea"*. We found that this approach does not improve our results, so we do not consider it further.

### 4.2 Background Knowledge Strategy

The use of background knowledge such as thesauri is a powerful approach since it can provide many linguistic relationships between words that are helpful to determine different relationships between concepts. Table 2 summarizes different linguistic relationships with typical examples as well as their associated kind of correspondence.

For preciseness, we briefly characterize the different linguistic relationships between words [10]. Two words $X \neq Y$ of a language are called *synonyms* if they refer to the same semantic concept, that is, if they are similar or equivalent in meaning. They are called *antonyms* if they are different in meaning (in the broad sense) or describe opposite or complementary things (in the narrow sense). $X$ is a *hypernym* of $Y$ if it describes something more general than $Y$. $Y$ is then called the *hyponym* of $X$. $X$ is a direct hypernym of $Y$ if there is no word $Z$ so that $Z$ is a hypernym of $Y$ and $X$ is a hypernym of $Z$.

**Table 2.** Typical linguistic and semantic relationships

| Linguistic relationship | Example | Corresp. type |
|---|---|---|
| Synonyms | river, stream | equal |
| Antonyms | valley, mountain | mismatch |
| Hypernyms | vehicle, car | inv. is-a |
| Hyponyms | apple, fruit | is-a |
| Holonyms | body, leg | has-a |
| Meronyms | roof, building | part-of |
| Cohyponyms | oak, maple, birch | related |

$X$ and $Y$ are cohyponyms if there is a concept $Z$ which is the direct hypernym of $X$ and $Y$. $X$ is a *holonym* of $Y$ if a "typical" $Y$ is usually part of a "typical" $X$. The expression "typical" is necessary to circumvent special cases, like cellar is part of house (there are houses without a cellar, and there are cellars without a house). $X$ is then called the *meronym* of $Y$.

In our current implementation of the Background Knowledge Strategy we use WordNet 3.0 [18] to determine the semantic relationship between two concepts of an input correspondence. We used the Java API for WordNet Searching (JAWS) [17] to retrieve information from WordNet, and implemented an interface to directly answer queries like "Is X a (direct) hypernym of Y?". We observed that WordNet is a very reliable source, which is able to detect many non-trivial relationships that cannot be detected by other strategies.

In case that an open compound $C$ matches a single word $W$, where $W$ is found in WordNet, yet $C$ is not, we gradually remove the modifiers of $C$ in order to detect the relationship. After each reduction, we check whether this form is in WordNet, and if not, proceed till we reach the head of $C$. For instance, we encountered correspondences such as ("US Vice President", "Person"), where "US Vice President" was not in the dictionary. However, "Vice President" is in the dictionary, so after the first modifier removal, WordNet could return the correct type (is-a).

### 4.3   Itemization Strategy

The itemization strategy is used if at least one of the two concepts in a correspondence is an itemization. We define an itemization as a list of items, where an item is a word or phrase that does not contain commas, slashes or the words "and" and "or". We call concepts containing only one item *simple concepts*, like "Red Wine", and concepts containing more than one item *complex concepts*, like "Champagne and Wine".

Itemizations need a different treatment than simple concepts, because they contain more information than a simple concept. Regarding itemizations also prevents us from detecting pseudo compounds, like "bikes and cars", which is not a specific form of cars, but something more general. Hence, there is in general an inverse is-a relationship between itemizations and the items they contain, e.g., between "cars and bikes" and cars resp. bikes. Two inv. is-a correspondences shown in Figure 1(right) are based on such itemizations (e.g., mouses and keyboards). Our itemization strategy is not restricted to such simple cases, but also checks whether there are is-a relationships between the items of an itemization. This is necessary to find out, for example, that "computers and laptops" is equivalent to a concept "computer", since laptop is just a subset of computer.

We now show how our approach determines the correspondence types between two concepts $C_1, C_2$ where at least one of the two concepts is an itemization with more than one item. Let $I_1$ be the item set of $C_1$ and $I_2$ the item set of $C_2$. Let $w_1, w_2$ be two words, with $w_1 \neq w_2$. Our approach works as follows:

1. In each set $I$ remove each $w_1 \in I$ which is a hyponym of $w_2 \in I$.
2. In each set $I$, replace a synonym pair ($w_1 \in I, w_2 \in I$) by $w_1$.
3. Remove each $w_1 \in I_1, w_2 \in I_2$ if there is a synonym pair ($w_1, w_2$).
4. Remove each $w_2 \in I_2$ which is a hyponym of $w_1 \in I_1$.

5. Determine the relation type:
   (a) If $I_1 = \emptyset, I_2 = \emptyset$: equal
   (b) If $I_1 = \emptyset, |I_2| \geq 1$: is-a
   (c) If $|I_1| \geq 1, I_2 = \emptyset$,: inverse is-a
   (d) If $|I_1| \geq 1, I_2 \geq 1$: undecided

The rationale behind this algorithm is that we remove items from the item sets as long as no information gets lost. Then we compare what is left in the two sets and come to the conclusions presented in step 5.

Let us consider the concept pair $C_1 =$ "books, ebooks, movies, films, cds" and $C_2 =$ "novels, cds". Our item sets are $I_1 = \{books, ebooks, movies, films, cds\}$, $I_2 = \{novels, cds\}$. First, we remove synonyms and hyponyms within each set, because this would cause no loss of information (steps 1+2). We remove $films$ in $I_1$ (because of the synonym $movies$) and $ebooks$ in $I_1$, because it is a hyponym of $books$. We have $I_1 = \{books, movies, cds\}, I_2 = \{novels, cds\}$. Now we remove synonym pairs between the two item sets, so we remove $cds$ in either set (step 3). Lastly, we remove a hyponym in $I_1$ if there is a hypernym in $I_2$ (step 4). We remove $novel$ in $I_2$, because it is a $book$. We have $I_1 = \{books, movies\}, I_2 = \emptyset$. Since $I_1$ still contains items, while $I_2$ is empty, we conclude that $I_1$ specifies something more general, i.e., it holds $C_1$ inverse is-a $C_2$.

If neither item set is empty, we return "undecided" because we cannot derive an equal or is-a relationship in this case.

## 4.4 Structure Strategy

The structure strategy takes the explicit structure of the ontologies into account. For a correspondence between concepts $Y$ and $Z$ we check whether we can derive a semantic relationship between a father concept $X$ of $Y$ and $Z$ (or vice versa). For an is-a relationship between $Y$ and $X$ we draw the following conclusions:

- $X$ equiv $Z \rightarrow Y$ is-a $Z$
- $X$ is-a $Z \rightarrow Y$ is-a $Z$

For a part-of relationship between $Y$ and $X$ we can analogously derive:

- $X$ equiv $Z \rightarrow Y$ part-of $Z$
- $X$ part-of $Z \rightarrow Y$ part-of $Z$

The approach obviously utilizes the semantics of the intra-ontology relationships to determine the correspondence types for pairs of concepts for which the semantic relationship cannot directly be determined.

For example, consider the correspondence ( vehicles.cars.convertibles, vehicles. cars). Let us assume that "convertibles" is not in the dictionary. No other strategy would trigger here. However, it can be seen that the leaf node "cars" of the second concept matches the father of the leaf node in the first concept. Since "convertibles" is a sub-concept of its father concept "cars", we can derive the is-a relationship for the correspondence.

To decide whether $X$ and $Z$ are equivalent or in an is-a or part-of relationship we exploit three methods: name equivalence (as in the example, cars = cars), WordNet and Compounding, thus exploiting the already implemented strategies.

### 4.5    Verification Step

We observed that the identification of is-a (subset) correspondences can fail when the concepts are differently organized within hierarchies in the input ontologies. Consider the correspondence ( "apparel.children_shoes", "clothing.children. shoes"). Based on the leaf concepts "children_shoes" and "shoes" both the Compound and Background strategies would suggest an "is-a" correspondence, because children shoes are obviously shoes. However, a closer look on the two paths reveals that both concepts are in fact equal.

To deal with such cases we implemented a verification step to post-process presumed is-a correspondences. For this purpose, we combine the leaf concept with the father concept and check whether the combination matches the opposite, unchanged leaf concept of a correspondence. For the above example, the combination of "children" and "shoes" on the target side leads to an equivalence match decision so that the is-a relationship is revoked.

This simple approach already leads to a significant improvement, but still needs extensions to deal with more complex situations such as:

1. The actual meaning is spread across multiple levels, like ( "children.footware. shoes", "children shoes").
2. The father node of a concept $A$ may not match the modifier of a corresponding concept $B$, like ( "kids.shoes", "children shoes"). Here, we would have to check whether the father node of $A$ ("kids") is a synonym to the modifier in $B$ ("children").

We plan to deal with such extensions in future work.

## 5    Evaluation

Evaluating our approach is more difficult than classic ontology matching techniques, because in many cases the true relationship cannot be unequivocally determined. For example, the correspondence type for (street, road) could be considered is-a (as suggested by WordNet) or equal. There might even be different relationships depending on the chosen domain or purpose of the ontologies. Consider the word *strawberry*, which biologically is not a berry but a nut. Thus, in a biological ontology, claiming *strawberry* is a *berry* would be wrong, whereas in a food ontology (of a supermarket etc.) it might be correct, since a customer would expect *strawberries* to be listed under the concept *berries*.

Another difficulty of evaluating the enrichment approach is its dependency on the match result of step 1 that might be incomplete and partially wrong. Furthermore, the evaluation may consider all relationship types or only the ones different from equality.

**Table 3.** Overview of the Gold Standards

| No | Domain | Lang. | #Corr. | equal | is-a | has-a | related |
|----|--------|-------|--------|-------|------|-------|---------|
| $G_1$ | Web Directories | DE | 340 | 278 | 52 | 5 | 5 |
| $G_2$ | Diseases | EN | 395 | 354 | 40 | 1 | 0 |
| $G_3$ | TM Taxonomies | EN | 762 | 70 | 692 | 0 | 0 |

For our evaluation, we use three test cases for which we manually determined the presumed perfect match result (Gold Standard) with semantic correspondence types. After the representation of these cases, we will first evaluate our approach under best-case conditions by providing it with all correspondences (without relationship types) of the Gold Standard. We then use the state-of-the art match tool (COMA 3.0) to determine an approximative initial match result for enrichment. We also report on how the 1-step semantic approach of S-match performed for our test cases. Finally, we summarize observations on the four individual strategies applied during enrichment.

### 5.1 Evaluation Scenarios

For our evaluation, we used three ontology matching scenarios of different domains and complexity with manually defined Gold Standards $G_1$, $G_2$, $G_3$. Table 3 provides key information about these standards, such as the domain, language (German, English) as well as the total number of correspondences and their distribution among the different semantic types (equal/is-a/has-a/related).

$G_1$ is a mapping between the Yahoo and Google Web taxonomies (product catalogs of shopping platforms), consisting of 340 correspondences. The ontologies are in German language, so WordNet has no impact on the result. This scenario contains many itemizations, which the other scenarios lack. $G_2$ is an extract of 395 correspondences between the diseases catalogs of Yahoo and dmoz. The ontologies are quite domain-specific (medical domain). $G_3$ is the largest mapping and based on the text mining (TM) taxonomies OpenCalais and AlchemyAPI. It was created and provided by SAP Research and consists of about 1,600 correspondences. In this scenario, half of the correspondences were of the type "related". We noticed significant problems with these correspondences, since many of them were actually of type is-a or has-a or even mismatches. We thus decided to ignore all "related" correspondences in $G_3$ leaving us with 762 correspondences of type equal, is-a and inverse is-a.

### 5.2 Tests against Gold Standards

We first evaluate our approach against the manually defined Gold Standards. The input was the Gold Standard containing only the correspondences, the output was the Gold Standard with a relation type annotation on each correspondence.

**Table 4.** Evaluation against benchmark

| (a) Non-equal types | | | (b) Equal-types | | | (c) Overall result | | |
| --- | --- | --- | --- | --- | --- | --- | --- | --- |
| r | p | f | r | p | f | r | p | f |
| $G_1$ .467 .690 .578 | | | $G_1$ .953 .889 .921 | | | $G_1$ .899 .864 .881 | | |
| $G_2$ .585 .800 .692 | | | $G_2$ .982 .947 .964 | | | $G_2$ .951 .941 .946 | | |
| $G_3$ .654 .977 .811 | | | $G_3$ .942 .213 .577 | | | $G_3$ .684 .675 .679 | | |

Table 4 shows the recall / precision and f-measure results for the three scenarios. Table **a)** only evaluates the non-equal types, which is of particular interest as such correspondences cannot be identified by standard matching approaches. It shows that the enrichment approach achieves a high f-measure of 58 to 81%, indicating a very good effectiveness. Precision was especially good (69 to 98%) while recall was somewhat limited.

Table **b)** only considers the equal-type. In the first and second scenario, the equal relationship dominates (about 90 % of all correspondences) and in these cases both recall and precision are very high. By contrast, in the third scenario we achieve only a poor precision and medium recall and f-measure. This is influenced by our policy that we denote the equal-type if no other type can be verified which is relatively often the case for the third scenario. We observe that the results for non-equal correspondences in **a)** and for equal correspondences in **b)** are inversely interrelated. For $G_3$, we achieved the best effectiveness for non-equal correspondences but the lowest for equal correspondences.

Finally, Table **c)** summarizes the overall results considering all correspondence types. F-measure values range from 68 to 95 %, indicating a high effectiveness of the proposed enrichment approach thus demonstrating its viability.

Running these tests took 2.18 s for $G_1$, 3.41 s for $G_2$ and 7.47 s for $G_3$. We thus observed an execution time of 5.5 ms per correspondences in the first scenario and approx. 10.0 ms per correspondence in the second and third scenario where WordNet was effectively used.

### 5.3   Tests with COMA 3.0

In the second set of experiments we apply the ontology matching tool COMA 3.0 [9] for the initial matching to determine a real, imperfect input mapping for the enrichment step. Since we had no ontologies for $G_3$, we could not generate a mapping for this scenario and were compelled to restrict the verification to the first two scenarios.

Table 5 shows the results for the COMA-based experiments. Table **a)** shows the quality results for the initial match result where we only checked the recall (completeness) and precision (correctness) of the correspondences generated by COMA (ignoring the correspondence type). The results are relatively low (f-measure between 61 and 72 %) underlining the hardness of the match scenarios.

**Table 5.** Evaluation with COMA match results

| (a) Quality of initial match result | | | (b) Results for non-equal types | | | (c) Results for equal type | | | (d) Overall results for enrichment | | |
|---|---|---|---|---|---|---|---|---|---|---|---|
| r | p | f | r | p | f | r | p | f | r | p | f |
| $G_1$ .702 .735 .718 | | | $G_1$ .145 .204 .174 | | | $G_1$ .762 .754 .758 | | | $G_1$ .669 .680 .674 | | |
| $G_2$ .673 .543 .608 | | | $G_2$ .365 .441 .403 | | | $G_2$ .703 .547 .625 | | | $G_2$ .670 .539 .604 | | |

Table 5b) shows the recall and precision for the detected non-equal types. We knew that the recall of b) must be lower than in table 4 a), because in the initial match result some typed correspondences were missing. Still, the recall for $G_1$ was surprisingly low. By analyzing the result we noticed that COMA aims at a high precision for equal results so that most non-equal results are not retained in its match results. In future work we plan to adjust the COMA settings to reduce this problem.

Table 5c) shows the results for the equal correspondences and eventually Table d) shows the overall results for all kinds of correspondences. Since most correspondences are of the equal type in $G_1$ and $G_2$, most correspondences were correctly typed, and therefore the result in d) is only slightly below the result in a).

## 5.4   Tests with S-Match

For this experiment we tried to apply the latest version of S-Match (s-match-20110422 from 2011) to our evaluation scenarios. It turned out to be very difficult comparing our approach with S-Match, because S-Match practically draws correspondences between each node pair since it also aims at determining mismatches.

For $G_1$, S-Match returned only 4 match correspondences, which have been all incorrect. This was very surprising despite the fact that the ontologies use German language. This is because there are several trivial correspondences with equal names which any matching tool should be able to detect.

By stark contrast, S-Match returned about 19,600 correspondences for scenario $G_2$. The root concept "Health" in the first ontology practically corresponded to any concept in the second ontology. This made it difficult to judge the result, because with 395 correspondences in the perfect result, the precision could be at most 2 %. There were 19,563 subset relations and 42 equivalence relations. According to our tests, 12 of the 42 equivalence relations were correct, but none of the subset relations. This would lead to a recall of 3.0 % and a precision being 0.06 %.

We saw that for the problem we addressed, to match taxonomies, S-Match is not a convenient tool. It may be helpful in smaller scenarios or apart from taxonomies, but in our tests it was of no use. While it only returned 4 correspondences in $G_1$, it returned 19,600 correspondences in $G_2$, which is hardly possible for any user to verify manually. Apparently S-Match does not rank correspondences to filter out only the most relevant ones.

## 5.5    Evaluating the Strategies

We also ran our test cases with each single strategy to reveal the strength and weaknesses of our strategies.

Compounding offers a good precision and practically works in all domains, even in different languages (Germanic Languages). Its recall is mostly limited because of the different possibilities how an is-a relation can be expressed. Background Knowledge (WordNet) turned out to be a very precise approach, allowing a precision close to 100 %. However, WordNet only works for the English language and has a limited recall, because of the limited vocabulary for specific domains. Addressing the recall problem would thus require the provision of additional dictionaries and thesauri. Itemization is able to derive the relation type between complex concepts, where the previous strategies invariably fail. However, itemization depends much on the Compound and WordNet strategy. In very complex concept names deriving the correct relation type is rather difficult, so both precision and recall are rather limited. Finally, the Structure Strategy is useful if all other strategies fail. It is able to slightly increase the recall and keep up the precision.

## 6    Conclusion and Outlook

We presented a new approach for semantic ontology matching that applies an enrichment step to extend correspondences determined with standard match approaches. We exploit linguistic knowledge in new ways to determine the semantic type of correspondences such as is-a and part-of relationships between ontologies. Knowing the intricacies and inconsistencies of natural languages, our approach delivered astonishingly good results in the three real benchmark scenarios. Even in the (German) scenario where background knowledge was practically of no help, we got a recall close to 50 %, and a considerably higher precision. We observed that our rather simple methods mostly achieve already a medium recall and good precision.

Our approach is largely generic and can deal with ontologies from different domains and even with different languages. The enrichment approach can reuse existing match tools, which is both an advantage but also a problem deserving further attention. Standard match tools only aim at finding equivalence correspondences so that many weaker correspondences may not be derivable from the initial match result. To reduce the problem we can use relaxed configuration settings for the initial matching or apply further enrichment strategies utilizing additional information from the ontology (as we have already started with the Structure strategy). We also plan to use additional, domain-specific background sources for improved effectiveness.

Furthermore, we intend to investigate how linguistic methods can be exploited to detect initially falsely detected correspondences, e.g., by taking antonyms or disproved compounds into account. Although this step will not add semantics to the mapping, it is potentially able to increase its precision.

**Acknowledgment.** This study was partly funded by the European Commission through Project "LinkedDesign" (No. 284613 FoF-ICT-2011.7.4).

# References

1. Aumueller, D., Do, H.H., Massmann, S., Rahm, E.: Schema and ontology matching with COMA++. In: Proc. SIGMOD Conf. (2005)
2. Bellahsene, Z., Bonifati, A., Rahm, E. (eds.): Schema Matching and Mapping. Springer (2011)
3. Bisetto, A., Scalise, S.: Classification of Compounds. University of Bologna. In: The Oxford Handbook of Compounding, pp. 49–82. Oxford University Press (2009)
4. Giunchiglia, F., Shvaiko, P., Yatskevich, M.: S-Match: An Algorithm and an Implementation of Semantic Matching. In: Bussler, C.J., Davies, J., Fensel, D., Studer, R. (eds.) ESWS 2004. LNCS, vol. 3053, pp. 61–75. Springer, Heidelberg (2004)
5. Giunchiglia, F., Autayeu, A., Pane, J.: S-Match: an open source framework for matching lightweight ontologies. Semantic Web 3(3), 307–317 (2012)
6. van Hage, W.R., Katrenko, S., Schreiber, G.: A Method to Combine Linguistic Ontology-Mapping Techniques. In: Gil, Y., Motta, E., Benjamins, V.R., Musen, M.A. (eds.) ISWC 2005. LNCS, vol. 3729, pp. 732–744. Springer, Heidelberg (2005)
7. Hamdi, F., Safar, B., Niraula, N.B., Reynaud, C.: TaxoMap alignment and refinement modules: Results for OAEI 2010. In: Proceedings of the ISWC Workshop, pp. 212–219 (2010)
8. Jiménez-Ruiz, E., Cuenca Grau, B.: LogMap: Logic-Based and Scalable Ontology Matching. In: Aroyo, L., Welty, C., Alani, H., Taylor, J., Bernstein, A., Kagal, L., Noy, N., Blomqvist, E. (eds.) ISWC 2011, Part I. LNCS, vol. 7031, pp. 273–288. Springer, Heidelberg (2011)
9. Massmann, S., Raunich, S., Aumueller, D., Arnold, P., Rahm, E.: Evolution of the COMA Match System. In: Proc. Sixth Intern. Workshop on Ontology Matching (2011)
10. Plag, I., Braun, M., Lappe, S., Schramm, M.: Introduction to English Linguistics, 2nd revised edn. Mounton de Gruyter (2007)
11. Raunich, S., Rahm, E.: ATOM: Automatic Target-driven Ontology Merging. In: Proc. Int. Conf. on Data Engineering (2011)
12. Reynaud, C., Safar, B.: Exploiting WordNet as Background Knowledge. In: Proc. Intern. ISWC 2007 Ontology Matching (OM 2007) Workshop (2007)
13. Sabou, M., d'Aquin, M., Motta, E.: Using the semantic web as background knowledge for ontology mapping. In: Proc. 1st Intern. Workshop on Ontology Matching (2006)
14. Shvaiko, P., Euzenat, J.: A Survey of Schema-based Matching Approaches. In: Spaccapietra, S. (ed.) Journal on Data Semantics IV. LNCS, vol. 3730, pp. 146–171. Springer, Heidelberg (2005)
15. Spiliopoulos, V., Vouros, G., Karkaletsis, V.: On the discovery of subsumption relations for the alignment of ontologies. Web Semantics: Science, Services and Agents on the World Wide Web 8, 69–88 (2010)
16. Zhang, X., Zhong, Q., Li, J., Xie, G., Li, H.: Rimom Results for OAEI 2008. In: Proc. 3rd Intern. Workshop on Ontology Matching OM 2008, pp. 182–189 (2008)
17. Java API for WordNet Searching (JAWS),
    http://lyle.smu.edu/~tspell/jaws/index.html
18. WordNet - A lexical database for English,
    http://wordnet.princeton.edu/wordnet/

# GEO-NASS: A Semantic Tagging Experience from Geographical Data on the Media*

Angel Luis Garrido, Maria G. Buey, Sergio Ilarri, and Eduardo Mena

IIS Department, University of Zaragoza, Zaragoza, Spain
{garrido,mgbuey,silarri,emena}@unizar.es

**Abstract.** From a documentary point of view, an important aspect when we are conducting a rigorous labeling is to consider the geographic locations related to each document. Although there exist tools and geographic databases, it is not easy to find an automated labeling system for multilingual texts specialized in this type of recognition and further adapted to a particular context.

This paper proposes a method that combines geographic location techniques with Natural Language Processing and statistical and semantic disambiguation tools to perform an appropriate labeling in a general way. The method can be configured and fine-tuned for a given context in order to optimize the results. The paper also details an experience of using the proposed method over a content management system in a real organization (a major Spanish newspaper). The experimental results obtained show an overall accuracy of around 80%, which shows the potential of the proposal.

**Keywords:** Geographic IR, gazetteer, semantic tagging, NLP, ontologies, text classification, media, news.

## 1 Introduction

The huge amount of digital information located in information systems in both public and private organizations is leading to a new problem: the difficulty to find accurate and relevant data among the vast amount of available data. The problem is aggravated daily because storage devices are becoming cheaper, the access to suitable document management software is becoming more common, and all these elements contribute to raise not only the total number of documents stored, but also its growth rate. In this situation, traditional indexers are not enough to provide the desired results when the search requires high precision and the volume of documents is large. This is why the interest in implementing expert labeling techniques for data catalogs increases every day, trying to to enable a situation where it is possible to find the desired information more easily.

The process of geotagging aims at extracting the places/locations where a text is framed. The complexity of this type of tagging depends on the tagging level

---

* This research work has been supported by the CICYT project TIN2010-21387-C02-02 and DGA-FSE.

B. Catania, G. Guerrini, and J. Pokorný (Eds.): ADBIS 2013, LNCS 8133, pp. 56–69, 2013.

to reach, the geography that must be covered, and the context of the text. One typical problem in this field is that we find names that are called like a location but are not. Therefore, we can identify three main problems in this field:

- *Geography - geography disambiguation*: it tries to find the exact location or geographic place that is being discussed in the text, by distinguishing among several locations with the same name.
- *Non geography - geography disambiguation*: it tries to detect correctly when the text is talking about a geographic place instead of about a different concept with the same name.
- *Selection of candidates*: when a certain location is mentioned in a text, references to other locations may also appear. However, not all the locations are always relevant, as they may not be related to the main story contained in the document. The objective is to establish a confidence threshold to decide if a tag is accepted or not.

In general, geolocation is a useful process commonly used in many information systems, but yet little known and rarely built in Content Management Systems (CMS). In a CMS it is typical to have indexing processes and search tools which can help to find quickly the documents that contain a particular word, but this kind of software has serious difficulties to face specific challenges when working with geographical issues. Therefore, integrating geolocation processes into a CMS can increase its capabilities of information management and enhance its usability from a documentary point of view.

There exist CMS that feature categorization professional tools like Drupal[1] or Athento[2], and other software products like OpenCalais[3], but it is usually not possible to customize such tools for a specific scenario. Therefore, working with those tools in very specialized multilingual environments may be impractical.

After analyzing different tools, we have not found any generic system able to automatically classify the information on a geographic basis that can be adapted to satisfy the particular specifications established by the organization that owns the documents, and of course, that takes into account all the typical difficulties for a geotagging system: to distinguish among places that share the same name, to identify a homonym of a place that is actually something else (a person, a verb, etc.) and to know how to geotag a text when it does not refer to explicit places.

Our proposal to solve this problem is to develope an automatic labeling system called GEO-NASS. The architecture of GEO-NASS includes different tools whose combination provides the desired results. These tools are a geographic database, Natural Language Processing (NLP) tools [1] and finally a disambiguation engine based on geographic, statistical and semantic techniques. The engine is designed to identify the location which is being told about in the case of homonyms, to detect whether a word is really a location or not, and finally to establish the

---

[1] http://drupal.org/
[2] http://www.athento.com/
[3] http://www.opencalais.com/

importance of each place/location related to the text for subsequent labeling. The system can also benefit from the context of the text to optimize its results. All of this is achieved through an open implementation which allows, on the one hand, adapting the geographical database to the context and, on the other hand, to fit the standard tagging system used in the specific organization considered. GEO-NASS has been tested with real data, specifically with about 10,000 news from one major Spanish newspaper. For comparison purposes, we considered news previously tagged manually by a professional documentation department of the mentioned newspaper. The experimental results obtained are very promising and show the interest of the proposal.

This paper is structured as follows. Section 2 briefly describes the state of the art. Section 3 explains the general architecture of our solution and presents the basic algorithmic details and some improvements introduced by using the context of the document and analyzing the semantics. Section 4 discusses the results of our experiments with real data. Finally, Section 5 provides our conclusions and some lines of future work.

## 2   Related Work

Although different techniques have been proposed to solve these problems of geotagging, we can find similar patterns in the most popular approaches [2]. They start applying a Named Entities Recognition (NER) and then they use a list of algorithms over these recognized entities. In this context, two important assumptions are considered: (1) that there is a single sense per discourse, that means that an ambiguous term is likely to mean only one of its senses when used multiple times, and (2) place names appearing in one context tend to show nearby locations. For example, if Florencia and Armenia appear in the same paragraph, then it is more likely that they indicate the two communities in Colombia rather than the larger and better known city in Italy and the country of Armenia, respectively.

Substantial effort has been applied to the most general task of Named Entity Recognition (NER) [3] concerned about in identifying proper names in a text. More recently, the specific task of determining which place is meant by a particular occurrence of a place name has been gaining attention. It requires general-world knowledge and cannot rely completely on information found in the text or even in a whole corpus. This general knowledge is provided in a gazetteer [4], which traditionally lists the names of all the places in an atlas. There are many of these free indexes that can be used for this purpose: Geonames, the NGA GEOnet Names Server, The World Gazetteer, the Alexandria Digital Gazetteer, etc.

Several disambiguation techniques have been studied for the task of geotagging. Some of them take into account the population of the location candidates as an important aspect to disambiguate places [2] or consider the context where the text is framed to establish a list of bonuses for certain regions [5]. Besides, many of these techniques usually construct an N-term window on the right and left of the entity considered to be a geographic term, as some words can contribute with a positive or negative modifier [6], or they try to find syntactic structures like "city, country" (e.g. "Sevilla, Spain") [7]. Other techniques focus on finding common points in the location hierarchy of each candidate that appears in the text and they consider if they share the same political father (province, community, country, etc.) because this is a relevant factor when a term is being disambiguated [8]. Moreover, ontologies have been also used in disambiguation algorithms for geotagging. They are formal and explicit specifications of shared conceptualizations [9] and, in recent years, they have been applied as substitutes or complements of gazetteers [10, 11] because they can provide a rich vocabulary of classes and relations that describe a particular area, in this case a geographical scope.

In order to build our proposal, we used the techniques we have found most interesting from this section and in addition we have incorporated new techniques. This is fully explained in the following sections.

## 3   Architecture and Methodology

The development of this system has been carried out based on the NASS System. *NASS* is the acronym of *News Annotation Semantic System*, a software designed to obtain tags from a particular thesaurus [12] by using semantic tools and information extraction technologies [13].

In order to do its work, NASS uses statistics, NLP tools, Support Vector Machines (SVM) [14], and ontologies. The process is as follow: NASS combines NLP with statistical tools to obtain keywords and performs a filtering process of texts through SVM and a detailed labeling using ontologies [15].

In this work, we have expanded the NASS system providing it with a new option: geotagging. NASS is a tool that supports the inclusion of independent tagging modules, so we have encapsulated the new geotagging functionality into a separate unit that is incorporated into the workflow of NASS. We have called GEO-NASS to the resulting module, which we detail below. We consider that there is an appropriate gazetteer for all the documents to be labeled, and associated with certain places in the gazetteer we have several items belonging to the thesaurus with which we intend to label texts. The architecture of GEO-NASS and the NASS integration can be appreciated in Figure 1. The engine of GEO-NASS uses the information provided from NASS (lemmas, keywords, and named entities), a database and a gazetteer to find possible locations, and also a set of its own context resources. With this information GEO-NASS finds the most probable locations the text is talking about, and extracts appropriate tags. Then the whole system merges the tags provided by NASS and the tags provided by GEO-NASS in one set.

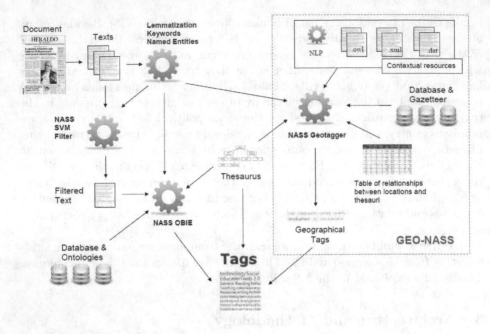

**Fig. 1.** General architecture of GEO-NASS and its integration with NASS

## 3.1  Basic GEO-NASS

As noted in Section 2, the problem of geographic information retrieval from a text
has been widely studied, so to implement GEO-NASS we have taken advantage
of the existing previous works.

The main functionalities considered in the initial version of GEO-NASS are:

- *Extraction of locations.* The aim is to extract from the text a list of proper
  nouns that correspond to a geographical location. For that purpose, our algorithm uses NASS functionalities: the text is analyzed by NASS using a
  Natural Language Processing software, and one of the tasks performed by
  this software is precisely to identify named entities [3]. Such entities may correspond to names of people, places, organizations, etc. Those named entities
  belonging to the gazetteer are candidates to be identified as places that must
  be labeled in the text. There are many gazetteers that can be used for this
  purpose. In our case, we have used *Geonames*[4] because it is an open source
  implementation, robust, comprehensive and frequently updated. We agree
  with other studies on the importance of a good adaptation of the gazetteer
  to the context [16].
- *Geographic disambiguation.* We will often find names of places with several
  entries in the gazetteer. For example, we can find in the gazetteer almost 20

---

[4] http://www.geonames.org/

different entries for cities and places called "Madrid". GEO-NASS assigns more weight to those locations with a common parent in the same document [17]. Thus, if there exist five cities that could either be from Italy or from different countries, then it is more likely that the text concerns Italy than five different countries. Furthermore, if the locations are contained one within the other and both have the same designation, we will always choose the most specific one [2, 7].

– *Detection of the main focus.* A text may contain many locations, but probably a piece of news occurs in only one place. To establish a single geographic focus for the story, different factors can be taken into account, such as the number of occurrences of the term in the text and the list of elements in common with the other terms [2]. GEO-NASS can set the focus on a place that is not reflected in the text, when the scores for the different locations do not reach the acceptance threshold established. That means that the text is talking about several places so unimportant that the focus of the document is actually the place which includes all of them (a province, a region, or a country).

So, the basic algorithm implies performing the following steps: 1) to locate the named entities, 2) to eliminate those that are not present in the gazetteer, 3) to disambiguate them in case there are several possibilities (applying the rule of focus and choosing the most specific place if there is one inside another), and 4) to assign weights to each place to sort the results. After the process, we have a sorted list of places and weights (scores), which will be enriched with the thesaurus descriptors related to each location.

## 3.2 Improvements to the Basic Approach of GEO-NASS

Based on the previous basic algorithm, we have analyzed several potential improvements that benefit from information about the context and the specific language of the text. This information can be exploited without effort by a human, but requires adding several additional resources and procedures to automate it with a computer. The new algorithm is detailed below in pseudo-code:

```
PROGRAM GEO-NASS(
    INPUT: text as string;context as ListOf(Attributes);
            min_weight, max_locations as real;
            g as Gazetteer; t as Thesaurus;
    OUTPUT: loc_list as ListOf(Location, weight, thesaurus_label);
BEGIN
    loc_list := Nothing; discard_list := Nothing;
    ne_list := Extraction_of_Named_Entites(text);
    ne_list := Remove_Stop_Words(ne_list);
    FOR EACH ne IN ne_list DO
        IF NOT(exist_in_gazetteer(g, ne)) THEN
            remove(ne, ne_list)
        ELSE
```

```
                loc_list_aux := Nothing;
                loc_list_aux := expand_locations(ne, g);
                loc_list_aux := filter_locations(loc_list_aux, g);
                loc_list.add(loc_list_aux);
        END IF
    END FOR
    FOR EACH loc IN loc_list DO
        w := calculate_weight(loc, text, context);
        IF w < min_weight THEN
                discard(loc_list, loc, w, discard_list)
            ELSE  assign_weight(loc_list, loc, w);
        END FOR
    IF loc_list = Nothing THEN
            IF discard_list <> Nothing THEN
                    loc_list := search_focus(discard_list)
                ELSE
                    loc_list := semantic_search(text, context);
        ELSE
            loc_list := filter_and_sort(loc_list, max_locations);

    loc_list := assign_thesaurus(loc_list, t, g);
    Output := loc_list;
  END.
```

Notice that we not only consider the text, but also other attributes of the document. These attributes can be the tittle, the date, a summary, the author, the language in which it is written, etc. These metadata, which add context to the document, will be used in certain parts of the algorithm to optimize the results. We add also as an input the minimum weight that a place must have to be considered and the maximum number of places to obtain. Both parameters are optional, but they could help to improve the results. The first parameter avoids labeling the text with tags related to locations of minor importance in the document. The second parameter is considered relevant because it can contribute, as discussed in Section 4, to reduce the noise caused by too many locations present in the text. The high-level functions that are used in the previous algorithm are explained as follow:

1. Extraction of named entities (*Extraction_of_Named_Entites*). Among the named entities we can find names of people and organizations. One way to minimize errors is to obviate those named entities that are clearly not places, which is not a trivial task. We propose to analyze the sentence in which the named entity is embedded. Thanks to NASS, the document has already been lemmatized and the system can check the type of each word by using a morphological analyzer and its function by using a syntactic parser. Penalties or bonuses are assigned, in case certain words that are acting with a certain function are found, to try to detect with reasonable accuracy if a named entity is or not a location. For example, if the named entity is acting

as subject of the sentence and the related verb is "to sing", that means that the named entity is probably a person and we can discard this occurrence. We also apply the syntactic structure detection method in this procedure.

2. Remove stop words (*Remove_Stop_Words*). Methods for obtaining named entities are not perfect and may cause noise. Nevertheless, as we have in the context the language as a parameter, we propose as an improvement the preparation (for each of the potential input languages) of a black list of words that, due to its high use frequency, may appear as named entities simply because they are capitalized. This simple measure speeds up the execution of the algorithm and avoids errors.

3. Expand and filter locations (*expand_locations, filter_locations*). As explained before in the basic algorithm, we first obtain all the places whose description matches the entity named in the text. Once expanded, the locations are filtered using the disambiguation mechanisms discussed in the previous section.

4. Calculate the weight (*calculate_weight*). In the basic algorithm described in Section 3.1, we use this formula to compute the weight of a location [2, 17].:

$$w = (wce + wp) * N$$

Where the meaning of each term in the formula is as follows:

- *wce (weight by common elements)* denotes that the weight is increased if there are other places with the same parent, thus applying the rule of "a single sense per discourse".
- *wp (weight per population)* denotes that the places with a higher population will be assigned a greater weight.
- *N (number of repetitions)* denotes the number of times the name of the place appears in the text.

We improved and extendes the previous formula by considering also the list of attributes of the document, the context, and some NLP and semantic aspects:

$$w = (wce + wp) * N * wpos * wk * wat$$

Where the meaning of the new terms is as follows:

- *Weight depending on the position (wpos, with $0 \leq wpos \leq 5$)*. We make an adjustment by the position of each location within the document. If it is located at the beginning we assign it a bonus, and if it is at the end we penalize it. The intuition is that in any text the most important keywords are usually at the beginning of the document [6].
- *Weight by keywords (wk, with $0 \leq wk \leq 2$)*. Another function provided by NASS is the detection of keywords by using statistical methods [18], which we have also taken advantage of to improve the results of GEO-NASS. We have observed, through specific cases, that locations appearing near the detected keywords are more important than those that are

farther away. We have developed a simple metric that consists of measuring the distance (in words) from each of the keywords to the locations. Those which are the closer receive a higher score, increasing the weight of that location.

— *Weight depending on the attributes and/or the type of place (wat, with $0 \leq wat \leq 10$)*. Finally, we use this parameter to improve the results in a practical and smart way. If we have simple but powerful common-sense context-dependent rules, this could be a good place to implement them using a rule table. If there are difficulties finding these rules, machine learning can be used to create the configuration table. For example, I could decide that documents with the value *International* in a field *topic* give a big bonus to locations of type *country*. This would make these places appear at the top of the list of results.

After computing the weight of each individual place, we can check if it has reached the minimum acceptance threshold required, which can be set according to the requirements of the specific context considered: a high threshold will increase the hit rate and reduce the errors (misplaced labels, i.e., labels that are used but should not be applied to the document), but will also increase the number of forgotten labels (places that are applicable but that are not used for labeling). All the places below the threshold are removed from the candidate list and transferred to a *list of places discarded*, that will be used in case no relevant place is found.

5. Search the focus (*search_focus*). In the basic algorithm the objective of this function is to try to find the focus of the news when at least one place is found relevant enough. Here the system also considers the list of discarded places for cases where no place is identified as relevant.

6. Semantic Search (*semantic_search*). If the list of places discarded is also empty, we are at a stage that we had not considered yet: What happens when we find nothing relevant in the gazetteer? Indeed, it is quite possible (about 18% of the times in our experimental tests). In fact, there are texts where the location is implicit. Our proposal to solve the problem is to follow a semantic approach by using ontologies. Specifically, we use ontologies including information about important aspects related to certain locations: for example, for a city we can populate the ontology with its most important streets, monuments and outstanding buildings, neighborhoods, etc. When a text has not explicit location identified, we will use the keywords extracted from the text (obtained through statistical analysis, based for example on the frequency of terms [18]) to query the ontologies. Then, we will choose the place whose ontology gives a higher number of correct answers for those queries. Ontologies are defined in OWL [19] or RDF and are interrogated using SPARQL [20]. Existing ontologies could be reused or could be designed from scratch.

7. Filter and sort (*filter_and_sort*). This function sorts the list of candidate locations according to their weight. Besides, if a maximum number of results has been set, then the list will be truncated to keep only the elements with the higher weights.

8. Assign a thesaurus (*assign_thesaurus*). We assume the existence of a related gazetteer that will be used for labeling. This function simply labels the document using the tags obtained from the thesaurus.

In summary, the system executes an algorithm that first extracts a set of named entities present in the text and then removes those words which may cause noise, such as articles or prepositions, that may appear as named entities simply because they are capitalized. For each Named Entity found, this algorithm searches if it is included in the gazetteer; if not, it removes it from the possible locations. Then, it expands and filters the location to obtain all the places whose description matches the named entity in the text using disambiguation mechanisms. Next, for each possible location, it calculates a weight using the formula described above; if it does not pass a minimum threshold value it is removed from the candidate list and transferred to a list of places discarded. In case it has not found any relevant place then it tries to find a focus from this list. If that list is empty, then the location may be implicit, and so it tries to solve the problem following a semantic approach by using ontologies. Otherwise, it finds a list of relevant places and sorts it according to their weight. Finally, it assigns to each location its appropriate thesaurus.

## 4   Experimental Results

This section discusses the results of experimental tests carried out to check the performance of the proposed algorithm. For testing in a real environment we have used a corpus of 9,520 news previously labeled by a professional documentation department of *Heraldo de Aragón*[5], a major Spanish media. The professional thesaurus used to label has more than 10,000 elements. The purpose is to approximate the automatic labeling of the locations to the labeling specified by that department. As the news in our data set were manually annotated by the professionals working in the documentation department of the newspaper, we can compare our automatic labeling with that performed by humans. Since the news are stored in a database, we can use the various fields of the table which are stored as context information. These fields are the list of attributes explained in Section 3.

Therefore, experiments have been performed with Spanish texts, so the input language is another context factor to consider. We had to run disambiguation algorithms in Spanish, with the additional difficulty that this implies [21–23]. We have used Freeling [24] as Spanish NLP tool, we have fixed a stop-word list, and we have prepared six ontologies populated with about a hundred words related with the most important cities in the region of the media.

We have performed four experiments. In the first one (experiment *E1*), we have used the basic algorithm presented in Section 3.1. In the second one (*E2*), we simply established a limit to the number of results. In the third one (*E3*), several improvements outlined in Section 3.2 were applied in a generic way for all

---

[5] http://www.heraldo.es/

the sections in the newspaper (sports, international, politics, etc.): the context window, the stop word list, the structure detection method and the weight *wpos* correction. Finally, in the fourth experiment (*E4*) the weights *wk* and *wat* were introduced, we added the semantic search option to solve cases in which no location appeared explicitly in the text, and the algorithm was improved to allow a different acceptance threshold for each specific section in the newspaper.

The results of the experiments are shown in Fig. 2. We analyze the following measures, commonly used in the Information Retrieval context [25]: the precision, the recall, and the F-Measure. The first three graphs refer to these measures applied on five major sections of the newspaper, and in the last graph we display the average values for all the sections.

Based on the results, we can make the following statements:

- The progression of the indicators from experiment 1 to 4 are pretty good: an 82% of recall, has been achieved, and the precision is improved from a modest 38% to an acceptable 58%, reaching an F-measure value of near 70%, which is acceptable to start using GEO-NASS in a real environment.
- In the first version of the algorithm, the number of labels used to annotate each document was not limited, yielding an average of 3.85 tags per document. As the average number of geotags per document used by the real documentation department was 1.89, that was one main reason for the low

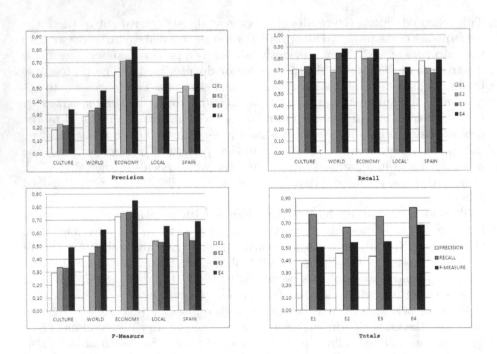

**Fig. 2.** Results of experiments with the news of a major Spanish media for the first two months of 2013

precision in experiment *E1*. So, the algorithm was generating excessive noise because it was labeling with all the places reaching the minimum threshold.

- In the second experiment we limited the number of results to three. After this there was an improvement in the recall, but obviously a setback for the precision, leading to the conclusion that the problem was on the scoring used to decide which places would be the first three, so we had to improve the adjustment of the weights.
- The results of the third experiment show an improvement in the recall but some decrease in the precision, which means that we have managed to improve the order of the results but we continue to generate many invalid tags.
- In the last experiment, the use of semantics to label items without explicit locations, combined with the *wat* adjustment, resulted in a substantial difference in the results (10%-15%). Adjusting the acceptance threshold per section contributed to a major reduction in the noise caused by excessive labeling.

Finally, if we look at the overall results for all the sections, we could say that the results of the precision could be considered to be not very good, but by analyzing samples of the results with the documentation department it was found that a significant proportion (around 50%) of the labels considered erroneous actually could be considered correct because their omission was due to redundancy or oversights of the person who tagged the news. In other words, if the documentation department had found those labels placed in advance then they would not have removed them (instead, they would have considered them as appropriate labels). Therefore, we can say that *the precision in practice is much better* than what is reflected in Fig. 2, reaching up to near the 80%.

## 5  Conclusions and Future Work

In this paper, we have presented a method to solve the problem of geotagging documents using specific vocabulary. The algorithm combines techniques well-known in the field of geographic labeling with other NLP tools and statistic methods. We also added the use of ontologies to solve certain problems that have been identified as potential loopholes in labeling standard procedures. Our main contributions are:

- Testing geotagging algorithms in a real environment, comparing experimental results with the actual labeling of a professional documentation department.
- Improving the results achieved by well known techniques using a new algorithm supported by NLP tools, ontologies and statistic methods.
- Measuring quantitatively the contribution of each geotagging technique to the final result.

The proposal has been tested in the real environment of a major Spanish newspaper. So, this paper also helps to increase the small number of experimental auto-tagging studies available for Spanish, which always implies a greater difficulty than English because of its great ambiguity and verb complexity [21–23]. In any case we believe that the algorithm used is independent of the language, question that we plan to test in a short-term.

In this work, we were able to take advantage of the fact that we have thousands of texts labeled by a professional team of archivists, which has allowed us to verify accurately the results of our tests and compare them with the actual labeling. The fact that it has been tested with professional archivists reduces the subjectivity that a manual labeling could have involved.

In conclusion we can say that, when using classic geotagging procedures over documents in real environments, the algorithms require certain adjustments related to language and context to provide acceptable results in a production environment, such as the case of applying geotagging over news in a media. Our proposed algorithm, incorporating semantic tools, NLP procedures and statistical functions, has responded adequately in such an environment, achieving 80% of precision and recall in practice. So, we are confident that it could also be used in other similar environments with guarantee of success.

As future work, we would like to automate the adaptation of the algorithms to real contexts, for which we think that the incorporation of machine learning techniques could be very helpful. Furthermore, we believe that it could be very interesting to expand the scope of the labeling, adapting these techniques for tagging specific places such as neighborhoods, streets, squares or buildings within a particular identified location.

# Reference

1. Smeaton, A.F.: Using NLP or NLP Resources for Information Retrieval Tasks. In: Natural Language Information Retrieval. Kluwer Academic Publishers (1999)
2. Amitay, E., Har'El, N., Sivan, R., Soffer, A.: Web-a-where: geotagging web content. In: 27th International Conference on Research and Development in Information Retrieval (SIGIR 2004), pp. 273–280. ACM (2004)
3. Sekine, S., Ranchhod, E.: Named Entities: Recognition, Classification and Use. John Benjamins (2009)
4. Hill, L.L.: Core elements of digital gazetteers: placenames, categories, and footprints. In: Borbinha, J.L., Baker, T. (eds.) ECDL 2000. LNCS, vol. 1923, pp. 280–290. Springer, Heidelberg (2000)
5. Quercini, G., Samet, H., Sankaranarayanan, J., Lieberman, M.D.: Determining the spatial reader scopes of news sources using local lexicons. In: 18th SIGSPATIAL International Conference on Advances in Geographic Information Systems, pp. 43–52. ACM (2010)
6. Rauch, E., Bukatin, M., Baker, K.: A confidence-based framework for disambiguating geographic terms. In: HLT-NAACL 2003 Workshop on Analysis of Geographic References, vol. 1, pp. 50–54. Association for Computational Linguistics (2003)
7. Li, H., Srihari, R.K., Niu, C., Li, W.: Location normalization for information extraction. In: 19th International Conference on Computational Linguistics, vol. 1, pp. 1–7. Association for Computational Linguistics (2002)

8. Pouliquen, B., Kimler, M., Steinberger, R., Ignat, C., Oellinger, T., Blackler, K., Fuart, F., Zaghouani, W., Widiger, A., Forslund, A.-C., Best, C.: Geocoding multilingual texts: Recognition, disambiguation and visualisation. The Computing Research Repository (CoRR) abs/cs/0609065 (2006)

9. Gruber, T.R.: A translation approach to portable ontology specifications. Knowledge Acquisition 5(2), 199–220 (1993)

10. Janowicz, K., Keßler, C.: The role of ontology in improving gazetteer interaction. International Journal of Geographical Information Science 22(10), 1129–1157 (2008)

11. Machado, I.M.R., de Alencar, R.O., de Oliveira Campos Jr., R., Davis Jr., C.A.: An ontological gazetteer and its application for place name disambiguation in text. Journal of the Brazilian Computer Society 17(4), 267–279 (2011)

12. Gilchrist, A.: Thesauri, taxonomies and ontologies - an etymological note. Journal of Documentation 59(1), 7–18 (2003)

13. Garrido, A., Gómez, O., Ilarri, S., Mena, E.: Nass: News Annotation Semantic System. In: 23rd International Conference on Tools with Artificial Intelligence, pp. 904–905. IEEE (2011)

14. Joachims, T.: Text categorization with support vector machines: learning with many relevant features. In: Nédellec, C., Rouveirol, C. (eds.) ECML 1998. LNCS, vol. 1398, pp. 137–142. Springer, Heidelberg (1998)

15. Garrido, A.L., Gómez, O., Ilarri, S., Mena, E.: An experience developing a semantic annotation system in a media group. In: Bouma, G., Ittoo, A., Métais, E., Wortmann, H. (eds.) NLDB 2012. LNCS, vol. 7337, pp. 333–338. Springer, Heidelberg (2012)

16. Lieberman, M.D., Samet, H., Sankaranarayanan, J.: Geotagging with local lexicons to build indexes for textually-specified spatial data. In: 2010 IEEE 26th International Conference on Data Engineering, pp. 201–212. IEEE (2010)

17. Gale, W.A., Church, K.W., Yarowsky, D.: One sense per discourse. In: Workshop on Speech and Natural Language (HLT 1991), pp. 233–237. Association for Computational Linguistics (1992)

18. Salton, G., Buckley, C.: Term-weighting approaches in automatic text retrieval. In: Information Processing and Management, vol. 24, pp. 513–523. Pergamon Press, Inc. (1988)

19. McGuinness, D.L., van Harmelen, F.: OWL Web Ontology Language Overview. W3C Recommendation (2004), http://www.w3.org/TR/owl-features/

20. Prudíhommeaux, E.: SPARQL Query Language for RDF. W3C Working Draft (2006), http://www.w3.org/TR/2006/WD-rdf-sparql-query-20061004/

21. Carrasco, R., Gelbukh, A.: Evaluation of TnT Tagger for Spanish. In: 4th Mexican International Conference on Computer Science, pp. 18–25. IEEE (2003)

22. Vallez, M., Pedraza-Jimenez, R.: Natural language processing in textual information retrieval and related topics. Hipertext.net (5) (2007)

23. Aguado de Cea G., Puch, J., Ramos, J.: Tagging spanish texts: The problem of 'se'. In: Sixth International Conference on Language Resources and Evaluation (LREC 2008), pp. 2321–2324 (2008)

24. Carreras, X., Chao, I., Padró, L., Padró, M.: Freeling: An open-source suite of language analyzers. In: 4th International Conference on Language Resources and Evaluation, pp. 239–242. European Language Resources Association (2004)

25. Manning, C.D., Raghavan, P., Schütze, H.: Introduction to information retrieval. Cambridge University Press, Cambridge (2008)

# Spatio-Temporal Keyword Queries in Social Networks

Vittoria Cozza[1], Antonio Messina[1], Danilo Montesi[1],
Luca Arietta[1], and Matteo Magnani[2]

[1] University of Bologna, Mura A. Zamboni 7, 40127 Bologna, IT
{vcozza,messina,montesi,arietta}@cs.unibo.it
[2] ISTI, CNR, via G. Moruzzi 1, 56124 Pisa, IT
matteo.magnani@isti.cnr.it

**Abstract.** Due to the large amount of social network data produced at an ever growing speed and their complex nature, recent works have addressed the problem of efficiently querying such data according to social, temporal or spatial dimensions. In this work we propose a data model that keeps into account all these dimensions and we compare different approaches for efficient query execution on a large real dataset using standard relational technologies.

## 1 Introduction

Nowadays it is common to use Twitter, Facebook and other on-line social networks (SNs) to share information and to learn about public and private events. Indeed, news are often reported and discussed in on-line SNs even earlier than in traditional mass media [9].

For example, at the end of August 2012 the event of the Hurricane Isaac that hit the Caribbean and the northern Gulf Coast of the United States was mentioned by a large number of tweets, mainly from the affected areas but also worldwide, first giving the news, then discussing the consequences of the event. Users were posting messages from different geographical *places* at different *times*, so each message was characterized by where and when it was posted, its topic (expressed as a short text) and the echoes it had in the network (message popularity). Since messages can include fake news or misleading information, the popularity of users also contributed to give reliability to the textual content.

SN messages are significantly different from other kinds of text documents, e.g., Web pages. They are created at a high rate and their information content can soon become outdated. This is the reason why the timestamp of these messages (e.g. freshness or recency) can be as important as their content. Several messages may repeat the same piece of information, therefore it is important to recognize when new information is added (diversity and coverage). Finally, SN messages are not independent pieces of text: they can be seen as dynamic contents coming from online platforms where interconnected users discuss topics and share news with friends. For this reason, as revealed by a survey conducted by Searchmetrics[1], number of shares and number of followers are

---

[1] http://www.searchmetrics.com/en/searchmetrics/press/ranking-factors/

B. Catania, G. Guerrini, and J. Pokorný (Eds.): ADBIS 2013, LNCS 8133, pp. 70–83, 2013.

used by Google to rank user status updates from Twitter. Indeed, traditional methods for searching the Web are not suitable for searching an event or an opinion in SN sites [12] and it typically makes sense to analyze SN messages only within their spatio-temporal and social context.

Let us consider the following search tasks. Where was Hurricane Isaac first reported? Was there a correspondence between the places Isaac hit and the origin of the comments about these places? That is, did the comments come from the places Isaac was hitting? Were there differences between topics under discussion about storms and hurricanes during the Isaac days in different locations?

In order to answer these queries, we need to be able to represent and manipulate the spatial, temporal and social features of the text messages. To this aim, in this work we define a data model and a ranking function based on popularity and spatio-temporal information retrieval ranking functions. Using our model, by expressing a spatio-temporal keyword query, we can retrieve messages available at query time involving people close to the query location, speaking about a certain topic (identified by the set of keywords). Popular messages by popular users are shown first.

It is worth noticing that despite extensive previous research on this general topic, existing works have not considered all these aspects in a unified model, e.g. in [2] the authors combine keyword search with temporal information, but not with spatial data while in [8, 11] time is not considered. In [10] an approach for searching SN messages is introduced, modeling users' and messages' interconnections and popularity but without considering spatial features and temporal marks. [5] has recently indicated the joint management of these aspects as one of the relevant open problems in the area.

In summary, to the best of our knowledge, no model to rank SN messages according to space, time, text and user/message popularity has been previously defined and studied with respect to its effectiveness and efficiency.

The contributions of this paper are as follows:

– We define a data model for spatio-temporal social interactions.
– We define a popularity-based spatio-temporal keyword ranking function.
– We experimentally compare alternative query processing approaches currently available on relational database management systems on a large real dataset.

The paper is organized as follows: in Section 2 we provide a description of related works both in the field of SN analysis and spatio-temporal information retrieval systems, in Section 3 we propose a model for SN interactions while in Section 4 we introduce a popularity-based spatio-temporal keyword ranking function. In Section 5 we introduce the Twitter dataset used in our experiments, then we highlight, by example, the potential richness of the information that can be extracted using the features incorporated in our model. Finally, in Section 6 we compare different approaches for top-k query processing and provide performance evaluations. The objective of the evaluation is to highlight the current support provided by relational database management systems for this kind of queries, and to identify critical bottlenecks requiring further research on efficient query execution.

## 2  Related Works

Several aspects of the problem of modelling spatio-temporal data have already been studied in the fields of spatial keyword queries, time-dependent text queries and in sensor networks.

In [14] the authors face the problem of retrieving web documents relevant to a keyword query within a pre-specified spatial region. In [4] a hybrid index structure to efficiently process spatial keyword queries with 'AND' semantics is introduced and experimentally evaluated. Cong et al. [8] also proposed to use a hybrid index: an inverted file is associated to each R-tree node such that both location information and text can be used to prune the search space at query time. In [11] the authors introduce a novel index called Spatial Inverted Index (S2I) to efficiently process top-k spatial keyword queries.

With regard to temporal ranking, related works have studied *time travel* or temporal range keyword queries, especially for versioned documents (web archives, blogs, wikis). Recent works have mainly focused on compression techniques for inverted indexes, to avoid the retrieval of relevant documents by traversing a whole index and later filtering documents out of temporal range. The time machine proposed in [2] allows to retrieve documents that existed at a specific time. To avoid processing redundant entities across multiple partitions, [1] uses a small synopsis for each partition. In [7] the focus is on optimisation through efficient index organisation.

Besides the problem of spatio-temporal keyword queries, in our work we focus on the context of microblogging sites and we consider information about the popularity of messages and users. Xi et al. [13] compare traditional Information Retrieval (IR) tasks and searching in newsgroups and they experimentally verify that including these features improves the search task, and in [10] the authors extend this idea to the context of microblog IR.

## 3  Data Model and Query Language

In this section we define a model for spatio-temporal interactions in SNs.

In a SN, like Twitter, the interaction dataset (a forest) is the set of all tweets and their interconnections through the *replyTo* relation. Each interaction involves a user being in a specific geographical place, performing a communicative act in the on-line platform at a precise time instant. In this work we focus on communicative acts expressed as textual interactions.

These actions, that we call *interactions*, remain as persistent communicative objects that can be later interpreted by the set of all the users to whom the object is available.

**Definition 1 (Interaction).** *An interaction is a tuple* $(t, l, u, m)$ *where t is the time and l is the spatial location (latitude and longitude) of the communicative act, u is the actor performing it, m is its textual message. If* $I = (t, l, u, m)$ *is an interaction, we will use the functions* $t(I) = t$ *(time of the interaction),* $loc(I) = l$ *(real place of interaction),* $post(I) = u$ *(poster) and* $msg(I) = m$ *(text message).*

Each interaction can refer to a previous one in the *replyTo* relation, constituting a tree-structure as in Figure 1 that refers to Table 1. An implicit property of the *replyTo* relation

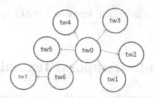

**Fig. 1.** Structure of Twitter interactions in Table 1

**Table 1.** Some Twitter interactions in our dataset, August 28

| id | uname, mentions | location | time | #replyTo | msg |
|----|-----------------|----------|------|----------|-----|
| tw0 | KSIOlajidebt, 97 | 0,0 | 14:05:35 | 6 | Why did Sahin choose liverpool :/l |
| tw1 | _olliehammond, 0 | 0,0 | 14:06:22 | 0 | @KSIOlajidebt cos he's a twat, that's why |
| tw2 | YuridSilmi, 7 | -6.9706917, 107.5776418 | 14:06:48 | 0 | @KSIOlajidebt because liverpool have more trophies than arsenal |
| tw3 | benncoxx, 1 | 53.49067936, -1.00937108 | 14:08:05 | 0 | @KSIOlajidebt because all the other Liverpool players are shit |
| tw4 | _Kidd_Shade, 0 | 0,0 | 14:08:10 | 0 | @KSIOlajidebt he obviously misread the carling cup as the champions league |
| tw5 | richmond4, 1 | 54.1787319, -4.47174819 | 14:08:39 | 0 | @KSIOlajidebt because Arsenal are finished without Song and RVP |
| tw6 | _dylP, 14 | 51.358484, 1.41081546 | 14:10:08 | 1 | @KSIOlajidebt Arsenal pulled out because there was no option to buy at the end of the loan! |
| tw7 | itsLeAnne17, 0 | 31.8840687, 34.81446363 | 14:11:11 | 0 | @Nikolesi23 may lovelife knb? haha |

is that a reply message has a higher time mark than its father. Given a set of spatio-temporal SN interactions $DB$ and a spatio-temporal keyword query, we retrieve the top-k interactions according to spatio-temporal text relevance with respect to query location, time and keywords. The most popular interactions by the most popular users are shown first. Spatio-temporal relevance is a wide topic not completely investigated in this work. According to our query model, we allow the user to specify a point of interest and to query by distance around it. As for the time we query by distance starting from a given timestamp and going backward. Moreover, it can also be useful to enrich queries with other spatial operators, e.g. to allow queries focused on specific political-geographical areas, and temporal operators.

**Definition 2 (Top-k spatio-temporal keyword query).** *A top-k spatio-temporal keyword query q is a tuple (l, t, d, k) where l is the searched conversation location, t is the searched conversation time instant, d is a set of query keywords and k is the number of expected results. In the following we will refer to the functions $l(q) = l$ (query place), $t(q) = t$ (query time), $d(q) = d$ (keywords).*

*We notate the result of the query applied to DB as $R_k$.*

*Given a list $R = \{I_1, I_2, ...\}$, $R \subseteq DB$, of temporal and text relevant interaction, for each $I_i \in R$, the score of $I_i$ is denoted as $RF(I_i, q)$.*

The top-k results of R are k results $R_k$ satisfying the following two conditions:

- $R_k \subseteq R$ and $|R_k| = min(k, |R_k|)$.
- For any $I_i \in R_k$ and $I_j \in R - R_k, RF(I_i, q) \geq RF(I_j, q)$

where RF is a popularity-spatio-temporal keyword ranking function, introduced in the next section.

## 4  Popularity-Based Spatio-Temporal Textual Ranking

In this section we define the properties and the constituents of the popularity-based spatio-temporal keyword ranking function used in the experiments.

We introduce basic functions for textual, temporal, spatial and popularity rank. We also exemplify these functions on some real data selected from the dataset that we use throughout the paper and that will be further described in Section 5.

Let us consider a dataset $DB$ containing $|DB|$ posts from a SN (Figure 1, Table 1) and the following example query:

$q_a$('53.408371 -2.991573', '2012/08/28 14:15:00', 'liverpool, arsenal, sahin', 3).

The textual rank of an interaction can be evaluated using boolean term frequency of searched keywords.

**Definition 3 (Interaction IR rank).** *Given a query description $d(q)$ containing $w_1 \ldots w_n$ keywords and an Interaction I, we define the textual rank $\delta_w$ as:*

$$\delta_w(q, I) = \frac{\sum_{i=1}^{n} p(w_i, msg(I))}{n} \tag{1}$$

*where $p(w_i, msg(I))$ equals 1 if the the word $w_i$ is present in the message $msg(I)$, 0 otherwise.*

Considering our working example, we are only interested in interactions containing at least one of the query keywords, and for those we compute the textual rank, e.g., $\delta_w(q_a, msg(tw0)) = \frac{1+0+1}{3} = 0.\bar{6}$ and $\delta_w(q_a, msg(tw3)) = \frac{1+0+0}{3} = 0.\bar{3}$.

With regard to the distance between query time and interaction time, we apply a linear transformation that scales this distance between 0 and 1 (let us call the result $x$). Then, temporal proximity might be expressed as simply as 1-$x$. However, in the case of SN data we often experience a large flow of multimedia and textual interactions per second, therefore we need a ranking function that is sensitive to small variations of time. Let us also call $T_{IN}$ the timestamp of the oldest interaction in $DB$, '2012/08/27 00:00:00' in our dataset. The temporal rank can be evaluated as in the following:

**Definition 4 (Time rank).** *The time rank $\delta_t$ is defined as:*

$$\delta_t(I, q) = 1 - \sqrt{2 \times x - x^2}, \text{ with } x = \frac{t(q) - t(I)}{t(q) - T_{IN}} \tag{2}$$

When no query time is specified, time is not considered in the ranking process. In our example, we have:

$\delta_t(tw0, q_a) = 0.723$ for $t(q) - t(I) = 9m25s$, and
$\delta_t(tw3, q_a) = 0.760$ for $t(q) - t(I) = 6m55s$.

Regarding space, we calculate the Haversine distance between the query location and the location of the interaction, then we apply a linear transformation that scales this distance between 0 and 1. Once the distance $x$ has been calculated, spatial proximity can be computed as $1$-$x$, as in [11], or we can use the same function applied to time, again to capture small variations. Spatial rank is defined as in the following:

**Definition 5 (Space rank).** *The space rank $\delta_s$ is:*

$$\delta_s(I, q) = \begin{cases} 1 - \sqrt{2 \times x - x^2}, \text{ with } x = \frac{H_a(loc(I), l(q))}{\pi \times r} & \text{if } loc(I) \neq null \\ 0 & \text{otherwise} \end{cases} \quad (3)$$

where $H_a$ is the Haversine distance[2].

In our working example, different interactions come from different locations, e.g., Indonesia (tw2), South Yorkshire UK (tw3), Isle of Man (tw5), Kent UK (tw6). We can e.g. compute: $\delta_s(tw_3, q_a) = 0.85$ for $H_a = 219.37$ Km. For a generic interaction $tw$ with unknown location we set $\delta_s(tw, q_a) = 0$.

The popularity of a user can be defined in different ways and inside a SN site we can simply compute its degree centrality, i.e., a function of the number of its followers. In this work we define the popularity of a user u, $pop(u)$, as its in-degree centrality expressed as the number of its mentions over the dataset. In the same way as we can use the popularity of the authors of a SN interaction to evaluate its rank, we can also consider the popularity of its content, that can be different from the popularity of its author. This can be usually computed in SNs counting the number of likes, sharings, replies or re-tweets received by the message. Replies show that you are consistently engaging your network with quality content, retweets increase your influence by exposing your content to extended follower networks. In this work we define the popularity of a message m, $pop(m)$, as its number of replies.

**Definition 6 (Popularity rank).** *The popularity function $\delta_P(I)$ is defined as a the combination of popularity of user and message.*

$$\delta_P(I) = \eta \times \delta_{P_U}(I) + (1 - \eta) \times \delta_{P_M}(I) \text{ with } 0 \leq \eta \leq 1 \quad (4)$$

*where $\delta_{P_U}(I) = \frac{pop(u)+1}{pop(u)+2}$ and $\delta_{P_M}(I) = \frac{pop(m)+1}{pop(m)+2}$ are the rank functions, each ranging between 0.5 and 1, of respectively user popularity ($P_U$) and message popularity ($P_M$).*

*For $\delta_{P_U}(I)$ and $\delta_{P_M}(I)$ it is also possible to apply a linear normalization of mentions and replies, as for spatial and temporal rank. The parameter $\eta$ regulates the importance of one rank over the other.*

As an example, $\eta$=0.5 means to assign the same relevance to user and message popularity:

$\delta_P(tw_0) = 0.5 \times ((97 + 1)/(97 + 2)) + 0.5 \times ((6 + 1)/(6 + 2)) = 0.932$
$\delta_P(tw_3) = 0.5 \times ((1 + 1)/(1 + 2)) + 0.5 \times ((0 + 1)/(0 + 2)) = 0.583$.

Text, time and space in a spatio-temporal information retrieval (ST-IR) query is defined as follows.

---

[2] Haversine formula calculates distances between points which are defined by geographical coordinates in terms of latitude and longitude.

**Definition 7 (ST-IR rank).** *Given an interaction I and a query q, the aggregation function $\tau$ returns a rank for I based on spatio-temporal textual relevance with respect to query location, query time and query keywords:*

$$\tau(I,q) = \alpha \times \delta_s(I,q) + \beta \times \delta_t(I,q) + (1 - \alpha - \beta) \times \delta_w(I,q) \qquad (5)$$

$$\text{with } 0 \le \alpha \le 1 \text{ and } 0 \le \beta \le 1 - \alpha.$$

$\delta_s(I,q)$ and $\delta_t(I,q)$ and $\delta_w(I,q)$, ranging between 0 and 1, represent three scores respectively for space, time and text.

The query parameters $\alpha$ and $\beta$ are used to regulate the importance of each rank with regard to the others. To give some examples, setting $\alpha = 0$ means not to give relevance to space rank; setting $\beta = 0$ means not to give relevance to temporal rank and reduces to a top-k spatial keyword search as in [11].

For example given $tw0$, $tw3$ and the rank functions as reported in this section, chosing $\alpha = \beta = \frac{1}{3}$ we have:

$\tau(tw0, q_a) = \frac{1}{3} \times 0 + \frac{1}{3} \times 0.723 + \frac{1}{3} \times 0.\overline{6} = 0.463$
$\tau(tw3, q_a) = \frac{1}{3} \times 0.85 + \frac{1}{3} \times 0.760 + \frac{1}{3} \times 0.\overline{3} = 0.648.$

Finally spatio-temporal textual rank will be weighted by popularity rank according to the following definition.

**Definition 8 (Ranking function RF).** *Given an interaction I and a query q, the ranking function RF is defined as:*

$$RF(I,q) = \delta_p(I) \times \tau(I,q) \qquad (6)$$

With respect to our working example, in the case of $tw0$ and $tw3$:

$RF(tw0, q_a) = 0.932 \times 0.463 = 0.431$ and $RF(tw3, q_a) = 0.583 \times 0.648 = 0.38.$

According to the formula and these settings, popular messages most relevant with respect to space, time and text and from popular users are ranked first.

## 5    Case Study

In this section we show how to retrieve opinions, breaking news and information about an event, e.g. a tropical storm, according to the ranking function presented in Section 4, from a Twitter dataset $TD$. Moreover we consider and analyze the quality of the function, that is the results it returns. $TD$ is a Twitter snapshot of about $10 \times 10^6$ tweets mainly from August 27 to August 31 2012 as shown in Figure 2b. We deal with $6 \times 10^6$ users from $2 \times 10^6$ places distributed as in Figure 2a. With respect to text, we had 60 000 different lexemes where the 17 most common words (lol, get, go, like, love, haha, day, good, other, time, know, con, one, tu, got, today) represent 16% of the total, following Zipf law. 'barackobama' appears 1 539 times, 'obama' 7 617, 'storm' 4 801, 'hurricane' 6 986, 'isaac 8 583' and 'love' is one of the most popular words with 196 208 occurrences. Lastly, we deal with popularity. In our dataset messages are posted by popular and less popular users in the SN as in Figure 3 and they have from

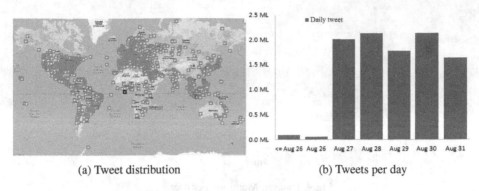

| (a) Tweet distribution | (b) Tweets per day |
|---|---|

**Fig. 2.** Dataset: space and time distribution

**Fig. 3.** Dataset: number of *ReplyTo* per tweet

0 to a large number of replies as in Figure 4. We notice 53% of records have 0 or 1 user_mentions and replies. 76% have 0 replies and 46% 0 user_mentions. Only 0.8% of the messages have replies and mentions 0 or 1 and not null geographic coordinates. This could be explained by the fact that very popular messages are typically from tv channels, newspapers or popular user fan pages. As an example the 3 most popular messages posted on August 29 afternoon are from Joseph Anthony 'Joey' Barton that is an English football player, actually a midfielder for Olympique de Marseille, Lewis Carl Davidson Hamilton, a British Formula One racing driver and from Chad Ochocinco, an American football wide receiver.

In the following we study several queries combining popularity, space, time and keywords. Throughout the examples we focus on the Hurricane Isaac, being it one of the most important events captured in our dataset. In our model the user can set 3 parameters that determine how each of the five features (space, time, text, user and message popularity) influences the ranking function. In the examples we set $\eta=0.5$ under the assumption that message and user popularity have the same importance. For time ranking, in the analysis we had $T_{IN}$ (the timestamp of the oldest interaction in $DB$) '2012/08/20 00:00:00'.

To perform a Popularity IR search one should only specify query keywords — this corresponds to setting $\alpha=\beta=0$.

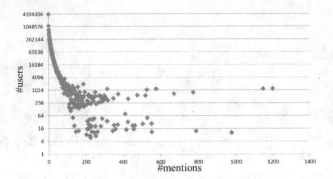

**Fig. 4.** Dataset: Mentions per user

*Example 1.* Query $Q_1$ (Popularity IR search): Extract popular messages about 'Obama' and 'Isaac'. $Q_1$=(null, null, 'hurricane isaac barackobama obama storm', '3')

The retrieved interactions are relevant to the query and they also include links to forecast images and a satellite video. The second message is from an authoritative user, Derek Brockway Met Office forecaster and BBC meteorologist.

To perform a temporal keyword search we need to specify query keywords and time. As an example we set $\beta$ to 0.5, that corresponds to giving the same relevance to time and text rank.

*Example 2.* Query $Q_2$ (Popularity Temporal Keyword search): Extract messages about Obama and Isaac until August 29.

$Q_2$=(null, '2012/08/29 20:20', 'isaac obama barackobama storm hurricane', '3')

Top 3 results are as in Table 2.

The news report that in August 29 the storm hit again, in fact messages contain feelings and links to pictures about the storm and the damages. The authors are very popular: Ian Somerhalder, actor and model, Ariana Grande, actress, singer and dancer, and Melissa Victoria Harris-Perry, author, television host and political commentator from New Orleans.

To perform a popularity-based spatio-temporal search the user will specify query time and a location expressed in geographical coordinates. Here we used $\alpha=\beta=1/2$ assuming that space and time have the same relevance.

**Table 2.** Top 3 results of Example 1

| Score: username - location [time] - text (#mentions, #replyTo) |
| --- |
| 0.488: ilchase_mkh [05:43:42 Aug 28 2012] Anyone that thinks #Isaac can still intensity to a Cat 3 hurricane is out of their mind. This storm looks like crap: http://t.co/gbjSrLj2 (1,23) |
| 0.472: DerekTheWeather [11:07:20 Aug 27 2012] Tropical storm #Isaac over the Gulf of Mexico moving towards Louisiana and becoming a hurricane http://t.co/Aj78BR1K (1,9) |
| 0.472: klorhy [13:50:50 Aug 27 2012] Why is 'Hurricane Isaac' trending? It's still a tropical storm not a hurricane o.O' (1,9) |

**Table 3.** Top 3 results of Example 2

| Score: username - location [time] - text (#mentions, #replyTo) |
| --- |
| 0.446: iansomerhalder [17:45:51 Aug 29 2012] Screw you Hurricane Isaac- leave my family and my home alone please... (13, 112) |
| 0.427: ArianaGrande [17:39:56 Aug 29 2012] I hope everybody is staying safe during Hurricane Isaac... Are you all ok? (5, 372) |
| 0.403: MHarrisPerry [16:42:32 Aug 29 2012] Feeling sad. #Isaac took the home JamesHPerry and I just bought. All safe. House was vacant except for my dreams http://t.co/h62AkGPB. (8,15) |

**Table 4.** Top 3 results of Example 3

| Score: username - location [time] - text (#mentions, #replyTo) |
| --- |
| 0.771: QuietSt0rm_ 7135 Westbranch Drive, Houston, TX 77072, USA [17:46:51 Aug 29 2012] YalSleep_iGrind: QuietSt0rm_ swiss_mocha FACT: YOU TRIED TO RUN OUT THE HAIR STORE WITH SYNTHETIC WIGS (19,7) |
| 0.736: GeorgeAdamWebb 80 Ross Road, Olive Branch, MS 38654, USA [18:18:18 Aug 29 2012] DaLJBeast fosho. You got Xbox live? (2,11) |
| 0.732: LukaLukaTuna 1405 Upper Kingston Road, Prattville, AL 36067, USA [17:58:10 Aug 29 2012] darrknight_ It's alright...Just....It snaking around me just startled me. Nuzzles her cheek against his, then lay her head in his neck. (1, 301) |

*Example 3.* Query $Q_3$ (Popularity Spatio-Temporal search): - Extract messages until 'August 29' from one of the most affected areas (Biloxi).

$Q_3$=('30.395833 -88.885278', '2013/08/29 20:20', null, '3')

With these setting, we retrieve 3 popularity-based space-time relevant messages, as in Table 3, from popular people. The same query, without including popularity, extracts messages exactly from Biloxi and mainly about the hurricane. This suggests that in emergency handling user popularity is not the central issue. In these cases a more general query model might be required.

Finally we perform a query including all the features. Here we used $\alpha=\beta=1/3$ assuming that space, time and text have the same importance.

*Example 4.* Query $Q_4$ (Popularity-based Spatio-temporal keyword search): - Extract messages about 'Isaac' and 'Obama' until 'August 29' from one of the most affected areas (Biloxi).

$Q_4$=('30.395833 -88.885278', '2013/08/29 20:20', 'hurricane isaac barackobama obama storm', '3')

With $Q_4$ we retrieve comments and pictures of the damages caused by the hurricane by people posting messages from the area close to the hurricane, as in Table 5.

**Table 5.** Top 3 results of Example 4

| Score: username - location [time] - text (#mentions, #replyTo) |
| --- |
| 0.525: missdreababy33 255 Richland Avenue, Baton Rouge, LA 70806, USA [17:30:25 Aug 29 2012] Barack Obama gave LOUISIANA supplies 5 days BEFORE Hurricane #Isaac even hit. Bush was 5 days LATE, AFTER Katrina hit, R-T for #Obama2012. (0,4) |
| 0.509: T_Jack_13 8801-8999 Knight Lane, Daphne, AL 36526, USA [15:04:27 Aug 29 2012] This what hurricane isaac did at my sis house http://t.co/ieETLLrz. (0,14) |
| 0.508: KalebHyde 517 Cliff Road, Montgomery, AL 36109, USA [18:16:35 Aug 29 2012] ?CloydRivers: That Hurricane Isaac don't want none. This is Merica, we ain't scared? (0,7) |

## 6    Experimental Evaluation

In the previous section through four sample queries and by using a Twitter dataset $TD$, we have shown the benefits of combining spatial, temporal, IR and social aspects and the richness of the query results. In the following, we briefly introduce how we built the dataset and how we implemented the top-k search. Three different search methods are shown and experimentally evaluated over the four sample queries. The objective of this evaluation is to assess the support for our query model currently available in standard relational database management systems — therefore, we limited the writing of custom code to the implementation of well known approaches, re-using built-in functions as much as possible.

### 6.1    Dataset

According to the Twitter Public API we can retrieve data by capturing a stream (Streaming API) or by searching and specifying a twitter identifier (Search API). We had a large original dataset with 7 029 259 tweets, collected using the Twitter Public Streaming API, then we enriched the dataset retrieving other related messages by id, using the *inreplytostatusid* field. Particularly, if a tweet is a child node, the *inreplytostatusid* field will contain the original Tweet's identifier. This step has led to the retrieval of 2 832 901 more tweets, while 705 394 calls by id remained unresolved.

Due to the rate limits of the Twitter Search API, we used a parallel architecture: 40 Pentium Pro Machines with GNU/Linux and 2 GB of RAM. The acquired data have been stored in a Postgres 9.1.6 database with Postgis 1.5 extension. To compute the rank of a given interaction we implemented ad hoc php scripts mainly based on sql queries that were needed to call Postgres/Postgis existing functions. For example, we used the Postgis implementation of Haversine distance[3]. As for the user popularity, we used the number of mentions over our testing dataset. Given a user with username *username*, this value can be computed by counting how many time @*username* appears in the dataset.

### 6.2    Evaluated Search Approaches

One naive way to answer our multi-feature queries is by sequentially scanning all database objects, computing the score of each object according to each feature, and

---

[3] Geographical distance: http://www.postgis.org/docs/ST_Distance.html

combining the scores into a total score for each object. Then the RF can be obtained by ordering the combined scores and returning the first k. We call this approach based on linear scanning '$M1$'. The second approach, that we call '$M2$', performs an index scan by using an index for the text. In particular, we use the 'tsvector' type to store pre-processed documents in the database and we index and search them by using a Generalized Inverted Index. In both approaches we need to analyze all the tuples and compute their score, despite being interested only in k results.

A well-known algorithm to efficiently find top-k elements [6] is the Threshold Algorithm (TA). This algorithm can be adapted to a relational scenario (R-TA), as in [3]. TA is based on an early-termination condition and can evaluate top-k queries without examining all the tuples. We call this approach '$M3$'. However, TA requires sorted access to the multiple attributes under consideration, therefore cannot be used for queries involving text.

When text is not involved, to sequentially and randomly access all other feature scoring lists we built a Generalized Search Tree (GIST) index for Location and a Gist for Time. Then, to sequentially access a score list we did a nearest neighbour search, that at each algorithm step takes a number of values, that we call '*limit*'. In Postgres this kind of query can be expressed using to the operator $< - >$ in the '*order by*' clause. To further speed up the spatial query, thanks to the STD_Within function it is possible to incrementally find nearest points. Likewise, for the time we search for an increasingly larger temporal range to find the nearest timestamps. For messages and user popularity we '*order by*' from more popular to less popular.

## 6.3   Performance Evaluation

In this Section we study the performance of the three aforementioned methods for the queries presented in our case study. Naturally, we observe that the approach $M1$ suffers from scalability problems with respect to database size and the number of features. This is indicated in Table 6. By using the textual index scan, to answer queries $q1$, $q2$ and $q4$ we need to scan 25 486 records. In the case of text search, based on the inverted index, the computation time is dependent only on the number and selectivity of keywords, as shown in Table 7, and not on the number $k$ of expected results. As summarized in Table 6, all the queries take around 40 seconds and this is mainly due to the time required to scan only and all the inverted index entries containing the searched keywords. To optimize space, time and popularity queries we can instead apply $M3$. $M3$ is an *instance*

**Table 6.** Linear scan vs Index scan

| Popularity-based query | M1 (sec.) | M2 (sec.) |
|---|---|---|
| q1 (text) | 324.79 | 28.85 |
| q2 (time-text) | 123.63 | 28.87 |
| q4 (space-time-text) | 140.89 | 31.73 |

**Table 7.** Keywords index search

| Keywords | Time (sec.) | #Records |
|---|---|---|
| isaac | 2.91 | 4 495 |
| isaac obama | 4.10 | 15 590 |
| isaac obama barackobama | 6.57 | 16 984 |
| isaac obama barackobama storm | 23.19 | 24 094 |
| isaac obama barackobama storm hurricane | 34.62 | 25 486 |

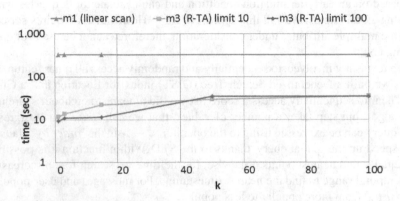

**Fig. 5.** Spatio-Temporal query (q3): Linear scan vs Relational Threshold algorithm

*dependent optimal* approach [3]. We scan *limit* values at each algorithm iteration, with $limit > k$, to find the best k records. If a large $limit$ value is chosen and feature distributions are homogeneous, we can find k best values with just one access to the database per feature, otherwise more accesses are required.

In Figure 5 we can see the $q3$ computation time according to $M1$, and how it decreases with $M3$ and different values of $limit$. In Figure 5 we show the performance of R-TA applied to the spatio-temporal query, but without considering popularity. If the popularity is included, the algorithm converges slowly. This happens because in the tested data the best result is only found after seven steps, after analyzing most of the data.

## 7   Discussion and Conclusion

In this work we propose a data and query model for spatio-temporal SN interactions. We show by example the benefits of using such a model to retrieve relevant messages. However, our analysis also emphasizes the potential negative effects of setting inappropriate weights, and the fact that weights may be query dependent when real SN data are involved. This tells us that even a generally effective setting of the weights may fail when applied to specific queries.

Our quantitative analysis has emphasized how the efficiency of multi-feature queries over a large dataset can be enhanced using indexes, but current relational database systems do not fully support this kind of queries even in cases where standard approaches like TA can be used. This highlights the need for extensions of methods such as [8, 11] to include temporal and SN-specific ranking parameters.

As a future direction, we aim to extend our data and query model by including other aspects of SNs not analyzed in this work (e.g. message graphs). As far as performance is concerned, we aim to optimize the search algorithms by using hybrid indexes.

**Acknowledgments.** We thank Anders Skovsgaard, Aarhus University, for providing the initial data used in the experiments. This work has been supported in part by the Italian Ministry of Education, Universities and Research PRIN project *Relazioni sociali ed identità in Rete: vissuti e narrazioni degli italiani nei siti di social network*, by FIRB project RBFR107725 and by EU FET-Open project DATASIM (FP7-ICT 270833).

# References

[1] Anand, A., Bedathur, S., Berberich, K., Schenkel, R.: Efficient Temporal Keyword Queries over Versioned Text. In: Proc. of ACM CIKM Conf. (2010)

[2] Berberich, K., Bedathur, S., Neumann, T., Weikum, G.: A time machine for text search. In: Proceedings of the 30th Annual International ACM SIGIR Conference on Research and Development in Information Retrieval - SIGIR 2007, p. 519 (2007)

[3] Bruno, N., Wang, H.: The threshold algorithm: From middleware systems to the relational engine. IEEE Trans. Knowl. Data Eng. 19(4), 523–537 (2007)

[4] Chen, Y.-Y., Markowetz, A.: Efficient Query Processing in Geographic Web Search Engines. In: Proc. of ACM Sigmod, pp. 277–288 (2006)

[5] Derczynski, L.R.A., Yang, B., Jensen, C.S.: Towards context-aware search and analysis on social media data. In: 16th Conference on Extending Database Technology. ACM (2013)

[6] Fagin, R., Lotem, A., Naor, M.: Optimal aggregation algorithms for middleware. J. Comput. Syst. Sci. 66(4), 614–656 (2003)

[7] He, J., Suel, T.: Faster temporal range queries over versioned text. In: Proceedings of the 34th International ACM SIGIR Conference on Research and Development in Information - SIGIR 2011, p. 565 (2011)

[8] Cong, G., Jensen, C.S., Wu, D.: Efficient Retrieval of the Top-k Most Relevant Spatial Web Objects. In: Int. Conf. on Very Large Data Bases (VLDB), pp. 337–348 (2009)

[9] Magnani, M., Montesi, D., Rossi, L.: Friendfeed breaking news: death of a public figure. In: 2010 IEEE Second International Conference on Social Computing (SocialCom), pp. 528–533. IEEE (2010)

[10] Magnani, M., Montesi, D., Rossi, L.: Conversation retrieval for microblogging sites. Inf. Retr. 15(3-4), 354–372 (2012)

[11] Rocha-Junior, J.B., Gkorgkas, O., Jonassen, S., Nørvåg, K.: Efficient Processing of Top-k Spatial Keyword Queries. In: Pfoser, D., Tao, Y., Mouratidis, K., Nascimento, M.A., Mokbel, M., Shekhar, S., Huang, Y. (eds.) SSTD 2011. LNCS, vol. 6849, pp. 205–222. Springer, Heidelberg (2011)

[12] Teevan, J., Ramage, D., Morris, M.R.: #twittersearch: a comparison of microblog search and web search. In: Proceedings of the Fourth ACM International Conference on Web search and Data Mining, WSDM 2011, pp. 35–44. ACM, New York (2011)

[13] Xi, W., Lind, J., Brill, E.: Learning effective ranking functions for newsgroup search. In: SIGIR, pp. 394–401 (2004)

[14] Zhou, Y., Xie, X., Wang, C., Gong, Y., Ma, W.-Y.: Hybrid Index Structures for Location-based Web Search. In: CIKM, vol. 49, pp. 155–162 (2005)

# GeoSocialRec: Explaining Recommendations in Location-Based Social Networks

Panagiotis Symeonidis, Antonis Krinis, and Yannis Manolopoulos*

Aristotle University, Department of Informatics, Thessaloniki 54124, Greece
{symeon,krinis,manolopo}@csd.auth.gr

**Abstract.** Social networks have evolved with the combination of geographical data, into location-based social networks (LBSNs). LBSNs give users the opportunity, not only to communicate with each other, but also to share images, videos, locations, and activities. In this paper, we have implemented an online recommender system for LBSNs, called GeoSocialRec, where users can get explanations along with the recommendations on friends, locations and activities. We have conducted a user study, which shows that users tend to prefer their friends opinion more than the overall users' opinion. Moreover, in friend recommendation, the users' favorite explanation style is the one that presents all human chains (i.e. pathways of more than length 2) that connect a person with his candidate friends.

## 1 Introduction

Over the past few years, social networks have attracted a huge attention after the widespread adoption of Web 2.0 technology. Social networks combined with geographical data, have evolved into location-based social networks (LBSNs). LBSNs such as Facebook Places, Foursquare.com, etc., which allow users with mobile phones to contribute valuable information, have increased both in popularity and size. These systems are considered to be the next big thing on the web [4].

LBSNs allow users to use their GPS-enabled device, to "check in" at various locations and record their experience. In particular, users submit ratings or personal comments for the location/activity they visited/performed. That is, they "check in" at various places, to publish their location online, and see where their friends are. Moreover, they can either comment on a friend's location or comment on their own. These LBSN systems, based on a user's "check in" profile, can also provide activity and location recommendations. For an activity recommendation, if a user plans to visit some place, the LBSN system can recommend an activity (i.e. dance, eat, etc.). For a location recommendation, if a user wants to do something, the LBSN system can recommend a place to go.

* This work has been partially funded by the Greek GSRT (project number 10TUR/4-3-3) and the Turkish TUBITAK (project number 109E282) national agencies as part of Greek-Turkey 2011-2012 bilateral scientific cooperation.

B. Catania, G. Guerrini, and J. Pokorný (Eds.): ADBIS 2013, LNCS 8133, pp. 84–97, 2013.
© Springer-Verlag Berlin Heidelberg 2013

Our prototype system GeoSocialRec is an online recommender system that relies on user check-ins to provide friend, location and activity recommendations. It provides also explanations along with the recommendations based on the democratic nature of users' voting. That is, GeoSocialRec interprets a rating by a user for an activity in a specific location, as a positive/negative vote for the "interestingness" of the location. Every registered user is presented with the option of checking in. The procedure involves selecting the location he is currently at, the activity he is performing there, and finally rating that activity. Based on the users' "check in" history and friendship network, GeoSocialRec provides friend, location and activity recommendations. Friends are recommended based on the FriendLink algorithm [7] and the average geographical distances between users' "check-ins", which are used as link weights. Users, locations and activities are also inserted into a 3-order tensor, which is then used to provide location and activity recommendations.

In this paper, we conduct a user study to measure the user satisfaction with different explanation styles. In particular, we have conducted a survey to measure user satisfaction against two styles of explanation. The first regards the "Peoples' Check-ins" style, which is based on all users' check ins, whereas the second is the "Friends' Check-ins" style, which relies only on the user's friends check-ins. As will be shown later, users tend to prefer the latter style. Moreover, in friend recommendation, the users' favorite explanation style is the one that presents all human chains (i.e. pathways of more than length 2) that connect a person with his candidate friends.

The remainder of this paper is organized as follows. Section 2 summarizes the related work. Section 3 describes the GeoSocialRec recommender system and its components. Section 4.1 presents experimental results for the evaluation of the accuracy of the recommendations. In Sections 4.2 and 4.3, we conduct two surveys to measure user satisfaction against the explanation styles in all three types of recommendations (friend, activity, location). Finally, Section 5 concludes the paper and proposes possible future work.

## 2 Related Work

Recently emerged LBSNs (i.e. Gowalla.com, Foursquare.com, Facebook Places etc.) provide to users activity or location recommendation. For example, in Gowalla.com a target user can provide to the system the activity he wants to do and the place he is (e.g. coffee in New York). Then, the system provides a map with coffee places which are nearby the user's location and were visited many times from people he knows. Moreover, Facebook Places allows users to see where their friends are and share their location in the real world.

There is a little research on the scientific field of LBSNs. Backstrom et al. [1] use user-supplied address data and the network of associations between members of the Facebook social network to measure the relationship between geography and friendship. Using these measurements, they can predict the location of an individual. Scellato et al. [10] proposed a graph analysis based approach to study social networks with geographic information. They also applied new geo-social metrics to four large-scale Online Social Network data sets (i.e. Liveljournal, Twitter, FourSquare, BrightKite). Quercia et al. [8] address the mobile cold-start problem when recommending social events to users without any location history.

Leung et al. [5] propose the Collaborative Location Recommendation (CLR) framework for location recommendation. The framework considers activities and different user classes to generate more precise and refined recommendations. The authors also incorporate a dynamic clustering algorithm, namely Community-based Agglomerative-Divisive Clustering (CADC), into the framework to cluster the trajectory data into groups of similar users, similar activities and similar locations. The algorithm can also be updated incrementally when new GPS trajectory data is available.

Ye et al. [11] believe that user preferences, social influence and geographical influence should be considered when providing Point of Interest recommendations. They study the geographical clustering phenomenon and propose a power-law probabilistic model to capture the geographical influence among Points of Interest. Finally, the authors evaluate their proposed method over the Foursquare and Whrrl datasets and discover, among others, that geographical influence is more important than social influence and that item similarity is not as accurate as user similarity due to a lack of user check-ins.

Moreover, there are tensor-based approaches. For example, Biancalana et al. [2] implemented a social recommender system based on a tensor that is able to identify user preferences and information needs and suggests personalized recommendations for possible points of interest (POI). Furthermore, Zheng et al. [13] proposed a method, where geographical data is combined with social data to provide location and activity recommendations. The authors used GPS location data, user ratings and user activities to propose location and activity recommendations to interested users and explain them accordingly. Moreover, Zheng et al. [12] proposed a User Collaborative Location and Activity Filtering (UCLAF) system, which is based on Tensor decomposition.

In contrast to the aforementioned tensor-based methods, our GeoSocialRec recommender system provides (i) location and activity recommendations (ii) friend recommendations by combining FriendLink algorithm [7] with the geographical distance between users. Moreover, our tensor method includes an incremental stage, where newly created data is inserted into the tensor by incremental solutions [9,3].

## 3   GeoSocialRec System Description

Our GeoSocialRec system consists of several components. The system's architecture is illustrated in Figure 1, where three main sub-systems are described: (i) the Web Site, (ii) the Database Profiles and (iii) the Recommendation Engine. In the following sections, we describe each sub-system of GeoSocialRec in detail.

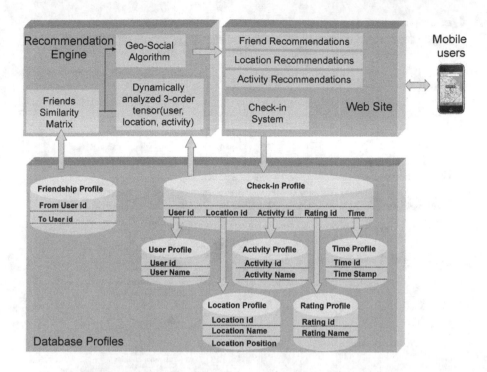

**Fig. 1.** Components of the GeoSocialRec recommender system

## 3.1   GeoSocialRec Web Site

The GeoSocialRec system uses a web site[1] to interact with the users. The web site consists of four subsystems: (i) the friend recommendation, (ii) the location recommendation, (iii) the activity recommendation, and (iv) the check-in subsystem. The friend recommendation subsystem is responsible for evaluating incoming data from the Recommendation Engine of GeoSocialRec and providing updated friend recommendations. To provide such recommendations, the web site subsystem implements the FriendLink algorithm [7] and also considers the geographical distance between users and check-in points. The same applies to the location and activity recommendation sub-systems, where new and updated location and activity recommendations are presented to the user as new check-ins are stored in the Database profiles. Finally, the check-in subsystem is responsible for passing the data inserted by the users to the respective Database profiles.

Figure 2 presents a scenario where the GeoSocialRec system recommends four possible friends to user *Panagiotis Symeonidis*. As shown, the first table recommends Anastasia Kalou and Ioanna Kontaki, who are connected to him with 2-hop paths. The results are ordered based on the second to last column of the

---

[1] http://delab.csd.auth.gr/geosocialrec

**EXPLANATION STYLE A:** We recommend the following users as possible 2-hop friends

| Name | Last Name | E-mail | Add as a friend | Picture | Number of common friends | Names of common friends |
|------|-----------|--------|-----------------|---------|--------------------------|-------------------------|
| Anastasia | Kalou | sasak2003@yahoo.gr | Add | | 3 | 1. Dimitrios Ntempos<br>2. Athina Giaouri<br>3. Foteini Vavitsa |
| Kontaki | Ioanna | gia_kodak@hotmail.com | Add | No Photo Available | 2 | 1. Dimitrios Ntempos<br>2. Athina Papagou |

**EXPLANATION STYLE B:** We recommend the following users as possible 3-hop friends!

| Name | Last Name | E-mail | Add as a friend | Picture | Paths | Number of found paths |
|------|-----------|--------|-----------------|---------|-------|-----------------------|
| Manolis | Daskalakis | emdaskalakis@gmail.com | Add | | Panagiotis Symeonidis--><br>Airam kortimanitsi--><br>Vasiliki-Eleni Provopoulou--><br>Manolis Daskalakis<br><br>Panagiotis Symeonidis --><br>TASSOS INGILIS--><br>Vasiliki-Eleni Provopoulou--><br>Manolis Daskalakis<br><br>Panagiotis Symeonidis --><br>foteini tzortzi--><br>Vasiliki-Eleni Provopoulou--><br>Manolis Daskalakis | 3 |
| George | Tsalikidis | tsalikgr@gmail.com | Add | No Photo Available | Panagiotis Symeonidis--><br>paulina marouda--><br>Christos Giannakis-Bompolis --><br>George Tsalikidis<br><br>Panagiotis Symeonidis --><br>TASSOS INGILIS--><br>Christos Giannakis-Bompolis --><br>George Tsalikidis | 2 |

**Fig. 2.** Friend recommendations provided by the GeoSocialRec system

table, which indicates the number of common friends that the target user shares with each possible friend. As shown in Figure 2, Anastasia Kalou is the top recommendation because she shares 3 common friends with the target user. The common friends are then presented in the last column of the table. The second table contains two users, namely Manolis Daskalakis and George Tsalikidis, who are connected to the target user via 3-hop paths. The last column of the second table indicates the number of found paths that connect the target user with the recommended friends. Manolis Daskalakis is now the top recommendation, because he is connected to Panagiotis Symeonidis via three 3-hop paths. It is obvious that the second explanation style is more analytical and detailed, since users can see, in a transparent way, the paths that connect them with the recommended friends.

Figure 3a shows a location recommendation, while Figure 3b depicts an activity recommendation. As shown in Figure 3a, the target user can provide to the system the activity she wants to do and the place she is (i.e. Bar in Athens). Then, the system provides a map with bar places (i.e. place A, place B, place C, etc.) along with a table, where these places are ranked based on the number of users' check-ins and their average rating. As shown in Figure 3a, the top recommended Bar is Mojo (i.e. place A), which is visited 3 times (from the target user's friends) and is rated highly (i.e. 5 stars). Regarding the activity recommendation, as shown in Figure 3b, the user selects a nearby city (i.e. Thessaloniki) and the system provides activities that she could perform. In this case, the top recommended activity is sightseeing the White Tower of Thessaloniki, because it is visited 14 times and has an average rating of 4.36.

We recommend the following Point(s) of Interest
for the Activity: Bar
in the city of:Athens

EXPLANATION STYLE A:

We recommend the following POI's (Point of Interest) based on total Check-ins!

| A/A | Point Of Interest | POI Address | Explanation Style A: Total Check-Ins | Average Rating from style A | Go To |
|-----|-------------------|-------------|------|------|-------|
| A | Mojo | Παπαδιαμαντοπούλου 8, Ζωγράφου 157 71, Ελλάδα | 3 | 5.0000 | Move! |
| B | A for Athens | Ερμού 82, Αθήνα 105 55, Ελλάδα | 3 | 3.6667 | Move! |
| C | Rox Box | Ειρήνης 2-10, Φιλαδέλφεια Χαλκηδόνα 143 41, Ελλάδα | 2 | 4.5000 | Move! |
| D | Holy Spirit | Λαοδίκης 18, Γλυφάδα 166 74, Ελλάδα | 2 | 4.0000 | Move! |
| E | Mo Better | Κωλέττη 28-42, Αθήνα, Ελλάδα | 2 | 4.0000 | Move! |
| F | Allo Bar | Thoukydidou 9-13, Chalandri 15232, Greece | 2 | 2.0000 | Move! |

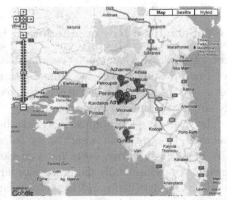

(a)

We recommend the following Activities
in the city of:Thessaloniki

EXPLANATION STYLE A:

We recommend the following activities based on total Check-ins!

| A/A | Activity | Point Of Interest | POI Address | Explanation Style A: Total Check-Ins | Average Rating from style A | Go To |
|-----|----------|-------------------|-------------|------|------|-------|
| A | Sight-seeing | White Tower | Nikis Avenue--Paralia Thessalonikis | 14 | 4.3571 | Move! |
| B | Education | Aristotle University of Thessaloniki | Egnatia & Kondriktonos--Aristotle Campus | 13 | 4.2308 | Move! |
| C | Sight-seeing | Aristotelous Square | Aristotle Square--City Center | 11 | 4.1818 | Move! |
| D | Museums | Archaeological Museum of Thessaloniki | Ανδρόνικου, Θεσσαλονίκη 54621, Ελλάς-- | 8 | 4.0000 | Move! |
| E | Cinema | Video Land | Ιθάκης 63-71, Εύοσμο 56224, Ελλάς-- | 8 | 2.6250 | Move! |
| F | Bar | BRISTOL | George Papandreou 24--Poseidonio | 7 | 4.2857 | Move! |

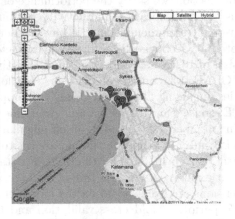

(b)

**Fig. 3.** Location and activity recommendations made by the Geo-social recommender system

### 3.2 GeoSocialRec Database Profiles

The database that supports the GeoSocialRec system is a MySQL (v.5.5.8)[2] database. MySQL is an established Database Management System (DBMS), which is widely used in on-line, dynamic, database driven websites.

The database profile sub-system contains five profiles where data about the users, locations, activities and their corresponding ratings are stored. As shown in Figure 1, this data are received by the Check-In profile and along with the Friendship profile, they provide the input for the Recommendation Engine sub-system. Each table field represents the respective data that is collected by the Check-In profile. User-id, Location-id and Activity-id refer to specific ids given to users, locations and activities respectively.

### 3.3 GeoSocialRec Recommendation Engine

The recommendation engine is responsible for collecting the data from the database and producing the recommendations, which will then be displayed on the web site. As shown in Figure 1, the recommendation engine constructs a friends similarity matrix by implementing the FriendLink algorithm proposed in [7]. The average geographical distances between users' check-ins are used as link weights. To obtain the weights, we calculate the average distance between all pairs of POIs that two users have checked-in. The recommendation engine also produces a dynamically analyzed 3-order tensor, which is firstly constructed by the HOSVD algorithm and is then updated using incremental methods [9], both of which are explained in later sections.

## 4    Experimental Results

In this section, we study the performance of FriendLink and ITR approaches in terms of friend, location and activity recommendations. To evaluate the aforementioned recommendations we have chosen two real data sets. The first one, denoted as GeoSocialRec data set, is extracted from the GeoSocialRec site [3]. It consists of 102 users, 46 locations and 18 activities. The second data set, denoted as UCLAF [13], consists of 164 users, 168 locations and 5 different types of activities, including "Food and Drink", "Shopping", "Movies and Shows", "Sports and Exercise", and "Tourism and Amusement".

The numbers $c_1, c_2$, and $c_3$ of left singular vectors of matrices $U^{(1)}, U^{(2)}, U^{(3)}$ for ITR, after appropriate tuning, are set to 25, 12 and 8 for the GeoSocialRec dataset, and to 40, 35, 5 for the UCLAF data set. Due to lack of space we do not present experiments for the tuning of $c_1, c_2$, and $c_3$ parameters. The core tensor dimensions are fixed, based on the aforementioned $c_1, c_2$, and $c_3$ values.

We perform 4-fold cross validation and the default size of the training set is 75% (we pick, for each user, 75% of his check-ins and friends randomly).

---

[2] http://www.mysql.com
[3] http://delab.csd.auth.gr/~symeon

The task of all three recommendation types (i.e. friend, location, activity) is to predict the friends/locations/activities of the user's 25% remaining check-ins and friends, respectively. As performance measures we use precision and recall, which are standard in such scenarios.

## 4.1   Comparison Results

In this section, we study the accuracy performance of ITR in terms of precision and recall. This reveals the robustness of ITR in attaining high recall with minimal losses in terms of precision. We examine the top-$N$ ranked list, which is recommended to a test user, starting from the top friend/location/activity. In this situation, the recall and precision vary as we proceed with the examination of the top-$N$ list. In Figure 4, we plot a precision versus recall curve.

As it can be seen, the ITR approach presents high accuracy. The reason is that we exploit altogether the information that concerns the three entities (friends, locations, and activities) and thus, we are able to provide accurate location/activity recommendations. Notice that activity recommendations are more accurate than location recommendations. A possible explanation could be the fact that the number of locations is bigger than the number of activities. That is, it is easier to predict accurately an activity than a location. Notice that for the task of friend recommendation, the performance of Friendlink is not so high. The main reason is data sparsity. In particular, the friendship network has average nodes' degree equal to 2.7 and average shortest distance between nodes 4.7, which means that the friendship network cannot be considered as a "small world" network and friend recommendations can not be so accurate.

For the UCLAF data set, as shown in Figure 5, the ITR algorithm attains analogous results. Notice that the recall for the activity recommendations, reaches 100% because the total number of activities is 5. Moreover, notice that in this diagram, we do not present results for the friend recommendation task, since there is no friendship network in the corresponding UCLAF data set.

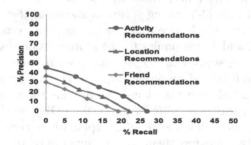

**Fig. 4.** Precision Recall diagram of ITR and FriendLink for activity, location and friend recommendations on the GeoSocialRec data set

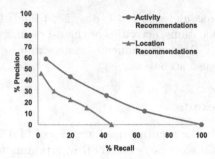

**Fig. 5.** Precision Recall diagram of ITR for activity and location recommendations on the UCLAF data set

## 4.2  User Study for Location and Activity Recommendations

We conducted a survey to measure user satisfaction against two styles of explanation. The first concerns the "Peoples' Check-ins" style (denoted as style $A$), and the second is the "Friends' Check-ins" style (denoted as style $B$). For the activity recommendation, Figure 6a shows the explanation style $A$ of the GeoSocialRec[4] site, while Figure 6b depicts the explanation style $B$.

Figure 6a depicts 3 recommended activities (Sightseeing, Education, Sightseeing) based on the explanation style $A$. As shown in the first row of Figure 6a, the first recommended activity to the target user is "sightseeing" to the monument of White Tower (the first and the second column). The explanation for this recommendation is the fact that White Tower has been visited by 14 different people and got an average rating of 4.3571 in [0-5] rating scale, as shown in the last two columns of the first row in Figure 6a.

Figure 6b depicts also a top-3 (Bar-Restaurant, Sightseeing, Transports) list of recommended activities. As shown in Figure 6b, the first recommended activity to the target user is eating to a bar-restaurant named Dishcotto (the first and the second column). The explanation for this recommendation is the fact that 6 check-ins in Dischotto have been made by the target user's friends and it got an average rating of 3 in [0-5] rating scale, as shown in the last two columns of the first row in Figure 6b. Notice that, for the location recommendation, the explanation styles $A$ and $B$ are similar to the aforementioned ones.

We designed the user study with 50 pre- and post-graduate students of Aristotle University, who filled out an on-line survey. The survey was conducted as follows: Firstly, we asked each target user to provide the system with ratings and comments for at least five point of interests (POIs), so that a decent recommendation along with some meaningful explanations could be provided by our system. Secondly, we asked them to rate separately, from 1 (dislike) to 5 (like), each recommended location/activity list based on the two different styles of explanations. In other words, we asked target users to rate separately each

---

[4] http://delab.csd.auth.gr/geosocialrec

EXPLANATION STYLE A:

We recommend the following activities based on total Check-ins!

| Activity | Point Of Interest | POI Address | Explanation Style A: Total Check-Ins | Average Rating from style A |
|---|---|---|---|---|
| Sight-seeing | White Tower | Nikis Avenue–Paralia Thessalonikis | 14 | 4.3571 |
| Education | Aristotle University of Thessaloniki | Egnatia & Kondriktonos–Aristotle Campus | 13 | 4.2308 |
| Sight-seeing | Aristotelous Square | Aristotle Square–City Center | 11 | 4.1818 |

(a)

EXPLANATION STYLE B:

We recommend the following activities based on the Check-Ins made by your friends!

| Activity | Point Of Interest | POI Address | Explanation Style B: Check-Ins made by your friends | Average Rating from style B |
|---|---|---|---|---|
| Bar-Restaurant | Dishcotto | Analipseos 6-20, Panorama 55236, Greece– | 6 | 3.0000 |
| Sight-seeing | Aristotelous Square | Aristotle Square–City Center | 4 | 3.7500 |
| Transports | International Airport 'Makedonia' (Thessaloniki) | Kalamaria Thessaloniki–Kalamaria | 4 | 2.7500 |

(b)

**Fig. 6.** Explaining recommendations based on (a) total peoples' check-ins, and (b) target user's friends' check-ins

explanation style to explicitly express their actual preference among the two styles.

We assume that, explanation style $B$ will be the users' favorite choice, since it relies on their friends' check-ins. Notice that according to homophily theory [6] (i.e., "love of the same") individuals tend to prefer the same things that similar other users do like.

Our results are illustrated in Table 1. The second and third columns contain for explanation style $A$, the mean $\mu_A$ and standard deviation $\sigma_A$ of the ratings provided by users for location and activity recommendations, respectively. As shown, the mean value of ratings $\mu_A$ for location recommendation is 3.77, whereas $\mu_A$ for activity recommendation is 3.63. The fact that the mean of ratings is higher than 2.5 in the [0-5] rating scale means that the quality of recommendations is good. The fourth and fifth columns contain for explanation style $B$, the mean $\mu_B$ and standard deviation $\sigma_B$ of the ratings provided by users. As shown, the mean value of ratings $\mu_B$ for location recommendation is 4.03, whereas $\mu_B$ for activity recommendation is 4.17. This is a clear support of the assumption that explanation style $B$ is the users' favorite choice.

Moreover, we computed the distribution of the difference between means of explanation styles $A$ and $B$, to verify that it is statistically significant. That is, the difference between ratings of style $A$ and $B$ should not be centered around 0. Thus, we measured the mean $\mu_d$ and standard deviation $\sigma_d$ of the differences between ratings of explanation style $A$ and ratings of explanation style $B$. These

**Table 1.** Results of the user survey for location/activity recommendations

| Recommendation Type | $\mu_A$ | $\sigma_A$ | $\mu_B$ | $\sigma_B$ | $\mu_d$ | $\sigma_d$ |
|---|---|---|---|---|---|---|
| Location | 3.77 | 1.13 | **4.03** | 1.40 | 0.26 | 0.32 |
| Activity | 3.63 | 0.96 | **4.17** | 1.44 | 0.54 | 0.31 |

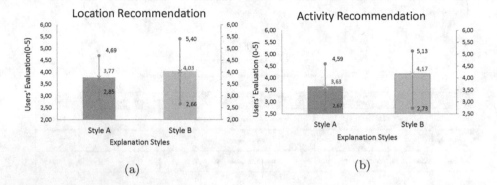

(a)                                          (b)

**Fig. 7.** Mean and standard deviation of users' ratings evaluating explanation styles $A$ and $B$ for (a) location recommendation, and (b) activity recommendation

values, for each recommendation type, are presented in the sixth and seventh columns of Table 1. We run paired t-tests with the null hypothesis $H_0(\mu_d = 0)$ for the two recommendation types (i.e. location and activity). We found that for both location and activity recommendations, $H_0(\mu_d = 0)$ is rejected at the 0.05 significance level. This verifies the assumption that explanation style $B$ is the users' favorite choice. Finally, Figures 7a and 7b show a visual representation of the mean and standard deviation of users' ratings, evaluating the explanation styles $A$ and $B$ for both location and activity recommendation, respectively. As expected, style $B$ outperforms $A$ in both recommendation types (i.e. location and activity recommendation). That is, likes of our friends have a greater impact in our own choices.

## 4.3   User Study for Friend Recommendations

We conducted a second survey to measure user satisfaction against the explanation styles in friend recommendation. We have also tested two styles of explanation. Explanation style $A$ justifies friend recommendations based on the number of common friends between the target user and his candidate friends. That is, explanation style $A$ considers only pathways of maximum length 2 between a target user and his candidate friends. Explanation style $B$ can provide more robust explanations, by presenting as explanation, all human chains (i.e. pathways of more than length 2) that connect a person with his candidate friends.

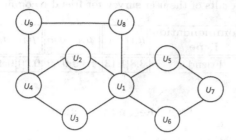

**Fig. 8.** Social Network Example

For instance, an example of a social network is shown in Figure 8. The explanation style $A$ for recommending new friends to a target user $U_1$ is as follows: "People you may know : (i) user $U_7$ because you have two common friends (user $U_5$ and user $U_6$) (ii) user $U_9$ because you have one common friend (user $U_8$) ...". The list of recommended friends is ranked based on the number of common friends each candidate friend has with the target user.

Based on explanation style $B$, a user can also get, along with a friend recommendation, a more robust explanation. This explanation contains all human chains that connect him with the recommended person. For instance, in our running example, $U_1$ would get as explanation for recommending to him $U_4$ the following human chains that connect them:

1. $U_1 \rightarrow U_2 \rightarrow U_4$
2. $U_1 \rightarrow U_3 \rightarrow U_4$
3. $U_1 \rightarrow U_8 \rightarrow U_9 \rightarrow U_4$

This means that $U_4$ is connected with $U_1$ with two pathways of length 2 and 1 pathway of length 3. It is obvious that explanation style $B$ is analytic and informative.

We designed the same user study with the one described in Section 4.2. We assumed that explanation style $B$ will be the users' favorite one, because it is more transparent and informative than explanation style $A$. Our results are illustrated in Table 2. As shown, the mean value of ratings $\mu_B$ of style $B$ is 4.10, whereas $\mu_A$ is 3.8. This is a first indication supporting our assumption that explanation style $B$ is the users' favorite choice. Moreover, we measured the mean $\mu_d$ and standard deviation $\sigma_d$ of the differences between means of explanation style $A$ and ratings of explanation style $B$. These values are presented in the sixth and seventh columns of Table 2. We run paired t-tests with the null hypothesis $H_0(\mu_d = 0)$. We found that the null hypothesis is rejected at the 0.05 significance level. This verifies our assumption that explanation style $B$ is the users' favorite choice.

Finally, Figure 9 shows a visual representation of the mean and standard deviation of users' ratings, evaluating the explanation styles $A$ and $B$ for friend recommendation. As expected, style $B$ outperforms $A$. That is, explanation style

**Table 2.** Results of the user survey for friend recommendations

| Recommendation Type | | $\mu_A$ | $\sigma_A$ | $\mu_B$ | $\sigma_B$ | $\mu_d$ | $\sigma_d$ |
|---|---|---|---|---|---|---|---|
| Friend | | 3.8 | 1.13 | **4.10** | 0.99 | 0.30 | 0.47 |

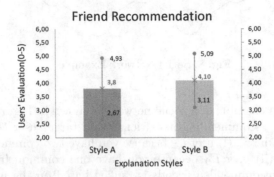

**Fig. 9.** Mean and standard deviation of users' ratings evaluating explanation styles $A$ and $B$ for friend recommendation

$B$ increases the acceptance of a recommender system, since users can understand the strengths and limitations of the recommendation process.

## 5   Conclusion and Future Work

In this paper, we have proposed the GeoSocialRec recommender system, which is capable of recommending friends, locations and activities and simultaneously provides explanations along with the recommendations. We have conducted a user study, which has shown that users tend to prefer their friends opinion more than the overall users' opinion. Moreover, in friend recommendation, the users' favorite explanation style is the one that uses all human chains (i.e. pathways of more than length 2) that connect a person with his candidate friends. As future work, we are planning on comparing our explanation styles with other hybrid explanation styles, which combine more features for justifying their recommendation.

## References

1. Backstrom, L., Sun, E., Marlow, C.: Find me if you can: improving geographical prediction with social and spatial proximity. In: Proceedings of the 19th International Conference on World Wide Web (WWW), New York, NY, pp. 61–70 (2010)
2. Biancalana, C., Gasparetti, F., Micarelli, A., Sansonetti, G.: Social tagging for personalized location-based services. In: Proceedings of the 2nd International Workshop on Social Recommender Systems (2011)

3. Brand, M.: Incremental singular value decomposition of uncertain data with missing values. In: Heyden, A., Sparr, G., Nielsen, M., Johansen, P. (eds.) ECCV 2002, Part I. LNCS, vol. 2350, pp. 707–720. Springer, Heidelberg (2002)
4. Economist. A world of connections: A special report on networking. The Economist: Editorial Team (2010)
5. Leung, K., Lee, D.L., Lee, W.C.: Clr: a collaborative location recommendation framework based on co-clustering. In: Proceedings of the 34th International ACM Conference on Research and Development in Information Retrieval (SIGIR), New York, NY, pp. 305–314 (2011)
6. Monge, P.R., Contractor, N.: Theories of communication networks. Oxford University Press (2003)
7. Papadimitriou, A., Symeonidis, P., Manolopoulos, Y.: Friendlink: Predicting links in social networks via bounded local path traversal. In: Proceedings of the 3rd Conference on Computational Aspects of Social Networks (CASON), Salamanca, Spain (2011)
8. Quercia, D., Lathia, N., Calabrese, F., Di Lorenzo, G., Crowcroft, J.: Recommending Social Events from Mobile Phone Location Data. In: Proceedings of the 10th IEEE International Conference on Data Mining (ICDM), Sydney, Australia, pp. 971–976 (2010)
9. Sarwar, B., Konstan, J., Riedl, J.: Incremental singular value decomposition algorithms for highly scalable recommender systems. In: International Conference on Computer and Information Science, Liverpool, UK (2002)
10. Scellato, S., Mascolo, C., Musolesi, M., Latora, V.: Distance Matters: Geo-social Metrics for Online Social Networks. In: Proceedings of the 3rd Workshop on Online Social Networks (WOSN), Boston, MA (2010)
11. Ye, M., Yin, P., Lee, W.-C., Lee, D.-L.: Exploiting geographical influence for collaborative point-of-interest recommendation. In: Proceedings of the 34th International ACM Conference on Research and Development in Information Retrieval (SIGIR), New York, NY, pp. 325–334 (2011)
12. Zheng, V.W., Cao, B., Zheng, Y., Xie, X., Yang, Q.: Collaborative filtering meets mobile recommendation: A user-centered approach. In: Proceedings of the AAAI Conference in Artificial Intelligence (AAAI), Atlanta, GA, pp. 236–241 (2010)
13. Zheng, W., Zheng, Y., Xie, X., Yang, Q.: Collaborative location and activity recommendations with GPS history data. In: Proceedings of the 19th International Conference on World Wide Web (WWW), New York, NY, pp. 1029–1038 (2010)

# Compact and Distinctive Visual Vocabularies for Efficient Multimedia Data Indexing

Dimitris Kastrinakis[1], Symeon Papadopoulos[2], and Athena Vakali[1]

[1] Department of Informatics, Aristotle University,
54124, Thessaloniki, Greece
{kastrind,avakali}@csd.auth.gr
[2] Information Technologies Institute, CERTH-ITI
57001, Thessaloniki, Greece
papadop@iti.gr

**Abstract.** Multimedia data indexing for content-based retrieval has attracted significant attention in recent years due to the commoditization of multimedia capturing equipment and the widespread adoption of social networking platforms as means for sharing media content online. Due to the very large amounts of multimedia content, notably images, produced and shared online by people, a very important requirement for multimedia indexing approaches pertains to their efficiency both in terms of computation and memory usage. A common approach to support query-by-example image search is based on the extraction of *visual words* from images and their indexing by means of inverted indices, a method proposed and popularized in the field of text retrieval.

The main challenge that visual word indexing systems currently face arises from the fact that it is necessary to build very large visual vocabularies (hundreds of thousands or even millions of words) to support sufficiently precise search. However, when the visual vocabulary is large, the image indexing process becomes computationally expensive due to the fact that the local image descriptors (e.g. SIFT) need to be quantized to the nearest visual words.

To this end, this paper proposes a novel method that significantly decreases the time required for the above quantization process. Instead of using hundreds of thousands of visual words for quantization, the proposed method manages to preserve retrieval quality by using a much smaller number of words for indexing. This is achieved by the concept of *composite words*, i.e. assigning multiple words to a local descriptor in ascending order of distance. We evaluate the proposed method in the Oxford and Paris buildings datasets to demonstrate the validity of the proposed approach.

**Keywords:** multimedia data indexing, local descriptors, visual word, composite visual word.

## 1 Introduction

Multimedia content is produced at unprecedented rates and is extensively used in both personal (e.g. holiday albums) and professional (e.g. stock image

B. Catania, G. Guerrini, and J. Pokorný (Eds.): ADBIS 2013, LNCS 8133, pp. 98–111, 2013.

collections) settings. As the size of media collections increases, the need for efficient content retrieval becomes more pronounced. One of the prevalent image search paradigms pertains to Content Based Image Retrieval (CBIR), which enables searching large image collections using their visual content to establish relevance. Typically, CBIR is implemented by means of a query-by-example application, which given an input query image returns the top $N$ most relevant results from the collection, assessing relevance on the basis of visual content alone.

In recent years, the performance of CBIR systems has substantially improved thanks to the development of rich image representations based on local descriptors (e.g. SIFT [8], SURF [2], etc.) and the use of full-text search technologies that formulate query-by-example as a text retrieval problem [11,15]. According to those, a set of local descriptors is extracted from each image and are subsequently quantized into *visual words*, leading to the so-called Bag of Words (BoW)[1] representation. The BoW representation is amenable to inverted indexing techniques, thus enabling indexing and efficient querying of very large image collections by use of robust full-text indexing implementations such as Lucene[2] and ImageTerrier[3].

Despite its success, the application of the BoW image indexing scheme is still considered a very challenging and computationally demanding task due to the fact that visual words are not natural entities (as are terms/tokens in the case of text documents). In fact, visual words are the result of a training process, whereby the local descriptors from a large collection of images are clustered around $k$ centres, and the corresponding centroids are considered as the words of the visual vocabulary. Having built such a vocabulary, new images are indexed by mapping their local descrptors to words of the vocabulary, i.e. for each image descriptor the most similar centroid (among the $k$ words of the vocabulary) is selected for use in the BoW representation. As the number of local descriptors per image typically lies in the range of some hundreds to a couple of thousands, it becomes obvious that deriving the BoW representation for an image may incur significant computational cost.

In fact, typical sizes for visual vocabularies range from hundreds of thousands to even millions of visual words (i.e. $k \sim 10^5 - 10^6$) according to related studies [12]. This creates two computational problems: (a) creating vocabularies of such sizes by means of clustering techniques becomes extremely expensive, (b) the indexing time for new images increases significantly due to the need for mapping each local descriptor of the image to its most similar visual word as explained above. While the first of these problems appears only at offline settings, and thus does not affect retrieval efficiency, the second problem incurs substantial overhead at indexing time. To this end, this paper proposes a new approach for visual word indexing that significantly reduces the number of visual words that are necessary to achieve satisfactory retrieval accuracy. This is achieved by

---

[1] In several works, the preferred abbreviation is BOV (Bag of Visual words).
[2] http://lucene.apache.org/
[3] http://www.imageterrier.org/

considering *composite visual words*, i.e. permutations of multiple visual words ordered according to their distance from the corresponding local descriptors. Even with a small visual vocabulary, our approach leads to a much more distinctive BoW representation. Our experimental study on two standard datasets, Oxford [12] and Paris buildings [13]) reveals that **as few as $k = 200$ visual words** can be utilized to match the retrieval performance of existing approaches using **two to three orders of magnitude more visual words**.

The paper is structured as follows: Section 2 offers the necessary background on the problem of BoW indexing, also covering important contributions in the area. Section 3 provides a description of the proposed approach. Next, we present an evaluation of the approach in Section 4 and we summarize our findings and discuss future steps in Section 5.

## 2   Background

### 2.1   Image Indexing Using Bag of Words Representations

We consider a collection of images $P = \{p_i\}$ to be indexed. For each image, we extract a set of local descriptors $F_i = \{f_{i,1}, ..., f_{i,|F_i|}\}$ where each descriptor is a feature vector, which is real-valued ($f_{i,x} \in \mathbb{R}^D$), e.g. in the case of SURF [2], or integer-valued ($f_{i,x} \in \mathbb{Z}^D$), e.g. in the case of SIFT [8]. Typical values for $|F_i|$, i.e. number of descriptors per image, range from a few hundreds to few thousands, while typical values for the dimensionality of the descriptor vectors are $D = \{64, 128\}$. To derive a BoW representation for an image, we need to discretize the set of local descriptors $F_i$ to end up with the BoW representation denoted as $W_i = \{w_{i,1}, ..., w_{i,k}\}$, where $w_{i,x}$ is the weight that visual word $x \in [1, k]$ has in the representation of image $p_i$, a process that is often called *feature quantization*. This presumes a visual vocabulary $V = \{v_1, ..., v_k\}$ where $v_x \in \mathbb{R}^D$ or $v_x \in \mathbb{Z}^D$ depending on the local descriptor of choice. Having derived the BoW vector for each image of the collection, indexing is typically implemented by means of inverted index structures and relevance is assessed on the basis of classic text retrieval schemes such as $tf * idf$ [15]. Table 1 summarizes the described notation.

The arising issue is the need of a rich and distinctive vocabulary $V$. To this end, a clustering process, e.g. k-means, must be carried out on a large number of images that act as the training or learning set for $V$. Since a satisfactory vocabulary may need to contain $> 10^6$ words, as evidenced by recent studies [12], it becomes clear that the feature quantization process may become a significant computational hurdle at both indexing and query time.

### 2.2   Related Work

The first popular attempt towards CBIR using a text retrieval approach was proposed by Sivic and Zisserman [15], who proposed the BoW representation for retrieving objects in video content. The descriptor vectors, which are computed

**Table 1.** Notation used in the paper

| Symbol | Description |
|---|---|
| $P = \{p_i\}$ | Collection of images, an image being denoted as $p_i$. |
| $n = |P|$ | Number of images in the collection to be indexed. |
| $f \in \mathbb{Z}^D$ | A $D$-dimensional local descriptor feature vector (integer-valued in the case of SIFT). |
| $F_i = \{f_{i,1}, ..., f_{i,|F_i|}\}$ | Set of local descriptor feature vectors for image $p_i$. |
| $|\bigcup_i F_i|$ | The set of all local descriptors in a collection (e.g. used to learn a vocabulary). |
| $V = \{v_1, ..., v_k\}$ | Visual vocabulary used for indexing. |
| $k = |V|$ | Size of visual vocabulary (number of visual words). |

for each frame, are quantized into visual words using k-means clustering. For each visual word, an entry is added to the index that stores all its occurrences in the video frames of the collection, thus building an inverted file. Text retrieval systems often promote documents where query keywords appear close together. This analogy is also adopted for BoW-based visual indexing [12,15], where it is required that matched regions in the retrieved frames of a video should have similar spatial arrangement to the regions of the query image.

To speed up the vocabulary construction step and the image query process, Nistér and Stewénius introduce a scheme in which local descriptors are hierarchically quantized in a vocabulary tree [11]. In particular, this tree is built by use of hierarchical k-means clustering (HKM) relying on a set of descriptor vectors for the unsupervised creation of the tree. An initial k-means process clusters the training data to groups, where each group consists of the descriptor vectors closest to its center. This process is recursively applied to each group, forming a specified maximum number of levels $L$. A descriptor vector is propagated down the tree by comparing the vector to the cluster centres that reside at each level.

Another clustering method often used for feature space quantization is the approximate k-means clustering (AKM) [12]. Typical k-means implementations fail to scale to large volumes of data, since the time complexity of each iteration is linear to the number of data, dimensionality of the data and the number of desired clusters. Instead of calculating the exact nearest neighbours between data points and cluster centres, an approximate nearest neighbour method can be applied to increase speed. In [12], a forest of eight randomized k-d trees [7] is used, which is built over the cluster centres at the beginning of each iteration. A forest of randomized k-d trees prevents points lying close to partition boundaries from being assigned to an incorrect nearest neighbour. This is especially important for the quantization of high dimensional features such as SIFT ($D = 128$).

In [13], Philbin et al. introduce an approach called *soft assignment*, where a high dimensional descriptor is mapped to a weighted combination of visual words, rather than a single visual word. The weight assigned to neighbouring clusters depends on the distance between the descriptor and the cluster centres. It was shown that soft assignment can boost the recall of a retrieval system and, if combined with spatial verification, precision could be increased too. However, this technique requires more space for the inverted index.

A similar work to the one proposed here introduces the concept of *visual phrases* [20]. The authors propose a method for mining visual word collocation patterns from large image collections and then use those for indexing. Compared to our approach, the concept of visual phrases does not take into account word ordering, which could harm precision, and in addition it requires an expensive phrase mining process to derive the phrase dictionary. A more sophisticated approach for deriving more descriptive visual vocabularies is presented in [21], in which a very large corpus of images is processed to extract a descriptive visual vocabulary consisting of both words and phrases. Though the reported retrieval accuracy is substantially improved, the approach of [21] presumes the existence of a very large image collection for training and the extraction of a vocabulary of considerable size ($k \sim 10^4 - 10^5$) that makes the approach considerably more demanding compared to ours.

In [22], an alternative visual vocabulary generation mechanism is proposed, wherein groups of visual words are extracted making sure that the spatial relations among the words of the same group are maintained. This could be considered as a generalization of the method proposed here since our compositve visual words maintain a Euclidean distance-induced ordering, but lose the spatial layout information. However, the vocabulary generation process is significantly more complex and carries the risk of leading to an incomplete vocabulary, i.e. not all possible spatial visual word configurations can be adequately represented in the vocabulary index. A similar approach, facing similar complications, is described in [19], where groups of spatially consistent local descriptors are called *bundled features* and are considered as the unit of visual indexing.

## 3   Proposed Framework

Our framework is described in two steps: (a) creating the visual vocabulary, (b) using composite visual words to index new image collections.

### 3.1   Visual Vocabulary Creation

Having stored the features of the collection, quantization of the feature space can take place, in order to produce the initial visual vocabulary. At a later step, this vocabulary will be enriched with composite visual words to improve retrieval.

**Clustering Large Numbers of Descriptor Vectors.** Feature discretization is performed by applying a clustering algorithm on the descriptors of the learning set. We used the VLFeat [17] implementation of Lloyd's k-means algorithm. To cluster the data, they need to be loaded in main memory, which is impossible in the case of millions of multidimensional features. Despite the fact that Lloyd's algorithm does not perform triangular inequality checks[4], it cannot run

---

[4] k-means with triangular equality checks requires $\frac{k \cdot (k-1)}{2}$ extra space to store the distances between centres [4].

in systems with limited main memory for large volumes of data, since it would require $n \cdot d$ space, $n$ being the number of vectors and $D$ their dimensionality.

To this end, we partition the input vectors and provide each partition to a streaming implementation of the k-means algorithm [10]. According to it, the set $F = F_i \cup \ldots \cup F_n$ of the extracted descriptors is divided into an appropriate number of subsets, to which k-means++ [1] is applied (such that an arbitrarily poor approximate solution of k-means is avoided). The union of the resulting cluster centres is then provided to a second execution of the algorithm, producing the set of visual words $V$.

**The Requirement for a Compact Vocabulary.** During the clustering step, Lloyd's k-means algorithm introduces significant time (and space) complexity requirements due to the large number of vectors to cluster and cluster centres to calculate. Simple k-means has $O(n \cdot D \cdot k \cdot I)$ complexity, where $n$ is the number of vectors to cluster, $D$ the dimensionality, $k$ the number of centres to find and $I$ the number of iterations of the algorithm. The above fact renders the application of k-means impractical in this context, when the size of the visual vocabulary becomes very large. As shall be shown below, this problem can be addressed at indexing time with the application of a technique that makes unnecessary the creation of a vocabulary of large size.

## 3.2   Indexing the Collection with Composite Visual Words

After the vocabulary creation step, a set $k$ of visual words is available for indexing. Then, each descriptor vector $f \in F_i$ of a new image $p_i$ is compared to the vocabulary visual words using the L2 distance. According to the BoW representation [15], in order to index the collection, each vector $f$ is assigned to the nearest visual word $v$ using the selected distance measure $d$:

$$v_f = \arg \min_{v \in V} d(f, v) \tag{1}$$

For each image in the collection, a document is created that contains the assigned visual words. Eventually, each such document is indexed using an inverted index of terms (visual words) pointing to documents.

**Composite Visual Words.** If the initial visual vocabulary is limited (due to the excessive computational requirements incurred by the clustering process when $k$ is too large), its distinctive capability will be limited, thus inflicting a decrease on the performance of the retrieval system. To this end, for each feature vector $f$, instead of indexing the nearest visual word only, we index the concatenation of the $B$ nearest words in ascending order using L2 distance. Figure 1 depicts the above idea: composite visual word $ACB$ corresponds to a feature vector lying nearest to center $A$, then $C$ and $B$. Similarly, visual word $BCA$ is created by a feature vector that lies nearest to $B$, $C$ and $A$.

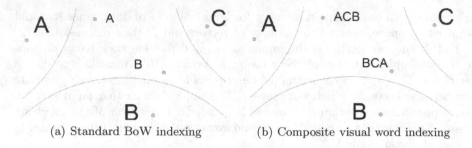

(a) Standard BoW indexing          (b) Composite visual word indexing

**Fig. 1.** Illustration of composite visual words

This approach implicitly exploits the inter-word relationships around neighbouring words so that the resulting composite visual word, as a permutation of relevant visual words, describes the corresponding descriptor $f$ more distinctively, even with a small number of initial visual words. Making composite visual words is somewhat similar to soft-assignment [13]; however, there is a key difference between the two techniques: in soft-assignment, a feature is assigned to several visual words (cluster centers) separately, whereas in making composite visual words a feature is assigned to their concatenation, thus effectively enriching the resulting visual vocabulary.

At this point, we need to point out the distinction between the set of visual words $V$ in the original visual vocabulary and the *effective visual vocabulary* resulting from the use of composite visual words. Since a composite word $w'$ is formed by a permutation of words in $V$, this method creates an extended vocabulary $V'$. The maximum number of possible words in $V'$ is: $|V|'_{max} = P(|V|, B)$, where $B$ is the user-defined number of words that will form a composite word, and $P(|V|, B)$ the number of $B$-permutations of $|V|$, i.e. $P(|V|, B) = \frac{|V|!}{(|V|-B)!}$.

For example, if $|V| = 100$ and $B = 3$, then the maximum number of words that $V'$ can possibly contain is $100 \cdot 99 \cdot 98 = 970,200$. For $B = 4$, we get $|V'| = 94,109,400$[5]. Of course, a vocabulary of such size might have a negative impact on retrieval, because a lot of terms would appear only once in the collection. For this reason, we devise a thresholding strategy based on the distances of features from candidate words, as explained below.

**Thresholding Strategy.** When assigning visual words to features, a restriction on distance would disqualify a word $v$ from the indexing of feature $f$ if:

$$d(v, f) > \beta \cdot \max_{v' \in V} d(v', f) \tag{2}$$

where $\beta \in (0, 1)$, and $\max_{v' \in V} d(v', f)$ is the maximum distance between feature $f$ and a visual word. We extend the above condition to include the $B$ nearest

---

[5] In practice, we expect somewhat smaller vocabularies than the aforementioned ones due to the fact that some word combinations would be highly improbable (depending also on the distribution of the local descriptor vectors of the images to be indexed).

candidate words, instead of the single nearest: instead of a fixed condition for all top $B$ words, we introduce a constraint that becomes progressively stricter for the next nearest word. The $i$-th nearest word, $1 \leq i \leq B$, is disqualified if:

$$d(v, f) > e^{-ai} \cdot \max_{v' \in V} d(v', f) \tag{3}$$

Increasing $a$ makes the condition stricter. In this case, notice that we do not consider a fixed number of words in the composite visual word; such a word may consist of many words as long as they satisfy the above condition. On the other hand, given a distance threshold and a user-defined maximum number of words $B$ that will form a composite word, there may be composite words formed by less than $B$ words. Apparently, constants $a$, $B$ and $|V|$ strongly affect the size of the composite vocabulary $V'$. In fact, the maximum number of words in $V'$ can be analytically determined as $|V|'_{max} = \sum_{i=1}^{B} P(|V|, B)$. This carries the risk of creating an even larger vocabulary $V'$ compared to the original one. However, this is highly unlikely if an appropriate condition is set, as will be shown in the experiments section. Apart from that, collections typically contain images that share a lot of visual patterns; therefore, we expect the vocabularies resulting from this process to not contain too many unique words.

Algorithm 1 specifies the process for extracting a composite visual word $v^+$ given a feature vector $f$ of an image, a set $V$ of visual words, a constant $a$ used in the distance condition and the number of words $B$ that can form $v^+$. At first, the algorithms calculates the distances of the visual words from the given vector and sorts them in ascending order. Then it concatenates the labels of the nearest $B$ words, as long as their distances are less or equal to the threshold. If a word $v_i$, $1 \leq i \leq B$ is disqualified, the process does not continue for $v_{i+1}$ as it would not satisfy the distance condition, since $d(v_{i+1}, f) \geq D(v_i, f)$ (words are already sorted with respect to $f$).

## 4 Experimental Evaluation

To test the validity of our approach, we developed a prototype implementation of the proposed framework and used the Oxford [12] and the Paris Buildings [13] datasets to assess the quality of retrieved results. As local descriptors, we made use of SIFT using one of the most popular implementations [16]. For building and maintaining the index, we use the Apache Solr full-text indexing framework. In the first dataset, we limited the extracted features per image to 2,000, whereas in the second, a maximum of 1,000 descriptors per image were extracted. Table 2 summarizes these basic statistics.

Each dataset contains manually created ground truth for 55 queries around 11 different landmarks, for which relevant and non-relevant images are known. Thus, each image in the dataset can be characterized by one of four possible states with respect to a given query:

- GOOD: a clear picture of the query object/building.
- OK: more than 25% of the query object is clearly visible.

**Algorithm 1.** Indexing with composite visual words

**Input:** $f, V, a, B$
**Output:** $v^+$ (composite visual words)
  **for all** $v \in V$ **do**
    $v.distance \leftarrow computeDistance(v, f)$
  **end for**
  $sort(V)$
  $d_{max} \leftarrow V.last().distance$
  $i \leftarrow 1$
  $v^+ \leftarrow \{\}$
  **while** $i \leq B$ **do**
    $d_i \leftarrow V[i].distance$
    $threshold \leftarrow e^{-ai}d_{max}$
    **if** $d_i \leq threshold$ **then**
      $v^+ \leftarrow v^+.concat(V[i])$
    **else**
      *break*
    **end if**
    $i \leftarrow i + 1$
  **end while**

**Table 2.** Basic dataset statistics

| Dataset | Oxford | Paris |
|---|---|---|
| $n = \lvert P \rvert$ | 5,063 | 6,412 |
| $\lvert \bigcup_i F_i \rvert$ | 9,731,989 | 6,314,776 |
| $avg(\lvert F_i \rvert)$ | 1,923 | 985 |

- JUNK: less than 25% of the query object is visible, or there are very high levels of occlusion or distortion.
- BAD: the query object is not present.

The calculation of mean Average Precision (mAP) considers as correct the images that fall under 'GOOD' or 'OK' states using the formula:

$$mAP = \frac{\sum_{q \in Q} AP(q)}{\lvert Q \rvert} \qquad (4)$$

where $Q$ is the set of queries, $q$ is an individual query and $AP(q)$ is the Average Precision for a specific query computed as:

$$AP(q) = \sum_{i=1}^{N} pr(i) \cdot \Delta rec(i) \qquad (5)$$

where $i$ is the rank in the list of top $N$ retrieved images, $pr(i)$ denotes the average precision at the $i$- and $(i - 1)$-th position and $\Delta rec(i)$ the difference of recall between the $i$- and $(i - 1)$-th position. We calculated the mAP for two values of $B$ (maximum number of words that may form a composite visual word). Obviously, a $B = 1$ setting does not generate any composite words, which allows us to compare the mAP of the proposed method against the standard BoW indexing scheme. We applied the distance condition of Equation 3 for an empirically selected parameter $a = 0.2$.

Figures 2(a) and 2(b) illustrate the mAP results for increasing size of the initial vocabulary $V$, i.e. increasing number of cluster centres (visual words). At this point, it should be noted that no supplementary mechanism for enhancing the quality of retrieved results was utilized (e.g. soft-assignment [13] or spatial verification [12]). It is noteworthy that the proposed approach achieves a mAP score of 0.383 for as few as 50 centres, whereas the baseline is limited to only 0.087 (Figure 2(a)). Performance appears to level off as $|V|$ rises, but this is natural since $|V|$ affects the size of the composite vocabulary $|V'|$: if $V'$ is too rich then probably different words are assigned to similar feature vectors, mitigating the performance benefits induced by the increased distinctive capability. Nevertheless, we can still build an image retrieval system with acceptable performance in less time, because we would have to quantize the feature space among 50 clusters only. For instance, in the case of k-means, with $O(n \cdot D \cdot k \cdot l)$ time complexity and $k = 50$, we quantize the feature space 10 times faster compared to the case of $k = 500$, incurring negligible loss in mAP (Figure 2(a)).

As illustrated in 2(b), for the Paris Buildings collection, there is still a considerable increase in mAP with our method, with a small decrease for 750 centers, as the baseline tends to outperform the $B = 3$ setting for larger initial vocabularies. Since the size of the composite vocabulary $V'$ depends on that of the initial vocabulary $V$ and constants $a$ and $B$, $V'$ could have been smaller for $B = 2$ or for larger values of $a$ (i.e. a stricter distance condition). In this way, constants $a$ and $B$ offer control over $|V'|$, given any $|V|$. Table 3 depicts the effective vocabulary $V'$ size dependence on the size of the original vocabulary, and for fixed values of $a$ and $B$ in the two datasets.

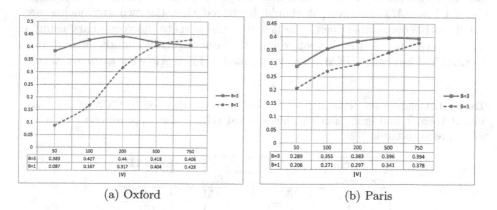

(a) Oxford                              (b) Paris

**Fig. 2.** Comparing standard BoW indexing ($B = 1$) with indexing based on composite visual words ($B = 3$). Retrieval accuracy is expressed with mAP.

**Fig. 3.** Top-3 results of the highest ranked queries for the Oxford and Paris buildings datasets (the query image is the first on the left)

**Table 3.** Size of extended vocabulary $V'$, given $|V|$, $B = 3$, and $a = 0.2$

| $|V|$ | $|V'|$ | |
|---|---|---|
| | Oxford | Paris |
| 50 | 3,688 | 3,220 |
| 100 | 26,978 | 20,613 |
| 200 | 120,950 | 104,810 |
| 500 | 599,683 | 477,239 |
| 750 | 980,644 | 778,499 |

Figure 3 illustrates the results of the proposed approach for two example queries (using $|V| = 200$). In addition, Table 4 presents a comparison between the proposed approach and three other indexing schemes based on the standard BoW model, as they were reported in [12]. It is remarkable that our approach attains a maximum mAP score of 0.44 for just $|V| = 200$ ($|V'| \approx 120K$), while Philbin et al. attained 0.355 mAP with $|V| = 10K$ and only slightly outperform our approach (mAP=0.464) when they use a vocabulary of significantly larger size ($|V| = 1M$). However, it should be noted that the two approaches are not directly comparable due to the fact that they create the vocabulary using 5 millions descriptors. It is also noteworthy that our approach slightly outperforms hierarchical k-means (HKM), which attained 0.439 mAP with as many as 1M

**Table 4.** Comparison between the proposed method, denoted as CVW (Composite Visual Words), and exact k-means, hierarchical (HKM) and approximate (AKM) k-means, as reported in [12]

| Approach | $|\bigcup_i F_i|$ | $|V|$ | mAP |
|---|---|---|---|
| k-means | 5M | 50K | 0.434 |
| HKM | 16.7M | 1M | 0.469 |
| AKM | 16.7M | 1M | 0.618 |
| CVW | 9.7M | $|V| = 200$, ($|V'| = 120K$) | 0.44 |

visual words generated from 16.7M descriptors. Approximate k-means (AKM) reaches 0.618 mAP with 1M cluster centers using a forest of eight randomized k-d trees, but this is not directly comparable to our approach as the authors submit only a subset of the original image as query taking into account only the actual object of interest.

## 5    Conclusions and Future Work

The paper proposed a novel approach for BoW-based indexing of images using a very compact visual vocabulary. The approach is based on the concept of composite visual words, i.e. sequences of visual words ordered based on their distance from the local descriptors of the images to be indexed. An appropriate thresholding strategy was devised to make the proposed indexing scheme effective by eliminating a large number of potentially spurious composite visual words, and an algorithm was described for extracting the composite words to be indexed by an incoming image.

Experiments on two standard datasets revealed that the proposed approach can accomplish decent retrieval results (as expressed by use of mean Average Precision). In particular, a setting of $B = 3$ nearest initial words from an initial vocabulary of 200 words attained 0.44 mAP, outperforming the standard BoW indexing scheme. Moreover, such a small initial vocabulary introduces significant performance gains on quantization and indexing; a vocabulary of 500 words with the standard method reaches comparable retrieval quality to an extended vocabulary (using composite visual words of maximum size $B = 3$) from an initial vocabulary of as few as 50 words, which offers a speedup of 10 for feature discretization. The parsimony of the produced vocabularies makes the proposal ideal for use in contexts with restricted computational resources, e.g. mobile applications.

**Future Work.** In the future, we consider experimenting on additional datasets of much larger scale $(n \sim 10^5 - 10^6)^6$ to investigate the scalability properties of the approach, both in terms of indexing time and in terms of robustness with respect to retrieval accuracy. In addition, we are interested in devising novel vocabulary learning methods that enable further gains in retrieval performance for even larger initial vocabulary sizes. More specifically, we will consider appropriate composite visual word ranking strategies to restrict the part of extended vocabulary $V'$ used for indexing.

As a more specific plan, it will be interesting to further investigate the retrieval performance of our approach in relation to parameters $a$ and $B$. We believe that these values are highly correlated with the size of the initial vocabulary $V$: a large vocabulary requires a strict distance condition, (i.e. a high $a$) and a small $B$, to avoid building an ineffective extended vocabulary $V'$. On the other hand,

---

[6] Or equivalently, we are going to use the ground truth of the two datasets used here with a large number of distractor images.

it would be wise to have a small $a$ and a high $B$ given a small initial vocabulary in order to produce as many useful words as possible. Developing a method for automatically adjusting these two parameters given the size of $V$ appears to offer considerable opportunities for future work.

**Acknowledgments.** Symeon Papadopoulos was supported by the SocialSensor project, partially funded by the European Commission, under contract number FP7-287975.

# References

1. Arthur, D., Vassilvitskii, S.: k-means++: The advantages of careful seeding. In: Proceedings of the Eighteenth Annual ACM-SIAM Symposium on Discrete Algorithms. Society for Industrial and Applied Mathematics (2007)
2. Bay, H., Tuytelaars, T., Van Gool, L.: Surf: Speeded up robust features. In: Leonardis, A., Bischof, H., Pinz, A. (eds.) ECCV 2006, Part I. LNCS, vol. 3951, pp. 404–417. Springer, Heidelberg (2006)
3. Csurka, G., Dance, C., Fan, L., Willamowski, J., Bray, C.: Visual categorization with bags of keypoints. In: ECCV International Workshop on Statistical Learning in Computer Vision, Prague (2004)
4. Elkan, C.: Using the triangle inequality to accelerate k-means. In: Proceedings of the 20th International Conference on Machine Learning (2003)
5. Harris, C., Stephens, M.: A combined corner and edge detector. In: Proceedings of the Alvey Vision Conference, pp. 147–151 (1988)
6. Jurie, F., Triggs, B.: Creating efficient codebooks for visual recognition. In: Tenth IEEE International Conference on Computer Vision, ICCV, vol. 1. IEEE (2005)
7. Lepetit, V., Lagger, P., Fua, P.: Randomized trees for real-time keypoint recognition. In: IEEE Computer Society Conference on Computer Vision and Pattern Recognition, vol. 2. IEEE (2005)
8. Lowe, D.G.: Distinctive image features from scale-invariant keypoints. International Journal of Computer Vision 60, 91–110 (2004)
9. Moosmann, F., Triggs, B., Jurie, F.: Fast discriminative visual codebooks using randomized clustering forests. In: Advances in Neural Information Processing Systems, vol. 19, pp. 985–992 (2007)
10. Nir, A., Jaiswal, R., Monteleoni, C.: Streaming k-means approximation. In: Advances in Neural Information Processing Systems, vol. 22, pp. 10–18 (2009)
11. Nister, D., Stewenius, H.: Scalable recognition with a vocabulary tree. In: IEEE Computer Society Conference on Computer Vision and Pattern Recognition, vol. 2 (2006)
12. Philbin, J., Chum, O., Isard, M., Sivic, J., Zisserman, A.: Object retrieval with large vocabularies and fast spatial matching. In: IEEE Conference on Computer Vision and Pattern Recognition, CVPR 2007 (2007)
13. Philbin, J., Chum, O., Isard, M., Sivic, J., Zisserman, A.: Lost in quantization: Improving particular object retrieval in large scale image databases. In: IEEE Conference on Computer Vision and Pattern Recognition, CVPR (2008)
14. Shotton, J., Johnson, M., Cipolla, R.: Semantic texton forests for image categorization and segmentation. In: IEEE Conference on Computer Vision and Pattern Recognition, CVPR (2008)

15. Sivic, J., Zisserman, A.: Video Google: A text retrieval approach to object matching in videos. In: Ninth IEEE International Conference on Computer Vision, ICCV (2003)
16. Van De Sande, K., Gevers, T., Snoek, C.: Evaluating color descriptors for object and scene recognition. IEEE Transactions on Pattern Analysis and Machine Intelligence 32(9), 1582–1596 (2010)
17. Vedaldi, A., Fulkerson, B.: VLFeat: An Open and Portable Library of Computer Vision Algorithms (2008), http://www.vlfeat.org/
18. Wang, C., Zhang, L., Zhang, H.: Learning to reduce the semantic gap in web image retrieval and annotation. In: Proceedings of the 31st Annual International ACM SIGIR Conference on Research and Development in Information Retrieval, pp. 355–362 (2008)
19. Wu, Z., Ke, Q., Isard, M., Sun, J.: Bundling features for large scale partial-duplicate web image search. In: IEEE Conference on Computer Vision and Pattern Recognition, CVPR (2009)
20. Yuan, J., Wu, Y., Yang, M.: Discovery of collocation patterns: from visual words to visual phrases. In: IEEE Conference on Computer Vision and Pattern Recognition, CVPR 2007, pp. 1–8. IEEE (2007)
21. Zhang, S., Tian, Q., Hua, G., Huang, Q., Li, S.: Descriptive visual words and visual phrases for image applications. In: Proceedings of the 17th ACM International Conference on Multimedia, pp. 75–84. ACM (2009)
22. Zhang, S., Huang, Q., Hua, G., Jiang, S., Gao, W., Tian, Q.: Building contextual visual vocabulary for large-scale image applications. In: Proceedings of the International Conference on Multimedia, pp. 501–510. ACM (2010)
23. Zhao, R., Grosky, W.: Bridging the semantic gap in image retrieval. In: Distributed Multimedia Databases: Techniques and Applications, pp. 14–36 (2002)

# A Probabilistic Index Structure
# for Querying Future Positions of Moving Objects

Philip Schmiegelt[1], Andreas Behrend[2], Bernhard Seeger[3], and Wolfgang Koch[1]

[1] Department SDF
Fraunhofer FKIE, Wachtberg, Germany
`{philip.schmiegelt,wolfgang.koch}@fkie.fraunhofer.de`
[2] Institute of Computer Science III
University of Bonn, Germany
`behrend@cs.uni-bonn.de`
[3] Department of Mathematics and Computer Science
Philipps-Universität Marburg, Germany
`seeger@informatik.uni-marburg.de`

**Abstract.** We are witnessing a tremendous increase in internet connected, geo-positioned mobile devices, e.g., smartphones and personal navigation devices. Therefore, location related services are becoming more and more important. This results in a very high load on both communication networks and server-side infrastructure. To avoid an overload we point out the beneficial effects of exploiting future routes for the early generation of the expected results of spatio-temporal queries. Probability density functions are employed to model the uncertain movement of objects. This kind of probable results is important for operative analytics in many applications like smart fleet management or intelligent logistics. An index structure is presented which allows for a fast maintenance of query results under continuous changes of mobile objects. We present a cost model to derive initialization parameters of the index and show that extensive parallelization is possible. A set of experiments based on realistic data shows the efficiency of our approach.

## 1 Introduction

Due to the advances in GPS-technology, navigational devices are available in almost all vehicles and mobile phones today. These devices are primarily used for the computation of (optimal) routes and for giving instructions to the driver using his/her actual position on a route. Moreover, web services where devices periodically transmit their actual position are very popular. Surprisingly, there is an obvious mismatch between the available information and the information used by these services.

We are convinced that the information about future positions is also very valuable in many application scenarios, and that all of the available knowledge should be exploited. It is very simple to transmit the routes being computed by the navigational devices to these web services and utilize that information.

B. Catania, G. Guerrini, and J. Pokorný (Eds.): ADBIS 2013, LNCS 8133, pp. 112–125, 2013.

In contrast to existing work, in this paper we focus on the efficient indexing of future positions for a high number of mobile objects given their preset trajectories and current positions. Our experimental evaluation confirms that a processing of this information in real-time is possible. Of course, the accuracy of the predicted position decreases over time. This can be modeled by a time variant probability density function, which can be efficiently handled by the methods proposed in this paper.

To illustrate the importance of using knowledge about future positions, let us consider a logistics company that manages a large fleet of trucks. Nowadays it is common that each parcels' position is known with high accuracy. Trucks delivering parcels have a rigid schedule, thus their future position is known with high accuracy. Still this knowledge is not yet exploited. A typical query would then be to select all trucks within a given region at the same time in the near future with a high probability:

```
select * from my_trucks where
position is within
    'my_company_headquarter'
[range (now + 1 hours)
    to (now + 2 hours)]
[probability 90%]
```

In order to optimize the delivery service of goods, it might be important to exchange goods among trucks dynamically. An optimal meeting point for a set of cooperating trucks has to be found:

```
select * from my_trucks t where
distance('truck_a', t) < 5 kilometers
[range 14:00 to 15:00]
[probability 75%]
```

Note that the results of these queries are not only important at the point in time when they will occur, but already earlier at the time when they have been computed. In fact, these "early" results are important for global planning and coordination of a fleet.

These examples show that the efficient determination of the future position of mobile objects represents a relevant problem. To this end, a framework is needed that supports continuous queries over a dynamic set of moving objects whose future travel routes are computed in advance. While the problem seems to be closely related to historical management of trajectories (like in [1]), the dynamic nature of the problem makes it substantially harder to address. Contrary to one-time queries, our problem is more related to continuous queries for the following reasons. First, mobile objects that start a new tour can influence the result sets of continuous queries. Second, traffic jams and other unforeseeable events will have a serious impact on traveling, and consequently also on continuous queries and their results. The difference to other approaches to continuous queries is that results already delivered to the user might become invalid later.

However, these early results are very important to the user for the purpose of planning. The key question of the paper is therefore how to manage the results of a set of continuous queries efficiently and how to update the results due to the occurrence of unforeseeable events. This paper is based on the previous work ([2] and [3]). However, we did not take into account the decreasing probability of the knowledge of an object position over time. This made the approach quite unrealistic for many scenarios. Furthermore, the parameters for the creation of the index had to be determined manually, whereas in this paper a cost model is proposed. We also show that an efficient parallelization of the algorithms introduced in the previous papers is possible. In sum, the additional contributions of this paper are the using of a probability density function (pdf) to approximate an objects position. Second, the introduction of a cost model to allow a precise determination of the parameters of the index structure. And third the added capability to parallelize the algorithms.

## 2 Preliminaries

In this section we introduce the formal definitions of moving objects, trajectories, and queries. We assume that the moving objects are bounded in a two-dimensional universe, e.g., the unit square. Our approach is based on a continuous timeline that allows us to compute all results, even if they are valid only between two discrete timestamps. This is a major difference to the discrete-time model commonly used in data stream management systems like [4].

### 2.1 Trajectories

Within the universe, a path can be specified as a sequence of two-dimensional points $P = (p_1, \ldots, p_n)$, termed waypoints throughout this paper. Associated to $P$ is a series of points in time $T = (t_1, \ldots, t_n)$. A trajectory $traj = (P, T)$ consists of a sequence of points and a sequence of time stamps with equal length $|P| = |T|$. Note that the trajectories of different objects do not need to and in general will not have the same length. Furthermore, if two consecutive waypoints are equal whereas the corresponding points in time are not, idleness is modeled.

For sake of simplicity, we assume that the movement between two waypoints $p_i$ and $p_{i+1}$ can be linearly interpolated. Note that our approach can be generalized to more advanced interpolation functions.

Note that in many scenarios trajectories are confined to road networks. This is, however, not necessary in our approach where arbitrary trajectories are handled.

### 2.2 Moving Objects

A moving object can be any locatable device, e.g., a GPS-enabled mobile phone or a trackable truck. With each object $o$, a trajectory $traj_o$ is associated. Similar to [5], we require that an object is aware of its current position, but also of its future positions and their certainty, not only its current position, direction, and speed. However, this only disqualifies very simple GPS loggers, but even cheap navigational devices have enough storage capacity and computational power to comply with this requirement.

The decreasing accuracy of the predicted position in the course of time is included in our design by using a probability density function (pdf), which allows us to model the uncertainty of the objects' position in a mathematically precise way. It is also possible to include uncertainty about the trajectory. However, the semantically more sound approach is to have a certain trajectory, where the objects current position is uncertain. This also allows for more flexibility, as the trajectories are independent of the type of moving object they are used by.

A moving object is therefore defined as a tuple $M = (traj, \mathcal{P})$, where $traj$ is the associated trajectory and $\mathcal{P}$ can be an arbitrary pdf.

In the remainder of the paper, we will discuss Gaussian functions with expectation $\mathbb{E}$ and covariance matrix $\mathbb{C}$ as pdfs only, for the sake of simplicity. However, any probability density function could be employed. In the two-dimensional case, $\mathbb{E}$ is a two-dimensional vector and $\mathbb{V}$ is a 2x2 matrix.

In the case of moving objects, the mean is defined by the trajectory, whereas the covariance matrix is a function growing with the time difference to the last reported accurate position. The uncertainty of the position of a moving object at time $t$ is therefore modeled by

$$\mathcal{N}\big(p_{uncertain}(t); p(t), \mathbb{V}(t - t_0)\big)$$

where $p_{uncertain}(t)$ is the random variable denoting the uncertain position of the object, $p(t)$ is the predicted position according to the given trajectory and $\mathbb{V}(t - t_0)$ is a covariance matrix, increasing with distance to time $t_0$ where the last certain position was reported. For the sake of simplicity we will assume that $\mathbb{V}(t_0)$ is a diagonal matrix obtained by multiplying the identity matrix with a scalar factor. By that the error of the expected value is the same regardless of the direction. For practical purposes we cut all probability values which are below a predefined threshold $\theta$.

An example is given in Figure 1, where an object moves on the road. The uncertainty at the first timestamp in the foreground of the figure is very small. This is indicated by a high peak of the pdf and a narrow base. On the planned path (further in the background) the uncertainty has increased, shown by a larger covariance and a smaller absolute value for the expected position. Note that the knowledge of a road network will have a great impact on the shape of the pdf. This is, however, beyond the scope of this paper and does not affect the methods used for indexing. Using such meta information will be addressed in future work.

Introducing probability in the context of moving objects introduces a broad variety of challenging tasks like tracking, filtering, or retrospection (see [6]). These problems are, however, beyond the scope of this paper.

**Fig. 1.** Object Moving with Uncertain Position

### 2.3   Queries

A continuous query $q$ over mobile objects is formally defined as

$$q = (pred(o), [t_{start}, t_{end}], \Theta_{(optional)}),$$

where $[t_{start}, t_{end}]$ denotes the time interval within the query result is of interest
to a user. $pred(o) \mapsto \{[s_0, e_0], \ldots [s_n, e_n]\}$ is a function returning for every moving
object $o$ the set of time intervals when $o$ qualifies for the query predicate $pred$.
These intervals are non-overlapping, and adjacent intervals are required to be
coalesced. Once a query is created, results are received continuously. $\Theta$ is an
optional probability threshold for this query. It is possible to specify that, e.g.,
only results having a probability greater then 95% should be reported. It is
worthwhile noting that this cannot be specified within the query predicate as
the predicate ultimately decides whether a moving object can be a result of the
query, e.g., if the object might be in the query region at any point in time.
However, the probability can dynamically change when after the transmission
of the expected trajectory the correct position is reported. The probability of
query result changes after the transmission, and previously discarded results
might have an updated probability which is higher than the query threshold $\Theta$.

## 3   A Probabilistic Index Structure

### 3.1   Index Structure

For indexing the spatial domain, a simple grid index [7] is used. The basic idea
of a grid index is to partition the spatial data space into partitions (cells) using
a two dimensional quadratic grid. Each cell of the grid index contains pointers

to the objects that intersect this cell. The advantage of such a simple structure is that the uncertainty of the position of a moving object can be included in an intuitive way. Pointers are not only generated for cells intersecting the (certain) trajectory, but for all cells the object might be within. The probability threshold $\theta$ is used to avoid having too many cells affected with a very low probability of the object actually being there. The computation of the overlap with the object's pdf is straightforward if a Gaussian pdf is used and the covariance matrix is monotonically increasing. The first is assumed in this paper for sake of simplicity, the second is a quite natural thought. This implies that the covariance matrix, dependent on time t, is defined as

$$\mathbb{C} = \begin{Bmatrix} a & 0 \\ 0 & a \end{Bmatrix} \cdot t = t \cdot a \cdot \mathbb{I}.$$

As the error ellipses are circles for sake of simplicity, as defined in the beginning of the paper, the radius $r$ has to be calculated as the only parameter of the circle. As in our two dimensional setting both spatial coordinates ($p_x$ and $p_y$) are normally distributed, the circle radius $r^2 = p_x^2 + p_y^2$ is $\chi^2$ distributed. Thus, for each desired confidence interval $p$, e.g. 95%, the value of the $\chi^2$ distribution can be conveniently looked up in a $\chi^2$ table, or be computed on the fly. The radius $r$ is therefore given by $r = \sqrt{\chi^2(p) \cdot a}$. In case of $p = 95\%$, a lookup in the $chi^2$ table shows that $chi^2(95\%) = 0.103$

As time passes, uncertainty on positions grows. In this case, it is sufficient to calculate the extent of the Gaussian pdf which is above the threshold $\theta$ for each waypoint. The resulting trapezoid is a superset of the region the object might be in. Note that a superset is sufficient for the index to function properly, as the actual intersection of the object's pdf and a query region always has to be calculated. Only performance might suffer if the superregion is too large. It is important to note, however, that the index is independent of the actual pdf used. Other distributions unlike the Gaussian could be employed. Only the calculation of the confidence region has then to be adapted, the index itself remaining unchanged.

The temporal index does not store any probability information, as time is assumed to be certain, i.e., all clocks are synchronized. An anomaly regarding time has to be taken into account, however. If the end time of the trajectory is closely before the start time of a query, and both the last waypoint and the border of the query are in close spatial vicinity, the query will be ignored. Due to the uncertainty in space, however, the query predicate might still be fulfilled with a high probability. This anomaly can be circumvented by adding 'dummy' elements to the trajectory, where the last waypoint is repeated. That is, the moving object is not immediately regarded as deleted, but treated as a non-moving object for some time, until the probability decreased below the threshold $\theta$. As this can be done programmatically, the described anomaly can automatically be handled and correct results are delivered.

In Figure 2, the interaction of the spatial and the temporal index is illustrated. A trajectory and a temporally bounded range query are stored in this example. The range query is shown in light green. As it touches four cells of the spatial grid, a reference to the query is stored in each cell. Also, references of the query are added to the temporal index in each cell overlapping with the lifetime of the query. The same is done for the trajectory. All cells where the probability of the object being in this cell is greater then $\theta$, a reference to the object is stored. These cells are marked in grey.

## 3.2   Cost-Based Parameters

This subsection introduces a cost model that can be used to set the grid resolution of the spatial grid indexes appropriately. This means that once the approximate lengths of the routes and the sizes of the queries are known, the optimal resolution of the index can be calculated with a simple formula. We assume here the case of a general update and a constant pdf for each object, which leads to a pessimistic model.

Let us consider the unit square to be the universe. The grid resolution $r$ is the number of slices in each dimension of the grid index. We assume that the resolution is constant for every dimension. However, we distinguish the resolution of the query index ($r_q$) and the resolution of the object index ($r_o$). Note that the side length of a grid cell is then $\frac{1}{r_o}$ and $\frac{1}{r_q}$, respectively. Overall, there are $|Q|$ queries and $|O|$ objects being distributed uniformly across the universe. We assume a constant number of objects and a constant number of queries. The average length of a trajectory is given by $l_{trajectory}$. We first examine range queries, all of them being rectangular with extension $l_{query}$ in every dimension.

In the worst case, the trajectory of an object crosses $l_{object} \cdot r_o$ grid cells. Analogously, $l_{query}^2 \cdot r_q^2$ cells are occupied by a single query. This means that a cell contains approximately $\frac{|O| \cdot (l_{object} \cdot r_o)}{r_o^2} = \frac{|O| \cdot l_{object}}{r_o}$ objects and $\frac{|Q| \cdot l_{query}^2 \cdot r_q}{r_q^2} = \frac{|Q| \cdot l_{query}}{r_q}$ queries.

The cost of an insertion of a moving object can therefore be estimated by

$$\underbrace{l_{object} \cdot r_o}_{\text{cells to consider}} \cdot \underbrace{\frac{|Q| \cdot l_{query}}{r_q}}_{\text{queries per cell}}$$

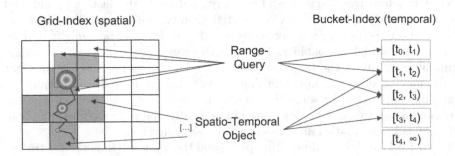

**Fig. 2.** Decoupled Indexes

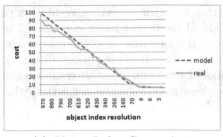
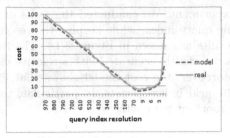

(a) Object Index Comparison        (b) Query Index Comparison

**Fig. 3.** Confirmation of Cost-Model

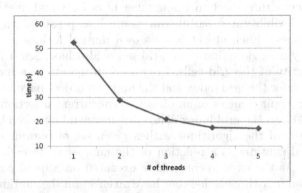

**Fig. 4.** Sizes of Thread Pool

$$= l_{object} \cdot |Q| \cdot l_{query} \cdot \frac{r_o}{r_q}$$

By using the same parameter set, we now examine the cost of a general update. First, the grid index has to be updated. The object has to be found and deleted, before the new trajectory can be inserted. $(r_o \cdot l_{object}) \cdot log(\frac{|O| \cdot l_{object}}{r_o})$. In the next step, the cells intersecting with the query have to be retrieved. $r_q \cdot l_{object} \cdot (\frac{|Q| \cdot l_{object} \cdot l_{query}}{r_q})$

Finally, new events have to be calculated for each qualifying query. Since this cost is constant for a given instance, it is not included in the cost model.

We also experimentally confirmed our model. The trajectories for the experiments were created synthetically to match the underlying assumptions of the cost model. However, we also found that the cost model is sufficiently accurate for realistic data sets, created by Brinkhoffs generator[8]. In order to demonstrate the accuracy we compared the runtimes with the cost model, too. This, however, is only possible by not using the pure numbers of operations as a cost measure, but to introduce an additional weighting factor for each operation to express the required CPU-time.

Figure 3(a) shows the time needed to access the object index as a function of its resolution. The resolution of the query index was kept constant in this experiment. We conducted the same experiments with a varying resolution of the query index, while keeping the resolution of the object index constant. The results are shown in Figure 3(b). The results demonstrate that the cost model is sufficiently accurate (with a relative error less than 10% for almost all cases). Moreover, the cost model shows that a global minimum for the resolution of the grid indexes exists. In particular, this observation is useful for setting this important parameter.

### 3.3 Parallelization

For the developed algorithms synchronization is essential due to the extremely high concurrency with respect to the number of objects and queries. Our goal was to provide flexibility and parallelism as much as possible, while still guaranteeing correctness. Each object has its own mutual lock, allowing multiple readers but only a single writer. The grid index also has such a lock, guaranteeing the integrity of the grid cells. In case of an update of a moving object, both a write-lock for the grid index and the updated object have to be acquired, preventing concurrent changes of an object or concurrent insertions of multiple queries (objects) into the grid index. The efficiency can be proven by measuring the total runtime of the algorithms with a given set of parameters. Figure 4 depicts the total runtime as a function of the number of threads available to the algorithms. As the experiments where executed on a quad core processor, having more than four threads does not have an effect on the runtime. The index structures and algorithms are also designed to be executed in a shared-nothing environment[9]. As this task is purely engineering due to the inherent ability of all methods to run in parallel it is not discussed in this paper.

## 4    Experiments

A set of experiments was conducted to validate the performance of our proposed approach to supporting continuous queries.

### 4.1 Dataset and Experimental Setup

Unfortunately, no real world dataset is available due to the novelty of our approach. Nevertheless, we tested the proposed data structures and algorithms on a real road network. Due to its popularity, we choose to use Brinkhoff's moving object generator[8] to simulate the moving objects. In our experiments, we restrict the discussion of the results to the network generated from the street map of the San Francisco Bay Area. The usage of other maps showed similar effects, and therefore, are omitted.

Our default setting of the generator results in a simulation of 10,000 moving objects and 1,000 randomly distributed range queries, each of them covered 1% of the data space as these numbers match with a typical application in the area of fleet management. Note that the algorithms are easily parallelizable and therefore using a larger experimental setting, e.g., a typical one in the domain of social networks, can be done. The experimental results for this are given in Figure 4. A more detailed description can be found in section 3.3 of this paper.

The algorithms presented here were implemented in JAVA and were performed on an Intel Dual Core Xeon 3.4 GHz with 6GB RAM running CentOS 5. The JAVA virtual machine occupied 3GB RAM.

## 4.2  Comparison

In the following, we compare our approach with the one presented in [5]. Let us first describe it briefly. Similar to our approach, the queries are stored in a grid index. There the moving objects are assumed to move along a straight line using the speed and the direction as reported at the last position. In fact, moving objects are required to report their position whenever their route or speed changes. This information is then used to compute the current results of each query. Note that future results are obtained by a simple extrapolation of the linear movement of the object. To limit the length of this predicted line, a *maximum-update-interval (MUI)* is defined. The period between two updates of a moving object must be within the MUI. An example is given in Figure 5. The trajectory is implied in the center of the figure, the necessary transmissions using a linear extrapolation approach above it, the single transmission of the trajectory approach beneath. The third transmission is necessary because the MUI has been reached, although the route did not change and the linear prediction is still correct.

We are aware that there are fundamental differences between this approach and ours, especially as it does not take probabilistic values into account. Recall that our approach delivers all results of a continuous query starting at the time the query is issued and results are updated when changes occur. Nonetheless we opted to use this competitive approach for our comparison, since it is still the one that is closest to our approach among those published so far. To keep the comparison as fair as possible, we compared the average total processing time of each object in both approaches. This means we added up the costs of all operations performed on a single object.

To model the dynamic changes of a trajectory, caused by a traffic jam, at each waypoint an object is updated with a given probability. This probability is called the *update rate*. We examined update rates between 0% and 15% in the experiments. In most real-life applications, this update rate will presumably be very small as we assume that a user follows the suggested track of the navigation system. The higher the update rate the higher the costs of our approach, while the update rate has no impact on the competitive approach.

Figure 6 depicts the average time for applying all updates to a moving object as a function of the MUI. It provides the costs of temporal updates for update rates 1%, 5%, 10% and 15% , In addition, the curve is plotted for the competitive approach. The costs of the competitive approach increase with increasing MUI, while the curves of our approach are obviously independent from MUI. Note that the range of the MUI is chosen in accordance with the recommendations

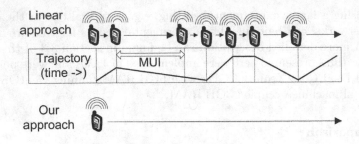

**Fig. 5.** Necessary Transmissions

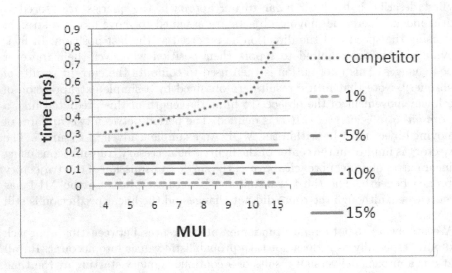

**Fig. 6.** Update Performance

in [5]. The results show that the performance of the competitive approach is superior to our approach only in case of general updates and small settings of MUI. For MUI > 6, our approach with general updates only is even superior to the competitive approach, If temporal updates are applicable, a large difference in the performance can be observed.

## 5   Related Work

There has been a large interest in supporting queries on moving objects, but only few papers address the problem of continuous queries on these objects. Most similar to our work is the QMOS (Query Moving Object Stream) system [5]. However, the fundamental difference to our work is that the knowledge about future trajectories is not exploited for query processing. Instead, the author predicts the future movement by only using a linear extrapolation. This requires that objects permanently transfer their current positions, an overhead that can be avoided in our approach. Moreover, the temporal horizon of continuous queries is very limited, while there is no limitation in our approach on how far a query is in the future.

Most related work (e.g. [10,11]) deals with indexing of historical data. In general however, these methods are not applicable to continuous queries as they lack the option to update predicted future trajectories. As emphasized before, the efficient processing of updates is of utmost importance in the context of continuous queries.

Using stream processing technology in the context of moving objects is suggested by different papers. The main idea of SCUBA [12] is to group objects into clusters. Moving objects are constrained to a road network and an update of the direction can occur when a crossroad is reached. PLACE [13] also offers the evaluation of moving queries on moving objects. The moving objects have to submit their position continuously to the system, which then decides whether the information is processed or discarded. Though queries are constantly reporting changing results, continuous queries are not supported. SINA (Scalable INcremental hash based Algorithm) [14] is similar to PLACE, but uses a different data structure.

The problem of managing trajectories led to the development of different kinds of index structures. Most are based on the TPR*-tree[15], an index structure of the R-tree family enhanced to better support the temporal domain. STRIPES [10] is based on dual transformations, but its limitation is that its efficiency is only acceptable in case of a small time-horizon, whereas a larger horizon leads to a rapid degeneration of query performance. The horizon has to be set in advance and cannot be adapted to the incoming data. As a consequence, moving objects have to report their position frequently, even if neither the path nor the speed has changed. Another approach for moving objects is to keep the objects in a one-dimensional index structure. Space filling curves are used by the authors of [11] for their index structures in combination with the popular B-Tree. In general, these kinds of adapted B-trees can handle updates more efficiently than methods derived from R-trees and are therefore the preferred choice in dynamic settings. However, as objects are still assumed to move in a linear fashion, the limitation of these approaches is that only the near future of moving objects can be indexed.

There are a few papers dealing with future trajectories. In [16], trajectories are considered uncertain within a given bound. Queries are defined on these trajectories, allowing the user to specify, for example, whether an object is possibly or definitely inside a given region. Based on a common relational database system, [17] proposes methods for bulk updating trajectories when a road obstruction occurs. The processing is based on triggers and relational database technology. This severely limits performance and does not scale well. [18] focuses on insertion and deletion of trajectories without considering the problem of updating results.

While there are indexing methods for historical trajectories, as for example [19,20], we are not aware of approaches indexing future trajectories. For indexing historical data, updates on past states are not allowed. The proposed methods are therefore not applicable to the problem of efficiently processing future trajectories, where updates frequently occur.

Related to our approach have been methods from data stream processing that are designed for processing predictive queries [21]. However, these methods are not designed for managing trajectories, but for supporting current and near future results only.

# 6   Conclusion and Future Work

In this paper we extended index structures and algorithms for processing the future trajectories of moving objects. The decreasing accuracy of the prediction was accounted for by using a probability density function. A cost model for determining the index parameters was introduced. It was also shown that the proposed methods can be efficiently parallelized.

In our present and future work, we address more advanced continuous queries like nearest neighbor queries. The predicates of these queries cannot be evaluated on the base of a single object, but require the inspection of multiple objects. We will also include the efficient processing of multi-modal pdf, where, e.g., at an intersection, many alternative routes are possible.

# References

1. Patroumpas, K., Sellis, T.K.: Managing trajectories of moving objects as data streams. In: STDBM 2004, pp. 41–48 (2004)
2. Schmiegelt, P., Seeger, B.: Querying the future of spatio-temporal objects. In: ACM GIS 2010, pp. 486–489 (2010)
3. Schmiegelt, P., Seeger, B., Behrend, A., Koch, W.: Continuous queries on trajectories of moving objects. In: IDEAS 2012, pp. 165–174 (2012)
4. Krämer, J., Seeger, B.: Semantics and implementation of continuous sliding window queries over data streams. TODS 34(1), 1–49 (2009)
5. Lin, D., Cui, B., Yang, D.: Optimizing moving queries over moving object data streams. In: Kotagiri, R., Radha Krishna, P., Mohania, M., Nantajeewarawat, E. (eds.) DASFAA 2007. LNCS, vol. 4443, pp. 563–575. Springer, Heidelberg (2007)
6. Koch, W.: On bayesian tracking and data fusion: A tutorial introduction with examples. IEEE AESS Magazine 25(7), 29–52
7. Samet, H.: The Design and Analysis of Spatial Data Structures (Addison-Wesley). Addison-Wesley Pub. (Sd)
8. Brinkhoff, T., Str, O.: A framework for generating network-based moving objects. Geoinformatica 6 (2002)
9. DeWitt, D., Gray, J.: Parallel database systems: the future of high performance database systems. Commun. ACM 35, 85–98 (1992)
10. Patel, J.M., Chen, Y., Chakka, V.P.: Stripes: An efficient index for predicted trajectories. In: SIGMOD 2004, pp. 637–646 (2004)
11. Jensen, C.S., Lin, D., Ooi, B.C.: Query and update efficient b + -tree based indexing of moving objects. In: VLDB 2004, pp. 768–779 (2004)

12. Nehme, R.V., Rundensteiner, E.A.: Scuba: Scalable cluster-based algorithm for evaluating continuous spatio-temporal queries on moving objects. In: Ioannidis, Y., Scholl, M.H., Schmidt, J.W., Matthes, F., Hatzopoulos, M., Böhm, K., Kemper, A., Grust, T., Böhm, C. (eds.) EDBT 2006. LNCS, vol. 3896, pp. 1001–1019. Springer, Heidelberg (2006)

13. Mokbel, M.F., Xiong, X., Hammad, M.A., Aref, W.G.: Continuous query processing of spatio-temporal data streams in place. Geoinformatica 9(4), 343–365 (2005)

14. Mokbel, M.F., Xiong, X., Aref, W.G.: Sina: Scalable incremental processing of continuous queries in spatio-temporal databases. In: SIGMOD 2004, pp. 623–634 (2004)

15. Tao, Y., Papadias, D., Sun, J.: The tpr*-tree: An optimized spatio-temporal access method for predictive queries. In: VLDB 2003, pp. 790–801 (2003)

16. Trajcevski, D., et al.: Managing uncertainty in moving objects databases. TODS 29(3), 463–507 (2004)

17. Ding, H., Trajcevski, G., Scheuermann, P.: Efficient maintenance of continuous queries for trajectories. Geoinformatica 12(3), 255–288 (2008)

18. Chon, H.D., Agrawal, D., El Abbadi, A.: Range and knn query processing for moving objects in grid model. Mob. Netw. Appl. 8(4), 401–412 (2003)

19. Hadjieleftheriou, M., Kollios, G., Tsotras, J., Gunopulos, D.: Indexing spatiotemporal archives. The VLDB Journal 15(2), 143–164 (2006)

20. De Almeida, V.T., Güting, R.H.: Indexing the trajectories of moving objects in networks. Geoinformatica 9(1), 33–60 (2005)

21. Dittrich, J., Blunschi, L., Vaz Salles, M.A.: Indexing moving objects using short-lived throwaway indexes. In: Mamoulis, N., Seidl, T., Pedersen, T.B., Torp, K., Assent, I. (eds.) SSTD 2009. LNCS, vol. 5644, pp. 189–207. Springer, Heidelberg (2009)

# An Adaptive Method for User Profile Learning

Rim Zghal Rebaï, Leila Ghorbel, Corinne Amel Zayani, and Ikram Amous

MIRACL-ISIMS, Sfax University, Tunis road Km 10, 3021 Sfax, Tunisia
rim_zghal@yahoo.fr, leila.ghorbel@gmail.com, zayani@irit.fr,
ikram.amous@isecs.rnu.tn

**Abstract.** The user profile is a key element in several systems which provide adapted result to the user. Thus, for a better quality of response and to satisfy the user, the profile's content must always be pertinent. So, the removal of irrelevant content is necessary. In this way, we propose in this paper a semi-supervised learning based method for automatically identifying irrelevant profile elements. The originality of this method is that it is based on a new co-training algorithm which is adapted to the content of any profile. For this, our method includes a preparation data step and a classification profile elements process. A comparative evaluation by the classical co-training algorithm shows that our method is better.

**Keywords:** user profile, navigation history, data preparation, semi-supe-rvised learning, co-training technique.

## 1 Introduction

In order to take into account of user's interests, navigation history, preferences etc. several systems, such as adaptive systems, use user profile. This latter must be automatically and frequently updated by the system after each user-system interaction. By the time, and especially after several addition operations, the profile can be overloaded and contains relevant and irrelevant elements. This can affect the result's relevance. For this, the solution is to remove irrelevant elements which are detected by an automatic classification.

In this way, several methods are proposed such as [15] and [10]. These methods use mainly a learning technique [16] which can be unsupervised or supervised. These techniques are used to identify the relevant new elements to be added to the profile or to identify the irrelevant already existing elements to be removed from the profile. The main objective of these methods is that the profile should always be not-overloaded and pertinent. But, by using one of these learning techniques the result is generally not adapted to any user's profile. Especially because each user has his specific interests, preferences, history, etc. and these techniques are applied in the same way whatever the profile's content.

Our contribution in this paper is to propose an automatic classification method adapted to the content of any user's profile. It is based on the semi-supervised

B. Catania, G. Guerrini, and J. Pokorný (Eds.): ADBIS 2013, LNCS 8133, pp. 126–134, 2013.

learning technique by using a new co-training algorithm. In fact, in the literature, several studies have proposed new version of co-training algorithm to make
it adapted to their addressed problem, among them we cite [8].

The remaining of this paper is presented as follows. Section 2 presents briefly
the required steps to apply the learning techniques and some user profile elements
classification works. In section 3, we present our method. Section 4 depicts the
evaluation of our proposal. Section 5 presents the conclusion and future work.

## 2   State of the Art

Automatic learning has been attracting a significant amount of research fields
such as information researches, image processing, etc. Before being applied, the
data preparation must be performed [5]. It is generally composed of three steps:
selection, preprocessing and transformation step. The first step allows to identify
the required data for the classification. Then, these data are copied and displayed
in a matrix which describes elements and their attributes. The second step deals
with cleaning the data in order to correct any inaccuracies or errors such as
duplicates, missing information, etc. The third step is to enrich, normalize and
code the data to apply the classification. The enrichment is done by adding
new attributes. Normalization and coding are done by regrouping or simplifying
attributes (coding of discrete attributes, changing type, etc.).

In the literature, we find several techniques of automatic learning. The most
used techniques are: the unsupervised, supervised and semi-supervised techniques. In this paper, we are interested in works that apply these techniques
on user's profiles either to classify the already existing profile elements in order
to remove the irrelevant ones such as [15], [10] and [3] or to classify the new
elements to be added to the profile such as [4], [11] and [14]. These latter can
provide pertinent profiles but after several updating operations, the profile can
be overloaded. Moreover, the profile cannot contain all the user's various interests for the reason that a new interest can be added only if there is a similarity
with the already existing interests. For these reasons, we are interested in the
methods that use learning techniques to classify the already existing profile elements. In [3] the profile elements are represented as hierarchical categories. Each
category represents the knowledge about a user interest and has an energy value.
This value increases when the user shows interest in the category and decreases
by a constant value for each period of time. Based on the energy value, the system classifies the categories: categories that have low-energy will be removed and
categories that have high-energy will persist. The proposed method in [15] allows
to classify the profile elements (here profile concepts) by using association rules
and Bayesian networks. The relevant concepts are maintained. In [10] authors
are based on the supervised learning technique by using the K-NN algorithm.
The classifier uses labeled users preferences pool to classify the preferences of
each user.

As we said at the beginning of this section, all these presented methods should
perform the data preparation before applying the most appropriate learning

technique to their contexts. By using these techniques, these methods [15], [10] and [3] are able to identify irrelevant profile elements. However, all these methods are based on the supervised learning technique which provides a prediction model that is not usually adapted on any user profile.

In this paper, we propose a method of profile elements classification adapted to the content of any profile. This method is based on a semi-supervised learning technique and uses a new co-training algorithm.

## 3    New Method Based on Co-training Algorithm

To classify profile elements into relevant/irrelevant, we propose an automatic method which can be applied to overloaded users' profiles. These profiles respect the profile model proposed in [20] and are composed of several parts. In this work, we are interested on the navigation history part which contains mainly the visited domains. Each visited domain is composed of the visited sub-domains, if exist, the visited documents and the visited links. These profiles are obtained after several navigation sessions in the INEX 2007[1] corpus which is part of the collection WIKIPEDIA XML. The 110000 XML documents in this corpus are related to one or more domains and interconnected by XLINK simple links. The used navigation method and the updating process are detailed in our previous work [19].

To apply the appropriate learning technique on these profiles, the data preparation is required. In our work, it consists of two steps: (i) selection, and (ii) transformation of data.

### 3.1    Data Preparation

The purpose of the data preparation is to provide a set of labeled user profiles on which the classification will be based. It consists of two steps: the selection of data related to each element (visited domains, visited sub-domains, visited documents and visited links) and the transformation of these data (cf. Fig. 1).

The data selection step consists in extracting the four user profile elements and their attributes. Table 1 presents all the extracted attributes.

These attributes describe mainly the different identifiers, the number of visits and clicks, the date of visits and clicks and the duration of visits.

After this step, we obtain four databases for each profile. These databases are the input of the data transformation step. This step allows to: (i) change the coding of some attributes, (ii) enrich the databases by adding new attributes and (iii) filter some not-discriminates attributes. The main objectives of this step are to facilitate the semi-automatic labeling of the profiles and improve the precision rates of the classification. For the coding of attributes, we have changed the coding of the attributes related to the date of visit in order to differentiate the recent dates and not-recent dates. So, each date will be replaced either by

---

[1] http://www-connex.lip6.fr/~denoyer/wikipediaXML

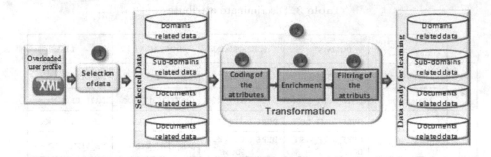

**Fig. 1.** Data preparation

**Table 1.** Selected elements and attributes from profile

| Element \ Attributes | DOMAIN | SUB-DOMAINE | DOCUMENT | LINK |
|---|---|---|---|---|
| Identifier | ID_DOM | ID_SDOM | ID_DOC | ID_LINK |
| Number of visit | NB_VISIT_DOM | NB_VISIT_SDOM | NB_VISIT_DOC | NB_VISIT_LINK |
| Date of visit | DATE_VISIT_DOM | DATE_VISIT_SDOM | DATE_VISIT_DOC | DATE_VISIT_LINK |
| Duration of visit | | | DURE_VISIT_DOC | |

**Table 2.** Added attributes of each element

| Element \ Attributes | DOMAIN | SUB-DOMAINE | DOCUMENT | LINK |
|---|---|---|---|---|
| Duration of visit | DURE_VISIT_DOM, DURE_MOY_VISIT_DOM (Domain average duration of visit) | DURE_VISIT_SDOM, DURE_MOY_VISIT-_SDOM (Sub-domain average duration of visit) | DURE_VISIT_DOC | DURE_VISIT_DOC |
| Number of visit | NB_SDOM, NB_DOC NB_LINK, NB_CLICK_TOT (Total number of the clicked links) | NB_DOC, NB_LINK, NB_CLICK_TOT (Total number of the clicked links), NB_VISIT_DOM (Domain Number of visit of sub-domain) | NB_VISIT_DOM, NB_LINK, NB_CLICK_TOT (Total number of the clicked links) | NB_VISIT_DOC, NB_VISIT_DOM |

"R" (ie. Recent date) or by "NR" (ie. Not Recent date). As for the enrichment of the databases, we firstly added to each element some attributes related to the other elements. Table 2 illustrates the added attributes of each element.

Secondly, we have semi-automatically labeled the profiles elements and added the labels "+" for the relevant elements and "-" for the irrelevant ones. The labeling starts with domains, sub-domains, documents and links. It is performed as follows; if an element is irrelevant then all its child elements are automatically labeled as irrelevant, otherwise all its child elements must be manually labeled.

**Table 3.** Discriminate attributes

| | Element\ Attributes | DOMAIN | SUB-DOMAINE | DOCUMENT | LINK |
|---|---|---|---|---|---|
| **SET 1** | Number of visit | NB_VISIT_DOM | NB_VISIT_SDOM | NB_ACCES | NB_CLIC |
| | Date of visit | DATE_VISIT_DOM | DATE_VISIT_SDOM | DATE_VISIT_DOC | DATE_CLIC |
| **SET 2** | Duration or number of visit | DURE_VISIT_DOM | DURE_VISIT_SDOM | DURE_VISIT_DOC | DURE_VISIT_DOC |
| | | DURE_MOY_VISIT_DOM | DURE_MOY_VISIT_SDOM | NB_VISIT_DOM | NB_VISIT_DOC |

Finally, based on the obtained labeled profiles, we proceeded to the filtering of the most discriminate attributes for classification. For this, we used ReliefF [9] algorithm. In the literature, there are many different algorithms for attributes selection such as Fisher filtering, Feature ranking, etc.; we are interested in ReliefF algorithm because it is unaffected by attributes interaction [9].The selected attributes are illustrated in table 3.

At the end of the data preparation, we obtain a set of labeled user profiles that represent labeled pool on which our proposed classification process will be based.

## 3.2  Profile Elements Classification

In our work, the data preparation and especially the semi-automatic labeling of a large number of users' profiles is a considerable work and the classification task must be adapted to the content of any user's profile. For these reasons, the semi-supervised learning technique has been chosen to classify automatically user's profile elements.

In the literature, several techniques for semi-supervised learning are proposed such as the self-training, the co-training, S3VM and T-SVM. As it is simple, able to provide better adapted result to classification problems of data, we are more interested in co-training.

Based on [17], the idea of co-training is that two separated classifiers are trained using the data of the Labeled Pool (LP) having two sub-attribute sets respectively. For the reason that co-training assumes that attributes for training must be split into two sets. Each sub-attribute set is sufficient to train a classifier and the two sets are conditionally independent given the class. Then, each classifier generates a prediction model. Based on this model, the classifier assigns labels to unlabelled data given as input. After that, the most confident predicted ones (obtained by the two classifiers), are selected and added to LP and the process repeats. When training is completed, after n rounds, the labels of the data to be classified are predicted. This process of co-training has been successfully applied to several classification fields. In our case, this process cannot robustly classify the elements of any profile. In fact, the content of any profile varies from one user to another according to the history of each one (durations of visits, total duration of sessions, etc.). So, two generated prediction models (by the two classifiers) based on one LP (the set of the labeled profiles result of the data preparation), which contains a mixture labeled data from several various users' profiles, cannot usually be applied to any profile and cannot provide good classification result. Therefore, we propose a new method based on a new co-training algorithm which can be adapted to the content of any user's profile. This method is based on N Adapted Labeled Pools (N-ALP). The initial content of N-ALP is similar to the content of the N-LP. Each LP consists in an overloaded labeled user profile. That means that we considered that each profile represents one LP.

To choose the best classifier (learning algorithm), we first made the choice of the supervised learning technique. In literature, there are several techniques of supervised learning. One of the criteria to compare these techniques is the comprehensibility of the generated prediction model. Based on this criterion, we choose the induction of decision trees technique [1]. We have applied to a set of overloaded profiles nine algorithms (ADTRee [18], C4.5 [12], DecisionStump [7], ID3 [13], RandomFoorest [2], and REPTree [13]) and we have obtained the best values of F-Measure and Classification by REPTree. So, we used two REPTree classifiers. Each classifier is based on a set of attributes, set 1 and set 2 (cf. Table 3). A classifier uses attributes related to date and number of visits and the other one uses the duration and number of visit related attributes. At the first round, these classifiers are trained based on the two attributes sets of N-LP and the unlabeled overloaded profile P (the input). So, 2*N prediction models are generated and the most confident labeled elements from each class (relevant/irrelevant) based on these models are added to N-ALP (elements from P). Then, the two classifiers are retrained n-1 rounds on N-ALP and P and after each round 2*N new prediction models are generated and the most confident labeled obtained elements are added to N-ALP. This process is applied to profile elements in the following order: domains, sub-domains, documents and links to obtain a labeled user profile. We only have a filtering step that eliminates the irrelevant elements to obtain a pertinent and not-overloaded user profile.

## 4   Evaluation

For the evaluation of our proposed method, 20 users have navigated for several sessions in the INEX 2007 corpus until we obtain 20 overloaded profiles. Based on these latter, we carried out a series of experiments. We begin by evaluating the classical co-training algorithm and our proposed co-training algorithm after applying them only to domain elements. Then, we finish by evaluating our algorithm after applying it to all profiles elements (domains, sub-domains, documents and links).

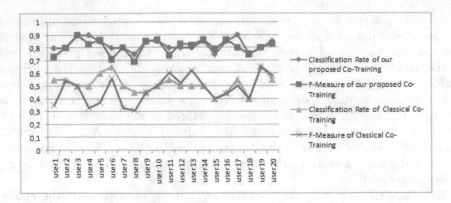

**Fig. 2.** Comparison between the classical co-training and our proposed co-training algorithm

So, for the first evaluation we proceed to the labeling of the profiles' domains elements: (i) manually by the 20 users themselves, (ii) based on the classical co-training algorithm and (iii) based on our proposed method. Then, we compare the profiles obtained by applying the classical and our co-training algorithm with the manual labeled profiles. For this, we used the F-Measure (FM=2*recall*precision/recall+precision) and the classification rates (CA=1-ER, ER is the error rate).

Figure 2 presents the result of the comparison between the classical co-training labeled profiles and our proposed co-training algorithm labeled profiles. We can notice that the best obtained CR and FM by the classical algorithm are 0.65. For example, for user 14 their respective values are: 0.5 and 0.5. In addition, this user labeled manually 11 domains as relevant, while classical co-training algorithm provides only 6 relevant domains. Whereas, by applying our algorithm the CR and FM are improved. For the same user 14, the values of CR and FM are respectively 0.85 and 0.86. Moreover, our method was labeled 9 relevant domains among 11.

In fact, the average CR increases from 0.515 to 0.8225 and the F-Measure increases from 0.4725 to 0.8055. For example, for the user 14 the CR was 0.40 and becomes 0.85. As for user 4 the FM was improved from 0.33 to 0.83. Thus, the obtained results in figure 2 prove the efficiency of our method.

For the second evaluation, we labeled all the profiles' elements: (i) manually by the 20 users and (iii) based on our proposed algorithm. Then we compare the obtained profiles by using the F-Measure and CA. Figure 3 depicts the obtained average values of the CR and F-Measure.

Based on the values illustrated in figure 3, we obtain 0.9156 as FM average value and 0.8263 as average CR. These values can confirm the effectiveness of our method.

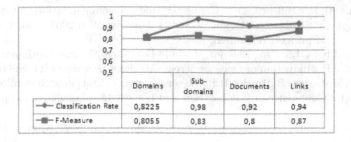

| | Domains | Sub-domains | Documents | Links |
|---|---|---|---|---|
| Classification Rate | 0,8225 | 0,98 | 0,92 | 0,94 |
| F-Measure | 0,8055 | 0,83 | 0,8 | 0,87 |

**Fig. 3.** Evaluation of our method on the four profile elements

## 5  Conclusion

In this paper, we presented the different steps which provide a pertinent user's profile from an overloaded one: the data preparation and the classification process. This latter is based on a new co-training algorithm adapted to any user's profile. With the comparison of set of users' profiles labeled by the classical co-training algorithm and those labeled by our proposed proves that our algorithm is better.

In the coming works we intend to implement our method in our adaptive navigation architecture [21] and evaluate its reliability on the navigation adaptation.

## References

1. Breiman, L., Friedman, J.H., Olshen, R.A., Stone, C.J.: Classification and Regression Trees. CRC Press (1984)
2. Breiman, L.: Random Forests. Machine Learning 45(1), 5–32 (2001)
3. Chen, C., Chen, M., Sun, Y.: A self-adaptive personal view agent. Journal of Intelligent Information Systems 18(2-3), 173–194 (2002)
4. Chunyan, L.: User profile for personalized web search. In: Fuzzy Systems and Knowledge Discovery FSKD, pp. 1847–1850 (2011)
5. Fayyad, M., Piatetsky-Shapiro, G., Smyth, P.: From data mining to knowledge discovery: an overview. In: Advances in Knowledge Discovery and Data Mining Book, pp. 1–34 (1996)
6. Guggenberger, A.: Semi-supervised Learning with Support Vector Machines. Technischen Universität Wien (2008)
7. Iba, W., Langley, P.: Induction of one-level decision trees. In: International Machine Learning Conference. Morgan Kaufmann (1992)
8. Jarraya, S.K., Boukhriss, R.R., Hammami, M., Ben-Abdallah, H.: Cast Shadow Detection Based on Semi-supervised Learning. In: Campilho, A., Kamel, M. (eds.) ICIAR 2012, Part I. LNCS, vol. 7324, pp. 19–26. Springer, Heidelberg (2012)
9. Kononenko, I.: Estimating attributes: Analysis and extensions of relief. In: Bergadano, F., De Raedt, L. (eds.) ECML 1994. LNCS, vol. 784, pp. 171–182. Springer, Heidelberg (1994)
10. Montaner, M., Lopez, B., De La Rosa, J.: A Taxonomy of Recommender Agents on the Internet. Artificial Intelligence Review 19(4), 285–330 (2003)

11. Pazzani, M.J., Billsus, D.: Content-Based Recommendation Systems. In: Brusilovsky, P., Kobsa, A., Nejdl, W. (eds.) Adaptive Web 2007. LNCS, vol. 4321, pp. 325–341. Springer, Heidelberg (2007)
12. Quinlan, J.R.: C4.5: Programs for Machine Learning. Morgan Kaufmann (1993)
13. Quinlan, J.R.: Induction of Decision Trees. Machine Learning 1(1), 81–106 (1986)
14. Sieg, A., Mobasher, B., Burke, R.: Web search personalization with ontological user profiles. In: Conference on Information and Knowledge Management, pp. 525–534 (2007)
15. Victoria, E., Analï, A.: Ontology-based user profile learning. Appl. Intell. 36(4), 857–869 (2012)
16. Witten, I.H., Frank, E.: Data Mining: Practical machine learning tools and techniques. Morgan Kaufmann (2005)
17. Xiaojin, Z.: Semi-Supervised Learning Literature Survey. Technical Report (2008)
18. Yoav, F., Llew, M.: The Alternating Decision Tree Learning Algorithm. In: ICML, pp. 124–133 (1999)
19. Zghal Rebaï, R., Zayani, C.A., Amous, I.: An adaptive navigation method in semi-structured data. In: Morzy, T., Härder, T., Wrembel, R. (eds.) Advances in Databases and Information Systems. AISC, vol. 186, pp. 207–215. Springer, Heidelberg (2013)
20. Zghal Rebaï, R., Zayani, C., Amous, I.: A new technology to adapt the navigation. In: International Conference on Internet and Web Applications and Services (in press, 2013)
21. Zghal Rebaï, R., Zayani, C., Amous, I.: MEDI-ADAPT: A distributed architecture for personalized access to heterogeneous semi-structured data. In: Web Information Systems and Technologies, pp. 259–263 (2012)

# Relative Reduct-Based Estimation of Relevance for Stylometric Features

Urszula Stańczyk

Institute of Informatics, Silesian University of Technology,
Akademicka 2A, 44-100 Gliwice, Poland

**Abstract.** In rough set theory characteristic features, which describe classified objects, correspond to conditional attributes. A relative reduct is such an irreducible subset of attributes that preserves the quality of approximation of a complete decision table. For a decision table a single reduct or many reducts may exist. In typical processing one reduct is selected for the subsequent generation of decision rules, while others can be discarded. Yet when the set of reducts is analysed as a whole, observations and conclusions drawn can be used to evaluate relevance of attributes, which in turn can be employed in reduction of features not only for rule-based but also connectionist classifiers. The paper describes the steps of such procedure applied in the domain of stylometric processing of literary texts.

**Keywords:** DRSA, Relative Reduct, Characteristic Feature, Relevance Measure, Decision Algorithm, ANN, Stylometry.

## 1 Introduction

Rough set theory has been proposed to deal with cases of incomplete and uncertain data [7]. Both Classic Rough Set Approach (CRSA) and Dominance-Based Rough Set Approach (DRSA) perceive objects of the universe through granules of knowledge. By indiscernibility relation CRSA recognises presence or absence of features thus the granules of knowledge are equivalence classes enabling nominal classification. DRSA uses dominance relation, observes preference orders in datasets, and sees granularity in unions of classes allowing also for ordinal classification [4,5]. The primary motivation behind invention of DRSA has been to support multi-criteria decision making problems.

One of the fundamental notions in rough set theory is that of a *relative reduct*, which is an inherent mechanism directed at dimensionality reduction. When a decision table containing training samples is analysed, it often happens that information stored there is found to be repetitive and excessive. To reduce it we can check whether all conditional attributes are indeed necessary for classification. Any irreducible subset of attributes that guarantees the same quality of approximation as the complete set corresponds to a relative reduct.

For the decision table a single or many reducts may exist. Usually, one reduct is chosen while others are disregarded, then basing on this choice the rough approximations are calculated and decision rules induced from the reduced table.

B. Catania, G. Guerrini, and J. Pokorný (Eds.): ADBIS 2013, LNCS 8133, pp. 135–147, 2013.
© Springer-Verlag Berlin Heidelberg 2013

However, since reducts vary in cardinalities and conditional attributes they include, the selection bears on the quality of the obtained solution, is far from trivial, and requires help from either domain expert knowledge or some supportive algorithm or dedicated procedure [6].

If the set of generated reducts is considered as a whole, it can be argued that it contains additional information on features retrieved from the decision table, which reflects their relevance for the classification task under study. Characteristics of reducts can be used to define their weights and measures of relevance for attributes that can be employed in the process of feature selection and reduction not only for rule-based, but also connectionist classifiers. In the research presented in the paper such methodology was applied to the task of authorship attribution from the field of stylometric analysis of texts.

Stylometry involves studies of writing styles and advocates that they can be uniquely described and expressed in numerical terms, allowing for author characterisation, comparison, and attribution [2]. Individual stylometric traits can be treated as patterns to be recognised and classified basing on their characteristic features, which here correspond to selected linguistic properties of text samples.

The paper is organised as follows. Section 2 gives basic notions and describes aims of stylometry, while section 3 provides brief overview of the employed data processing techniques. Details of the experimental setup and definitions of weights and relevance measure used in research are explained in section 4. Section 5 specifies results of conducted experiments. Section 6 contains concluding remarks.

## 2    Aims of Stylometry

When we read, a preference of some book genre over other is usually easily explained. One simply likes suspense and an intricate plot of a thriller or crime fiction, another enjoys unlimited imagination of fantasy or science fiction, while yet someone else yearns for emotions and reflections brought by classic romance or poetry. But why do we fancy David Baldacci and put away Lee Child? Why John R.R. Tolkien and not Christopher Paolini? Why do we remember some authors and their books in detail and forget others? The answer to these questions lies in a concept that we grasp intuitively, but have trouble with a precise definition. The answer is in a writing *style*.

When we write, we put down on paper (in the present times of electronic ways and means this expression should be treated rather metaphorically than literally) the uniqueness of our minds, our way of thinking, forming phrases and sentences. Our social and cultural backgrounds, education history, lifetime experiences, and both conscious and subconscious linguistic preferences accumulate to individualities of writing styles.

The first and foremost aim of stylometry (or computational stylistics, if we want to emphasise the role of computations involved in knowledge discovery) is to study complexities and subtleties of styles. By comparative analysis, given sufficient number of representative writing samples, it is possible to characterise one style as opposed to others, some similarities and differences between authors can be detected, and also authorship attribution is executed [8].

To work its magic, stylometry requires three elements, complementing each other. The first is a sufficiently wide corpus of texts, because to observe patterns we need a base from which conclusions could be drawn, the second — a set of linguistic features reflecting characteristics for considered writing samples, such as lexical and syntactic descriptors [1], while the third is some data processing technique allowing for pattern recognition and classification. Typically there are employed either computer-aided statistic-oriented computations or mechanisms from artificial intelligence domain [14].

## 3    Data Mining Techniques Employed

In the research two distinctively different approaches to classification were used: rule-based and connectionist, namely Dominance-Based Rough Set Approach (DRSA) and Artificial Neural Networks (ANN), briefly presented below.

### 3.1    DRSA Processing

Rough set methodology enables construction of decision algorithms by induction of rules of *IF...THEN...* type from the training samples. Depending on values of observed conditional attributes the rules classify objects to decision classes.

Classical Rough Set Approach (CRSA), as defined by Z. Pawlak in 1980ties [7], uses equivalence relation to see objects as either discernible or indiscernible with respect to considered features. With indiscernibility relation only nominal classification is possible. To rectify that and support multi-criteria decision making, Dominance-Based Rough Set Approach (DRSA) has been proposed [9].

DRSA exploits orderings present in datasets and defines their preferences. The attributes are considered as either *gain* features for which higher or increasing values are preferred, or *cost* features where lower or decreasing values are preferred. With dominance relation objects are regarded as: equal to or more preferred, or equal to or less preferred than other objects. If for two considered objects $x$ and $y$, all features of $x$ have values equal to or more preferred than those for $y$, then $x$ dominates $y$ (or $y$ is dominated by $x$). A set of objects dominated by $x$ is denoted as $D^-(x)$, and objects dominating $x$ as $D^+(x)$.

The starting point for induction of decision rules is reduction of excessive information contained in the decision table by selection of a reduct and calculation of lower and upper approximations of upward and downward unions of classes $Cl_t$ (dominance cones):

$$\underline{P}(\bigcup_{s \geq t} Cl_s) = \{x \in U : D_P^+ \subseteq \bigcup_{s \geq t} Cl_s\} \tag{1}$$

$$\overline{P}(\bigcup_{s \geq t} Cl_s) = \{x \in U : D_P^- \cap \bigcup_{s \geq t} Cl_s \neq \emptyset\}$$

where $P$ is a subset of conditional attributes $C$, and $t = 1, \ldots, n$. The differences between the upper and lower approximations define the boundary regions between dominance cones.

Each irreducible subset of conditional attributes which preserves the quality of approximation of the complete set of attributes is called a relative reduct and their intersection a core. Attributes absent in all reducts can be disregarded, those present in the core are essential and necessary for classification, and these included only in some reducts present optional choices. Instead of making this choice we can also treat the set of all reducts and their characteristics as a source of additional information on the considered features and use it in processing [12].

When no object from the decision table remains unclassified, a set of constructed decision rules is complete. When it is also irredundant, it is minimal.

One of the main advantages of using a rule-based classifier lies in its explicitness — the decision rules are easily understandable — meet the conditions specified in the premise part of a rule and you know the class, but it also has a disadvantage of involved computational costs. Decision algorithms that provide a minimal cover of the learning samples are relatively quickly calculated but they can be (and usually are) significantly outperformed by other algorithms. It is a popularly used approach to generate some higher number (even all) rules from examples and then discard some of them by imposing hard constraints, such as minimal support (which gives a number of learning samples for which a rule is valid) or maximal length (corresponding to the number of conditions in the premise part of the rule) required, by some assumed or calculated weights, or presence or absence of some specific features [10].

## 3.2   ANN Classifier

Artificial Neural Networks constitute an alternative to rule-based approaches to classification. A unidirectional feedforward network with Multilayer Perceptron (MLP) topology is a popular solution with good generalisation properties. The neurons that form such network are grouped into some number of layers. Depending on this number the network can deal with different classes of problems. When ANN has only the input and output layers and no hidden layer, it can solve only linearly separable problems. With one hidden layer it can classify simplexes, and with two layers it allows for recognition of any object [3].

Apart from the number of layers, the specification of a neural network comprises definitions of inputs and outputs, number of neurons in hidden layers, the training rule, and activation function which determines when neurons fire. In classification tasks sigmoid is the most widely employed activation function.

The number of network inputs corresponds to the number of considered characteristic features, while the output nodes usually reflect recognised classes of objects. The number of hidden neurons bears on the learning process — with just few nodes the network can run into trouble converging, while when there are too many of them the network learns quickly yet performs poorly for unknown samples. The actual numbers are typically established through tests and the hidden layers are adapted to other layers to form a decreasing tendency: inputs the most numerous, then in subsequent layers fewer and fewer neurons.

The training rule describes how weights of interconnections between neurons are modified within the learning phase to arrive at the highest classification accuracy. In backpropagation algorithm, often used for MLP, the error on the output, which corresponds to the difference between the expected and actual output value, is minimised for all outputs and all training facts.

Since the initiation of interconnection weights strongly influences the learning results, multi-starting procedure can be employed, when each network is trained several times, each time with randomisation of weights before the training starts, and calculating minimal, median, and maximal classification accuracy.

## 4    Experimental Setup

The experiments described in this paper started with selection of texts for stylometric analysis and features to be considered. Next, there was proposed a methodology for evaluation of attribute relevance basing on performance of rule- and ANN-based classifiers, as given below. The results of conducted tests were then analysed, providing a basis for some conclusions.

### 4.1    Stylometric Characteristic Features

In order to capture stylistic individualities, distinguishing features cannot be generally applicable. For any pair or group of authors the one and only, always valid and always efficient set of characteristics does not exist. Descriptors can base on the same type of linguistic properties, yet they have to be considered in the limited context of specific writers, their works perceived through available samples, and processing techniques employed.

In authorship attribution studies two groups of textual markers are most often exploited, namely lexical and syntactic. The first provides statistics such as average word lengths, average sentence lengths, frequencies of usage for words and phrases, distributions of these averages and frequencies. Syntactic markers describe composition of a text in terms of sentences either simple or complex, and organisation into paragraphs and sections, by referring to punctuation marks. These descriptors have to be based on sufficiently high number of text samples. If samples were few, the results could be unreliable. Unfortunately, there is no universal rule defining exactly how many samples provide the reliable base.

In the research conducted these examples were constructed basing on literary works of Jane Austen and Edith Wharton, available for on-line reading and download thanks to Project Gutenberg (http://www.gutenberg.org). Texts of selected novels, listed in Table 1, were divided into smaller parts, of comparable lengths, typically corresponding to chapters. In the learning dataset there were 100, and in the testing set 90 samples per author. Each text sample was a few pages long, and contained about 4000 words.

For both learning and testing samples the frequencies of usage were calculated for 25 selected stylometric features (17 lexical and 8 syntactic): but, and, not, in,

**Table 1.** Literary works submitted to stylometric processing

| | Datasets | |
| | Learning | Testing |
|---|---|---|
| Jane Austen | "Emma" | "Northanger Abbey" |
| | "Mansfield Park" | "Sense and Sensibility" |
| | "Pride and Prejudice" | "Lady Susan" |
| | "Persuasion" | |
| Edith Wharton | "A Backward Glance" | "Certain People" |
| | "The Age of Innocence" | "House of Mirth" |
| | "The Reef" | "Summer" |
| | "The Glimpse of the Moon" | |

with, on, at, of, this, as, that, what, from, by, for, to, if, a fullstop, a comma, a question mark, an exclamation mark, a semicolon, a colon, a bracket, a hyphen, which were successfully used in the past authorship attribution studies. Lexical descriptors were chosen basing on the list of the most popular English words.

As calculated frequencies are continuous real values, observation of orderings in datasets for all conditional attributes is trivial. Yet to proceed with DRSA procedures preference orders need to be assumed. In this context preference ties certain, lower or higher, frequencies of features with certain authors and such information in the stylometric domain is unavailable. In such cases preference orders can be assumed arbitrarily, moving the problem of evaluation of attribute relevance and optimisation of the obtained solution to the field of the processing techniques employed [11,13].

### 4.2 Proposed Framework

To observe the relevance of conditional attributes in the considered classification task 3-step methodology was proposed:

1. 1st step - preliminary assessment of importance for characteristic features
   (a) Finding all relative reducts
   (b) Calculating values for defined reduct weights and attribute relevance measures based on reducts and ordering attributes accordingly
   (c) Induction of DRSA decision algorithm providing a minimal cover and testing without and with hard constraints upon rules
   (d) Induction of all rules on examples and testing without and with hard constraints on rules
2. 2nd step - backward reduction of attributes following their previously obtained orderings
   (a) Rule-based classifiers - modifications of algorithms by discarding decision rules referring to selected attributes, testing without and with constraints
   (b) Connectionist classifiers - construction of networks for classification with the whole and reduced set of characteristic features

3. 3rd step - analysis of results and drawing conclusions

In experiments data mining with DRSA methodology involved using 4eMka software [5,9], while artificial neural networks were simulated with California Scientific Brainmaker. All networks constructed had two hidden layers, the number of inputs equal to the number of considered features, and two outputs corresponding to two recognised authors. Each network was trained 20 times and only average performance is presented.

## 4.3   Relative Reduct-Based Relevance Measures

For the constructed decision table as many as 2887 relative reducts were found (the core was empty), with cardinalities ranging from 2 to 10. Their characteristics in relation to conditional attributes included are listed in Table 2.

The weights of relative reducts specified in 6 columns in the right part of the table were defined as follows:

1. Attributes included were substituted with the percentage of reducts in which they appeared, expressed as fractional numbers
2. For each reduct as its base weight ($RedBW$) there was calculated a product of fractions introduced in the previous step with changing the scale by multiplying the outcome by 1000
3. Weights of reducts ($RedW$) were based on $RedBW$ by their discretisation accordingly to:
   $RedW = 1$ if $0.001 > RedBW$
   $RedW = 2$ if $0.001 \leq RedBW < 0.01$
   $RedW = 3$ if $0.01 \leq RedBW < 0.1$
   $RedW = 4$ if $0.1 \leq RedBW < 1$
   $RedW = 5$ if $1 \leq RedBW < 10$
   $RedW = 6$ if $10 \leq RedBW$
4. For a conditional attribute $a$ its relevance measure, denoted $MRA(a)$, is calculated as:

$$MRA(a) = \sum_{i=MinW}^{MaxW} RedW_i \cdot Nr(a, RedW_i) \tag{2}$$

   where $Nr(a, RedW_i)$ returns the number of reducts with weight equal $RedW_i$ including attribute $a$, and $MinW$ and $MaxW$ are respectively the minimal and maximal reduct weights (in this particular case equal to 1 and 6)
5. For each reduct its average weight $ARedW$ is obtained by dividing $RedW$ by reduct cardinality

In simple words, reducts have higher values of assumed weights $RedW$ when they include relatively few attributes appearing in high percentage of reducts. And further, attributes have higher relevance measures $MRA$ if they occur more often in many reducts with higher weights. The values of $MRA$ gave base to ordering of considered features presented as A series in Table 3.

Average reduct weights $ARedW$ ranged from 0.1 to 3. Following the line of reasoning that higher weights could be considered as more informative with

**Table 2.** Analysis of characteristic features based on relative reducts

| Cond | Number of reducts of specific cardinality and calculated weight | | | | | | | | | | | | | | | |
| | Reduct cardinalities | | | | | | | | | | Reduct weights | | | | | |
| Attr | 2 | 3 | 4 | 5 | 6 | 7 | 8 | 9 | 10 | Total | 1 | 2 | 3 | 4 | 5 | 6 |
|---|---|---|---|---|---|---|---|---|---|---|---|---|---|---|---|---|
| but | 0 | 2 | 55 | 118 | 406 | 321 | 37 | 4 | 0 | 943 | 0 | 7 | 66 | 543 | 326 | 1 |
| and | 0 | 6 | 49 | 105 | 468 | 410 | 71 | 10 | 0 | 1119 | 0 | 8 | 140 | 612 | 358 | 1 |
| not | 1 | 30 | 45 | 23 | 0 | 0 | 0 | 0 | 0 | 99 | 0 | 0 | 1 | 43 | 55 | 0 |
| in | 0 | 1 | 42 | 63 | 344 | 507 | 153 | 39 | 3 | 1152 | 7 | 47 | 227 | 629 | 242 | 0 |
| with | 0 | 4 | 31 | 65 | 338 | 462 | 99 | 21 | 0 | 1020 | 2 | 23 | 166 | 614 | 215 | 0 |
| on | 0 | 0 | 18 | 28 | 71 | 176 | 84 | 23 | 1 | 401 | 4 | 31 | 149 | 209 | 8 | 0 |
| at | 0 | 1 | 26 | 39 | 328 | 440 | 109 | 35 | 3 | 981 | 7 | 39 | 184 | 520 | 231 | 0 |
| of | 0 | 12 | 58 | 97 | 508 | 420 | 83 | 17 | 0 | 1195 | 0 | 13 | 138 | 634 | 408 | 2 |
| this | 0 | 11 | 31 | 100 | 391 | 342 | 47 | 1 | 0 | 923 | 0 | 4 | 120 | 495 | 303 | 1 |
| as | 0 | 4 | 33 | 90 | 427 | 356 | 66 | 29 | 3 | 1008 | 7 | 35 | 143 | 544 | 278 | 1 |
| that | 0 | 4 | 45 | 31 | 241 | 412 | 100 | 29 | 3 | 865 | 3 | 32 | 168 | 521 | 141 | 0 |
| what | 0 | 1 | 14 | 59 | 270 | 265 | 73 | 8 | 0 | 690 | 0 | 15 | 107 | 434 | 134 | 0 |
| from | 0 | 7 | 43 | 68 | 383 | 379 | 77 | 24 | 3 | 984 | 7 | 31 | 148 | 500 | 297 | 1 |
| by | 1 | 19 | 8 | 120 | 398 | 254 | 28 | 3 | 0 | 831 | 0 | 11 | 110 | 480 | 228 | 2 |
| for | 0 | 8 | 38 | 122 | 442 | 283 | 43 | 7 | 0 | 943 | 0 | 10 | 92 | 525 | 315 | 1 |
| to | 0 | 7 | 15 | 58 | 347 | 342 | 76 | 16 | 1 | 862 | 1 | 20 | 126 | 513 | 202 | 0 |
| if | 0 | 0 | 16 | 19 | 245 | 301 | 83 | 32 | 3 | 699 | 6 | 34 | 134 | 420 | 105 | 0 |
| . | 0 | 4 | 22 | 87 | 248 | 206 | 50 | 10 | 0 | 627 | 2 | 10 | 96 | 375 | 143 | 1 |
| , | 1 | 22 | 106 | 26 | 17 | 6 | 0 | 0 | 0 | 178 | 0 | 3 | 20 | 47 | 108 | 0 |
| ? | 0 | 0 | 11 | 18 | 150 | 205 | 49 | 16 | 1 | 450 | 2 | 10 | 112 | 308 | 18 | 0 |
| ! | 0 | 5 | 26 | 91 | 326 | 219 | 58 | 21 | 0 | 746 | 0 | 26 | 101 | 406 | 213 | 0 |
| : | 0 | 0 | 0 | 1 | 60 | 122 | 44 | 16 | 3 | 246 | 7 | 25 | 140 | 74 | 0 | 0 |
| ; | 3 | 5 | 51 | 72 | 77 | 10 | 0 | 0 | 0 | 218 | 0 | 0 | 22 | 129 | 65 | 2 |
| ( | 0 | 0 | 0 | 0 | 17 | 57 | 39 | 25 | 3 | 141 | 7 | 41 | 77 | 16 | 0 | 0 |
| - | 0 | 0 | 13 | 30 | 164 | 344 | 67 | 19 | 3 | 640 | 4 | 19 | 157 | 390 | 70 | 0 |
| Total | 3 | 51 | 199 | 306 | 1111 | 977 | 192 | 45 | 3 | 2887 | 7 | 58 | 400 | 1561 | 857 | 4 |

regard to included conditional attributes, only those satisfying the condition $0.7 \leq ARedW \leq 3$ were next studied in more detail and are given as B series in Table 3. The characteristic features are ordered with respect to the frequencies with with they appear in reducts of decreasing average weights.

Both series were next used in the process of reduction of features for both DRSA- and ANN-based classifiers, by which the relevance of attributes can be observed in improved or worsened performance.

## 5   Tests Performed

Generation of firstly a minimal cover, then all rules on examples decision algorithm resulted in performance with maximum of 76% and 86% of correctly recognised samples, respectively, when hard constraints upon rule supports are imposed. These results, although certainly can be considered as satisfactory, say nothing specific about relevance of individual features, their importance for this

**Table 3.** Ordering of characteristic features based on an analysis of relative reducts in which the features are included, defined weights and measures. L series means reduction of conditional attributes with lowest values of considered elements, while M series corresponds to elimination of features with highest values of these elements.

A Series

| Attr | MRA | |
|---|---|---|
| ( | 384 | |
| not | 450 | M15 |
| : | 773 L1 | |
| , | 794 | M14 |
| ; | 919 L2 | M13 |
| on | 1389 L3 | M12 |
| ? | 1680 L4 | M11 |
| - | 2423 L5 | M10 |
| . | 2531 L6 | M9 |
| if | 2681 L7 | |
| what | 2757 | M8 |
| ! | 3044 L8 | M7 |
| that | 3360 L9 | M6 |
| by | 3424 L10 | |
| to | 3481 | M5 |
| this | 3869 L11 | |
| at | 3872 | M4 |
| for | 3977 L12 | |
| from | 4004 | |
| but | 4020 | M3 |
| with | 4077 L13 | |
| as | 4078 | M2 |
| in | 4508 L14 | M1 |
| and | 4680 L15 | |
| of | 5028 | |

B Series

| Attr | 3 | 2.5 | 2 | 1.7 | 1.5 | 1.3 | 1 | 0.8 | 0.7 | |
|---|---|---|---|---|---|---|---|---|---|---|
| ( | 0 | 0 | 0 | 0 | 0 | 0 | 0 | 0 | 16 | |
| : | 0 | 0 | 0 | 0 | 0 | 0 | 0 | 1 | 55 L1 | M12 |
| on | 0 | 0 | 0 | 0 | 0 | 4 | 17 | 25 | 52 L2 | |
| ? | 0 | 0 | 0 | 0 | 0 | 9 | 9 | 13 | 138 | M11 |
| - | 0 | 0 | 0 | 0 | 0 | 12 | 20 | 50 | 122 L3 | |
| if | 0 | 0 | 0 | 0 | 0 | 15 | 13 | 85 | 153 | M10 |
| what | 0 | 0 | 0 | 1 | 0 | 13 | 31 | 119 | 165 L4 | |
| at | 0 | 0 | 0 | 1 | 0 | 25 | 13 | 220 | 132 | |
| in | 0 | 0 | 0 | 1 | 0 | 38 | 26 | 222 | 158 | M9 |
| but | 0 | 0 | 0 | 2 | 1 | 48 | 85 | 235 | 196 L5 | |
| with | 0 | 0 | 0 | 3 | 0 | 30 | 28 | 196 | 176 | M8 |
| that | 0 | 0 | 0 | 4 | 0 | 42 | 13 | 106 | 154 L6 | M7 |
| . | 0 | 0 | 0 | 4 | 1 | 19 | 65 | 81 | 183 L7 | |
| as | 0 | 0 | 0 | 4 | 1 | 30 | 63 | 212 | 241 | M6 |
| ! | 0 | 0 | 0 | 5 | 0 | 21 | 68 | 152 | 194 L8 | |
| and | 0 | 0 | 0 | 5 | 1 | 48 | 96 | 221 | 244 | M5 |
| to | 0 | 0 | 0 | 7 | 0 | 13 | 41 | 162 | 203 L9 | |
| for | 0 | 0 | 0 | 7 | 1 | 28 | 85 | 253 | 226 | M4 |
| this | 0 | 0 | 0 | 10 | 1 | 29 | 74 | 221 | 198 L10 | M3 |
| from | 0 | 0 | 1 | 6 | 0 | 40 | 58 | 209 | 181 L11 | |
| of | 0 | 0 | 1 | 10 | 1 | 56 | 94 | 256 | 257 | M2 |
| , | 0 | 1 | 0 | 16 | 0 | 97 | 14 | 27 | 1 L12 | |
| not | 0 | 1 | 0 | 24 | 0 | 36 | 14 | 24 | 0 | M1 |
| by | 1 | 0 | 0 | 18 | 1 | 7 | 120 | 86 | 312 L13 | |
| ; | 1 | 2 | 1 | 4 | 0 | 50 | 10 | 63 | 65 | |

task. Evaluation of relevance can be executed through elimination of attributes and making observations on how this influences the classification accuracy.

Experiments involved construction of new decision algorithms by constricting the previously inferred all rules on examples. From the complete set of decision rules there are eliminated these which in the premise parts contain conditions on discarded attributes. The attributes to be disregarded are indicated by established orderings specified as A and B series in Table 3. Detailed performance of these new constricted algorithms is given in Table 4.

Decision algorithms from A test series show clearly distinctive trends. Reduction of rules with features characterised by lower values of $MRA$ relevance measure decreases the classifier performance instantly, even when just few conditional attributes are discarded and with them relatively low numbers of rules disregarded. On the other hand, removal of attributes with higher values of $MRA$ brings very quick and significant decrease in the number of remaining decision

**Table 4.** Performance of DRSA classifiers. Columns present parameters of reduced decision algorithms: a) number of decision rules without any constraints, b) number of conditional attributes considered, c) maximal classification accuracy [%] specifying only correctly classified samples, without taking into account ambiguous cases of no rules matching or contradicting decisions, d) maximal support required of rules to achieve the highest classification ratio, e) number of rules meeting hard constraints.

| | | a) | b) | c) | d) | e) | | a) | b) | c) | d) | e) |
|---|---|---|---|---|---|---|---|---|---|---|---|---|
| | L01 | 50433 | 23 | 80 | 48 | 49 | M01 | 34888 | 23 | 86 | 66 | 17 |
| | L02 | 40814 | 21 | 74 | 32 | 152 | M02 | 25162 | 22 | 86 | 66 | 17 |
| | L03 | 36892 | 20 | 62 | 20 | 81 | M03 | 14475 | 20 | 86 | 66 | 16 |
| | L04 | 33350 | 19 | 62 | 20 | 80 | M04 | 5662 | 17 | 86 | 63 | 18 |
| | L05 | 28104 | 18 | 62 | 20 | 70 | M05 | 3023 | 15 | 86 | 61 | 21 |
| | L06 | 23623 | 17 | 64 | 20 | 64 | M06 | 1138 | 13 | 86 | 61 | 20 |
| A Series | L07 | 17341 | 16 | 66 | 19 | 72 | M07 | 831 | 12 | 86 | 61 | 20 |
| | L08 | 9345 | 14 | 66 | 19 | 50 | M08 | 626 | 11 | 86 | 55 | 27 |
| | L09 | 5963 | 13 | 70 | 14 | 160 | M09 | 292 | 9 | 86 | 55 | 27 |
| | L10 | 4330 | 12 | 70 | 14 | 158 | M10 | 202 | 8 | 86 | 55 | 27 |
| | L11 | 2153 | 10 | 64 | 10 | 141 | M11 | 172 | 7 | 86 | 55 | 27 |
| | L12 | 828 | 8 | 56 | 6 | 193 | M12 | 133 | 6 | 86 | 55 | 27 |
| | L13 | 80 | 5 | 18 | | | M13 | 94 | 5 | 86 | 55 | 27 |
| | L14 | 19 | 3 | 6 | 4 | 12 | M14 | 39 | 4 | 85 | 46 | 20 |
| | L15 | 9 | 2 | 5 | 4 | 4 | M15 | 17 | 2 | 65 | 46 | 9 |
| | L01 | 51321 | 24 | 86 | 66 | 17 | M01 | 44733 | 23 | 86 | 52 | 52 |
| | L02 | 47709 | 23 | 86 | 66 | 17 | M02 | 38349 | 21 | 62 | 20 | 58 |
| | L03 | 36604 | 21 | 86 | 66 | 17 | M03 | 19924 | 19 | 64 | 19 | 75 |
| | L04 | 24360 | 19 | 86 | 66 | 17 | M04 | 14496 | 18 | 63 | 17 | 86 |
| | L05 | 10772 | 16 | 86 | 66 | 17 | M05 | 6773 | 16 | 50 | 10 | 117 |
| | L06 | 6384 | 14 | 86 | 66 | 17 | M06 | 2975 | 14 | 52 | 5 | 359 |
| B Series | L07 | 5180 | 13 | 86 | 66 | 17 | M07 | 1239 | 12 | 37 | 3 | 266 |
| | L08 | 2397 | 11 | 86 | 66 | 16 | M08 | 887 | 11 | 36 | 3 | 221 |
| | L09 | 1100 | 9 | 86 | 66 | 15 | M09 | 222 | 9 | 16 | | |
| | L10 | 451 | 7 | 86 | 63 | 18 | M10 | 34 | 6 | 2 | | |
| | L11 | 295 | 6 | 86 | 62 | 17 | M11 | 6 | 4 | 2 | | |
| | L12 | 131 | 4 | 86 | 55 | 28 | | | | | | |
| | L13 | 30 | 2 | 61 | 7 | 28 | | | | | | |
| Minimal cover DA | | 30 | 25 | 76 | 6 | 6 | | | | | | |
| Full DA | | 62383 | 25 | 86 | 66 | 17 | | | | | | |

rules while maintaining the high classification accuracy almost to the end, when there are just few features left for consideration.

These observed trends are reversed for algorithms built within B test series. Here elimination of decision rules containing conditional attributes that appear most often in relative reducts of highest average weights causes worsened classification ratio, but when rules referring to features less often present in such reducts are removed, the recognition is decreased only for the very last algorithm, which operates on just two features.

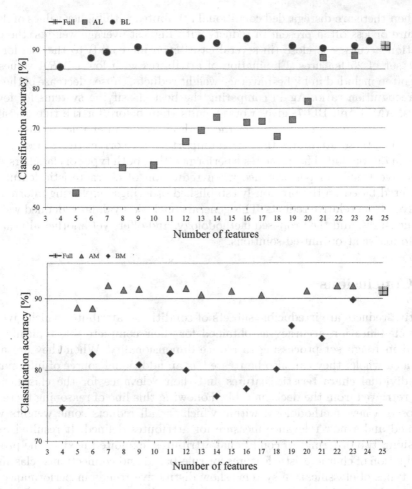

**Fig. 1.** ANN classification accuracy in relation to the number of features, with their reduction based on orderings specified by A and B series

When AM13 and BL12, the two shortest decision algorithms which maintain the maximal classification accuracy of all rules on examples inferred, are compared, along with conditional attributes they operate on, it turns out that they have three features in common, namely: a comma, a colon, not. It means that these elements show the highest relevance for this task, although just by themselves they are insufficient to always guarantee the high recognition ratio.

The power of these rule-based classification systems and respective relevance of characteristic features observed can be compared against that of ANN classifier. Such combination of elements from two approaches gives a hybrid DRSA-ANN solution. The performance is plotted in Fig. 1. The two graphs displaying the classification accuracy in relation to the number of attributes left, show noticeable resemblance to the results given before in Table 4.

When there are disregarded conditional attributes with higher values of $MRA$ measure or less often present in reducts with highest average weights, the classification accuracy is close (in several cases even increased) to the one for the whole set of 25 features. Elimination of attributes with lower $MRA$ values, or more often included in highest average weight reducts, causes decrease of the correct recognition ratio. Again comparing the best classifying systems for fewest inputs, AM11 and BL07, gives worse results than before for the rule classifiers (they operate on more inputs), yet the methodology still enables to distinguish and reduce these attributes that are unimportant for correct recognition.

It can be concluded from the tests performed that both types of classifiers show distinctive trends in performance when reduction of characteristic features is conducted based on the previously established orderings, exploiting information retrieved from relative reducts. This general conclusion validates defined weights and measures, and the proposed methodology, and offers yet another alternative way to arrive at optimised solutions.

## 6   Conclusions

Relative reducts are irreducible subsets of conditional attributes which give the same classification accuracy as obtained for their complete set, typically employed in rough set processing to reduce dimensionality. When they are analysed as a whole, they can also be treated as an additional source of information on individual characteristic features and their relevance for the classification task, retrieved from the decision table. Following this line of reasoning there was proposed a new methodology, within which for all reducts some weights were assumed and a new relevance measure for attributes defined. It resulted in establishing two orderings of conditional attributes, exploited next in the process of reduction of characteristic features for rule-based and connectionist classifiers. Both types of classification systems show distinctive trends in performance observed, while attribute elimination is executed. By the proposed methodology less relevant features are detected and discarded while keeping high classification accuracy, which validates the usefulness of the presented research framework.

**Acknowledgments.** 4eMka Software used in search for relative reducts and induction of decision rules [5,4] was developed at the Laboratory of Intelligent Decision Support Systems, (http://www-idss.cs.put.poznan.pl/), Poznan University of Technology, Poland.

## References

1. Burrows, J.: Textual analysis. In: Schreibman, S., Siemens, R., Unsworth, J. (eds.) A Companion to Digital Humanities, ch. 23. Blackwell, Oxford (2004)
2. Craig, H.: Stylistic analysis and authorship studies. In: Schreibman, S., Siemens, R., Unsworth, J. (eds.) A Companion to Digital Humanities. Blackwell, Oxford (2004)

3. Fiesler, E., Beale, R.: Handbook of neural computation. Oxford University Press (1997)
4. Greco, S., Matarazzo, B., Sowiński, R.: The use of rough sets and fuzzy sets in Multi Criteria Decision Making. In: Gal, T., Hanne, T., Stewart, T. (eds.) Advances in Multiple Criteria Decision Making, ch. 14, pp. 14.1–14.59. Kluwer Academic Publishers, Dordrecht (1999)
5. Greco, S., Matarazzo, B., Sowiński, R.: Dominance-based rough set approach as a proper way of handling graduality in rough set theory. Transactions on Rough Sets 7, 36–52 (2007)
6. Moshkov, M.J., Skowron, A., Suraj, Z.: On covering attribute sets by reducts. In: Kryszkiewicz, M., Peters, J.F., Rybiński, H., Skowron, A. (eds.) RSEISP 2007. LNCS (LNAI), vol. 4585, pp. 175–180. Springer, Heidelberg (2007)
7. Pawlak, Z.: Rough sets and intelligent data analysis. Information Sciences 147, 1–12 (2002)
8. Peng, R., Hengartner, H.: Quantitative analysis of literary styles. The American Statistician 56(3), 15–38 (2002)
9. Słowiński, R., Greco, S., Matarazzo, B.: Dominance-based rough set approach to reasoning about ordinal data. In: Kryszkiewicz, M., Peters, J.F., Rybiński, H., Skowron, A. (eds.) RSEISP 2007. LNCS (LNAI), vol. 4585, pp. 5–11. Springer, Heidelberg (2007)
10. Stańczyk, U.: DRSA decision algorithm analysis in stylometric processing of literary texts. In: Szczuka, M., Kryszkiewicz, M., Ramanna, S., Jensen, R., Hu, Q. (eds.) RSCTC 2010. LNCS (LNAI), vol. 6086, pp. 600–609. Springer, Heidelberg (2010)
11. Stańczyk, U.: Application of DRSA-ANN classifier in computational stylistics. In: Kryszkiewicz, M., Rybinski, H., Skowron, A., Raś, Z.W. (eds.) ISMIS 2011. LNCS (LNAI), vol. 6804, pp. 695–704. Springer, Heidelberg (2011)
12. Stańczyk, U.: Reduct-based analysis of decision algorithms: Application in computational stylistics. In: Corchado, E., Kurzyński, M., Woźniak, M. (eds.) HAIS 2011, Part II. LNCS (LNAI), vol. 6679, pp. 295–302. Springer, Heidelberg (2011)
13. Stańczyk, U.: Rule-based approach to computational stylistics. In: Bouvry, P., Kłopotek, M.A., Leprévost, F., Marciniak, M., Mykowiecka, A., Rybiński, H. (eds.) SIIS 2011. LNCS (LNAI), vol. 7053, pp. 168–179. Springer, Heidelberg (2012)
14. Waugh, S., Adams, A., Twedeedie, F.: Computational stylistics using artificial neural networks. Literary and Linguistic Computing 15(2), 187–198 (2000)

# Spatial Itemset Mining: A Framework to Explore Itemsets in Geographic Space

Christian Sengstock and Michael Gertz

Database Systems Research Group, Heidelberg University, Germany
{sengstock,gertz}@informatik.uni-heidelberg.de

**Abstract.** Driven by the major adoption of mobile devices, user contributed geographic information has become ubiquitous. A typical example is georeferenced and tagged social media, linking a location to a set of features or attributes. Mining frequent sets of discrete attributes to discover interesting patterns and rules of attribute usage in such data sets is an important data mining task.

In this work we extend the frequent itemset mining framework to model the spatial distribution of itemsets and association rules. For this, we expect the input transactions to have an associated spatial attribute, as, for example, present in georeferenced tag sets. Using the framework, we formulate interestingness measures that are based on the underlying spatial distribution of the input transactions, namely area, spatial support, location-conditional support, and spatial confidence. We show that describing the spatial characteristics of itemsets cannot be handled by existing approaches to mine association rules with numeric attributes, and that the problem is different from co-location pattern mining and spatial association rules mining. We demonstrate the usefulness of our proposed extension by different mining tasks using a real-world data set from Flickr.

## 1 Introduction

The past couple of years have witnessed a huge increase in spatial data sets. Examples include georeferenced social media, status information of cell phones (like active services and reachable networks), and Web-click or shopping transactions occurring at a particular location. Mining of interesting patterns and correlations from such spatial data sets has become an important body of research like place and event discovery [1], extraction of trajectories [2], and geographic topic discovery [3]. For this, the approaches leverage and combine different models to extract spatial knowledge, like parametric and non-parametric distributions of individual attributes, and bag-of-words representations.

In this work we introduce a general model to extract geographic knowledge from such data sets based on an extension of frequent itemset mining [4]. For that, we expect the spatial data to be represented by transactions that are distributed in geographic space. The resulting itemsets and derived association rules thus have an inherent spatial distribution, based on the spatial distribution

B. Catania, G. Guerrini, and J. Pokorný (Eds.): ADBIS 2013, LNCS 8133, pp. 148–161, 2013.

of the underlying *geo-transactions*. In addition to frequent patterns emerging from co-occurring items in the transactions, this allows to filter, compare, and visualize itemsets and association rules based on their spatial characteristics, like local or global patterns, or patterns having similar spatial distributions.

Traditionally, mining spatial datasets has focused on mining spatial association rules, e.g., [5], and finding co-location patterns, e.g., [6]. However, these approaches focus on different problems. Spatial association rules try to unveil geographic knowledge by introducing items with spatial semantics. However, the resulting itemsets and rules have no underlying spatial distribution of the input transactions. The aim of co-location pattern mining is to find sets of object classes that frequently co-occur in geographic space. Thereby the input consists of labeled spatial objects and not transactions. Also, the resulting patterns have no underlying spatial distribution. Since both approaches can be formulated in a frequent itemset mining framework, they can benefit from our extension to mine such spatial association rules and co-location patterns that exhibit certain spatial characteristics.

Our work can be summarized as follows: (1) We propose an extension to frequent itemset mining where the input transactions have an associated spatial attribute. We show that existing approaches to handle numeric attributes in transactional data are not sufficient to model such spatial attributes. Moreover, we detail that spatial itemset mining is different from existing spatial association rule mining and co-location pattern mining approaches. (2) We introduce various spatial interestingness measures based on the spatial characteristics of itemsets and association rules. The measures allow to filter, compare, and visualize frequent itemsets in geographic space. (3) We discuss computational issues and outline a basic strategy to implement geo-transactional extensions. (4) We demonstrate the usefulness of our framework by performing different data mining tasks on a real-world dataset taken from Flickr.

The remainder of the paper is structured as follows. First, we discuss related work. In Section 3, we introduce the spatial itemset mining model. In Section 4, we discuss computational considerations and outline a basic strategy to implement our extensions. In Section 5, we detail our experiments. We summarize the paper and outline ongoing work in Section 6.

## 2   Related Work

Frequent itemset mining is a major domain in data mining, and several extensions to the core model [4] have been developed in the past two decades. Since we view a spatial object as a transaction having an associated spatial data type in continuous space, our work is heavily related to *frequent itemset mining with numeric attributes*. Techniques in this area can be classified into (1) discretization approaches to discover intervals of numeric attributes leading to high support and/or confidence in association rules [7–10], and (2) approaches that capture the distributional characteristics of numeric attributes in statistics in order to represent them in association rules [11, 12].

In our work we are not interested in varying discretizations of geographic space. Instead, we treat the bandwidth of a density estimate of the spatial data points as a parameter defining the scale-level of interest. We then discover interesting itemsets and association rules based on an appropriately chosen bandwidth, omitting the need for an explicit discretization.

Our work is, however, influenced by previous work on describing numeric attributes by statistics. In [11] the concept of quantitative association rules was introduced. There, numeric attributes are summarized by statistics that can occur on the LHS or the RHS of an association rule. In [12] an approach to find quantitative association rules where the statistic on the RHS is significant was proposed. Both works use simple statistics like mean and variance to describe the numeric attribute. However, we expect the numeric attribute of a transaction to be a spatial location. Since this distribution is multi-modal and heterogeneous, such simple statistics are not applicable. In this work we propose an extension of frequent itemset mining based on representing the spatial attribute by a non-parametric density estimation. This allows for a flexible representation of the spatial distribution of transactions. However, it increases the amount of parameters we need to capture in order to model the distribution. We present two ways to model the non-parametric distribution, namely a tree-based approach and a sparse grid-based approach.

From a data mining perspective our work tries to reveal interesting patterns from spatial data. Prominent techniques in the domain of geographic data mining and knowledge discovery are co-location pattern mining, e.g., [6, 13, 14] and spatial association rule mining, e.g., [5, 15]. Co-location patterns are sets of object or event classes that frequently co-occur in geographic space. The input of co-location pattern mining are spatial objects labeled with a single class. By transforming classes that occur in close spatial proximity to co-location instances, frequent patterns are found by mining frequent co-location instances using a participation ratio threshold [13]. This is different from the problem we are facing. We expect as input a set of transactions having an associated spatial data type. The resulting itemsets describe co-occurrences of items in transactions having a certain characteristic in geographic space. By transforming a transaction into a set of spatial objects labeled by the respective items, we are able to transform the transactional input to a co-location pattern mining input. However, then the results will represent spatial proximity patterns instead of co-occurring itemsets in transactions.

Spatial association rules (SAR) are association rules that have a spatial predicate in the LHS and/or in the RHS side [5, 15]. In general, any frequent itemset mining approach can be used to mine SARs. For this, the transaction database has to contain categorical items having spatial semantics (e.g., locations, areas, spatial relationships). In contrast, our approach does not expect a prior transformation of the spatial attribute. Instead, we want to model the spatial distribution of resulting itemsets and association rules.

Both, co-location pattern mining and spatial association rule mining can benefit from an extension of frequent itemsets mining to filter, compare, and visualize the itemsets and association rules based on their spatial distribution. For this, the input transactions need to have an associated spatial data type. Such a spatial location is inherent to a co-location instance [16]. For a transaction with items having geographic semantics, as expected for SAR mining, an associated spatial data type will also appear naturally in most cases.

Regional association rule mining was proposed by Ding et al. [17]. The approach allows to mine association rules that are valid only in small regions. A reward-based region discovery algorithm is used to determine regions for association rules. There is a similarity between regional association rule mining and our concept of spatial confidence introduced in this paper. A spatial confidence expresses the probability that an association rule occurs at a given location in space, which in turn allows to determine regions where an association rule is valid. Different from the approach introduced by Ding et al., the focus of our framework, however, is a general extension to model the spatial characteristics of itemsets and association rules, with region discovery being a particular application.

# 3 Spatial Itemsets

To discover interesting itemsets in geographic space we first introduce an extension to the frequent itemset mining framework. Based on this, we are able to describe spatial characteristics of itemsets that allow us to formulate constraints and define spatial interestingness measures.

## 3.1 Geo-transactions and Spatial Itemsets

We assume a set $\mathcal{O}$ of spatial objects. Each object $o = (id, p, I) \in \mathcal{O}$ has an identifier $o.id$, a point-based location $o.p \in \mathbb{R}^2$, and a set $o.I \subseteq \mathcal{A}$ of items taken from a *base itemset* $\mathcal{A}$. A typical example of such a set of objects are Flickr photos, each of which has an id, a location and a set $I = \{a_1, \ldots, a_k\}$ of tags.

Following the traditional itemset mining framework [4], the object-identifier $o.id$ in combination with the attributes $o.I$ of each object can be considered a transaction. For an itemset $X \subseteq \mathcal{A}$, the set $O_X$ of transactions supporting $X$ is defined as $O_X = \{o \mid o \in \mathcal{O} : X \subseteq o.I\}$, $|O_X|$ is the support count, and $sup(X) = |O_X|/|\mathcal{O}|$ is the support of itemset $X$. The task of frequent itemset mining is to find those itemsets in $\mathcal{O}$ having a minimum support $\delta$.

We call our input objects $o \in \mathcal{O}$ *geo-transactions*, meaning that they are transactions having an associated spatial point. The set of points associated with an itemset $X$ is defined as $P_X = \{o.p \mid o \in O_X\}$. We call $(X, P_X)$ a *spatial itemset*.

We use the points $P_X$ of an itemset $X$ to determine and explore different *spatial characteristics* of an itemset. These characteristics include aspects such as spatial coverage or clustering, most generally described by a density distribution. Based on the spatial characteristics, spatial itemsets can be filtered, compared, and visualized, allowing to gather more insights into patterns of itemsets in geographic space.

### 3.2 Spatial Characteristics

To discover interesting itemsets in geographic space, we now describe itemsets on the basis of their spatial characteristics. Taken most generally, the spatial characteristics of itemset $X$ are described by the spatial density distribution of its points $P_X$. However, we do not always need a full description of the distribution. Often, a simple statistics of the point set is enough to determine interesting characteristics, like coverage, inclusion, or the mean. In the following, we first describe the density distribution of itemset points and derive spatial support and confidence measures. Then, we define statistics for spatial itemsets, in particular to describe the spatial coverage.

**Spatial Density Distribution.** The point set $P_X$ can be represented by a spatial density distribution $f_{P_X}(u) \in [0,1], u \in \mathbb{R}^2$, describing the probability to find transactions supporting $X$ around a location $u$ in geographic space $\mathbb{R}^2$.

The density $f_{P_X}(u)$ is obtained by an estimator $\hat{f}$. Estimators can be parametric (e.g., Gaussian distribution, Gaussian Mixture Models) where the density $f_{P_X}(u; \Theta)$ depends on parameters $\Theta$ inferred from the data. Non-parametric estimators $f_{P_X}(u; \theta)$ assume no underlying model, but smoothing conditions on the density function, specified by a smoothing bandwidth $\theta$ [18]. A large bandwidth results in a smooth distribution representing large-scale variation, while a small bandwidth results in a peaky distribution representing small-scale variation.

To compare, filter, and visualize spatial characteristics, we need a density distribution for each spatial itemset. Using a simple parametric model, like a Gaussian with parameters $\Theta = (mean, variance)$, we can easily augment the itemsets by statistics capturing the necessary information, similar to quantitative association rule mining approaches [11, 12]. However, for spatial attributes representing geographic data, simple parametric models like a Gaussian distribution will not adequately describe the underlying distribution. Hence, complex parametric models (like Gaussian Mixture Models with a huge number of components) or non-parametric estimators need to be chosen, to describe the multi-modal and heterogeneous distribution of geo-transactions. Since the required number of parameters to calculate an estimate may be large, choosing an appropriate estimator is crucial to allow for efficient mining of spatial itemsets. Moreover, parametric models like Gaussian Mixture Models have expensive runtime complexity to fit the model. Hence, in Section 4 we propose two promising non-parametric estimators, a histogram estimator using a sparse-grid, and an estimator based on a CF-tree [19], which can be used to augment itemsets easily and that has reasonable runtime and space requirements for real-world geographic data sets.

In Section 5, we compare spatial itemsets on the basis of their spatial density distributions. For this, we define the following distance function. Let $F_X \sim f_{P_X}$ and $F_Y \sim f_{P_Y}$ denote the density distributions of itemsets $X$ and $Y$, discretized over the set of points $u_1, \ldots, u_n \subseteq \mathbb{R}^2$. We use the symmetric Kullback-Leibler divergence

$$D_{sym}(F_X, F_Y) := D_{KL}(F_X \| F_Y) + D_{KL}(F_Y \| F_X) \tag{1}$$

with

$$D_{KL}(F_X||F_Y) := \sum_{i=1}^{n} \ln \frac{F_X(i)}{F_Y(i)} F_X(i) \tag{2}$$

to obtain a measure of similarity between the two distributions.

**Spatial Support and Confidence.** Following the traditional frequent itemset mining framework, the support of $X$ is defined as the probability to find a transaction supporting $X$ in the database $\mathcal{O}$

$$supp(X) := P(X) = \frac{|O_X|}{|\mathcal{O}|}. \tag{3}$$

Given two disjoint itemsets $X$ and $Y$, the confidence $conf(Y \Rightarrow X)$ is the probability to find transactions supporting $X$ among those transactions supporting $Y$

$$conf(Y \Rightarrow X) := P(X|Y) = \frac{P(X \cap Y)}{P(Y)} = \frac{|O_{X \cap Y}|}{|O_Y|}. \tag{4}$$

Given the spatial density distribution $f_{P_X}$ of itemset $X$, we can calculate support and confidence at a certain location. Note that $f_{P_X}$ is just based on those transactions supporting $X$. Hence, by treating $f_{P_X}(u)$ as a probability in the overall set of itemsets in $\mathcal{O}$, it equals the itemset-conditional probability

$$P(u|X) := f_{P_X}(u). \tag{5}$$

The *spatial support* of itemset $X$ in $\mathcal{O}$ at location $u$ is then

$$supp(X, u) := P(X, u) = P(u|X)P(X). \tag{6}$$

Note that the spatial support equals the joint probability to find an itemset $X$ in the transaction database at location $u$ in geographic space. Since the itemsets are mined in the itemset space $2^A$ and the transactions are distributed in geographic space $\mathbb{R}^2$ we see that mining spatial itemsets is based on observations in the joint sample space $2^A \times \mathbb{R}^2$, with $P(X, u)$ being a probability measure on that space.

Given a location, we can set a spatial support threshold $\delta$ similar to the support threshold in frequent itemset mining, to mine those itemsets having a certain frequency in the database. To obtain those itemsets that are mostly unique at a given location $u$, we define the *location-conditional support* measure

$$P(X|u) = \frac{P(X, u)}{P(u)} = \frac{f_{P_X}(u)}{f_{P_{\{\}}}(u)} P(X), \tag{7}$$

where $f_{P_{\{\}}}$ is the density of all input transactions. $P(X|u)$ will be high if the transactions supporting $X$ have a high probability to occur only around $u$.

Similar to the definitions of support, we define the *spatial confidence* of itemset $X$ given $Y$ at location $u$ as

$$conf(Y, u \Rightarrow X) := P(X|Y, u) = \frac{P(X \cap Y, u)}{P(Y, u)} = \frac{f_{P_{X \cap Y}}(u)}{f_{P_Y}(u)} \frac{|O_{X \cap Y}|}{|O_Y|}, \tag{8}$$

describing how likely the transactions supporting itemset $Y$ around location $u$ will also support $X$.

Holding $X$ and $Y$ fixed, the spatial characteristics of itemsets are described by the *spatial interestingness measures* $P(X, u)$, $P(X|u)$, and $P(X|Y, u)$. Depending on the mining task, the measure either returns the value at a particular location $u$ or the distribution over geographic space. In Section 5, we use the measures to explore spatial itemsets and association rules.

**Spatial Point Set Statistics.** If only a certain spatial characteristic is of interest, we may not need to model the spatial distribution $f_{P_X}$. For that, we define a measure $M$ to be any function $M(P_X)$ of $P_X$, and we use it to describe spatial characteristics of an itemset point set $P_X$. For example, let $BB_X$ be the bounding box surrounding the points $P_X$ and $area(BB_X)$ be the size of the covered area. Then

$$M(P_X) := area(BB_X) \tag{9}$$

is a valid *spatial interestingness measure* of a spatial itemset $X$. Based on the area measure $M(X)$, an itemset $X$ can then be considered a *local itemset* (if $M(X)$ is small), or a *global itemset* (if $M(X)$ is large). Instead of calculating the area on the basis of a minimum bounding box, it can also be calculated using the convex hull of $P_X$, or on the basis of the confidence region of a Gaussian distribution fitted by the points $P_X$. The necessary information about the point set $P_X$ needed to calculate $M$ is called the statistics $T_M$.

Extracting the statistics $T_M$ is similar to the approach proposed in [11]. Instead of using traditional measures like mean and variance, geometric measures like area or shape are useful interestingness measures in the presence of a spatial attribute. In Section 5, we mine itemsets whose area is smaller and/or greater then given maximum and/or minimum thresholds.

**Spatial Itemset Mining.** Given our model of geo-transactions and spatial itemsets, we define *spatial itemset mining* as the process to discover itemsets $X \subseteq \mathcal{A}$ and derived association rules $Y \Rightarrow X$ in a database of geo-transactions $\mathcal{O}$. The spatial itemsets $(X, P_X)$ exhibit spatial characteristics in geographic space that can be filtered, compared, or visualized in the exploration process. Spatial characteristics of itemset $X$ are determined by a spatial interestingness measure $M$, derived from the distribution $f_{P_X}$, or from a particular statistic $T_M$ of the points set $P_X$.

## 4   Computational Considerations

In this paper our focus is on the extension of the theoretical itemset mining framework by spatial attributes. In the following, we give a short outline of the major implications for an implementation of the proposed modifications and present our FP-growth extension.

## 4.1 Density Estimation

A simple non-parametric estimator is based on a sparse grid [20]. For this, the points are discretized by a grid using a bandwidth $\theta$. The density estimate $\hat{f}_{P_X}(u)$ is then just the normalized number of points falling in the cell covering $u$. Since the distribution of geographic locations can be assumed to be sparse (only a fraction of cells contain points), the space complexity is reasonable. The grid can be updated iteratively having constant insert and estimate operations.

The estimates of the histogram estimator are discrete with respect to the chosen bandwidth. To obtain continuous estimates a non-parametric approach with reasonable space requirements is based on a CF-tree [21], known from the BIRCH clustering algorithm [19]. In a CF-tree, the points are iteratively summarized by sufficient statistics of Gaussian distributions. Thereby, each point corresponds to one Gaussian and the partitioning takes place during the insert operation. The tree is parametrized by a branching factor $b$ and a distance threshold $t$. While the branching factor $b$ affects the height of the tree and has no effect on the resulting density estimates, the distance threshold $t$ affects the number of resulting leaf nodes and is equivalent to the smoothing bandwidth $\theta$. Given a CF tree with leaf nodes $q_1, \ldots, q_k$, the estimator calculates the estimate $\hat{f}_{P_X}(u)$ based on a mixture of Gaussians $\mathcal{N}(u; q_1), \ldots, \mathcal{N}(u; q_k)$. The insert and estimation complexity is $O(\log N)$, with $N$ being the total number of transactions.

## 4.2 Itemset Augmentation and Mining

Using any frequent itemset mining algorithm with a minimum support threshold $\delta$, the density estimates $f_{P_X}$ or the statistics $T_M$ can be obtained post-hoc. For this, the resulting itemsets are post-processed to aggregate the necessary information (e.g., a CF-tree or a bounding box). If the itemsets have no index about their supporting transactions, this operation is of complexity $O(RN)$, where $R$ is the number of resulting itemsets and $N$ is the number of transactions.

In our work we augment the data structures of an FP-tree [22] to capture the necessary information during the mining process in synopsis data structures. For this, each node in the tree has a synopsis representing the aggregated spatial distribution or statistics, in addition to the transaction counts. To allow the FP-growth algorithm to work, the synopsis must support a merge operation between two synopses. Given two density distributions, the merged distribution is simply their normalized sum, hence this operation is supported naturally. For particular statistics, this operation will almost always hold, since a statistic is itself an aggregate. For example, the merge operation of two bounding boxes consist only of choosing the most extreme lower-left and upper-right points.

If the synopsis data structure is of fixed size and has constant insert and merge operations, the space complexity overhead and the runtime overhead are constant. This is true for the bounding box example. However, the space and runtime overhead for synopses holding information about density estimations (sparse grid, CF-tree) will increase with the number of transactions. The increase of the CF-tree approach is expected to be log-linear, as the space complexity and the insert and estimate operation runtime complexity are in the

**Table 1.** Comparison of FPB- and FPC-trees to the FP-tree in terms of tree size, tree generation runtime (build rt), and FP-growth runtime (query rt) with respect to the number of input transactions using a minimum support count $s = 60$ and a CF distance threshold $t = 1.0$

| n | size | | build rt | | query rt | |
|---|---|---|---|---|---|---|
| | FPB | FPC | FPB | FPC | FPB | FPC |
| 5K | 3.09 | 14.19 | 2.57 | 24.99 | 1.45 | 7.26 |
| 10K | 3.10 | 14.49 | 2.41 | 24.25 | 1.36 | 7.68 |
| 20K | 3.10 | 14.71 | 2.53 | 27.69 | 1.38 | 9.87 |
| 30K | 3.10 | 14.76 | 2.56 | 28.66 | 1.36 | 10.14 |
| 40K | 3.11 | 14.67 | 2.52 | 30.11 | 1.32 | 10.15 |
| 50K | 3.10 | 14.81 | 2.47 | 29.13 | 1.30 | 10.05 |

order of the logged transactions. We implemented the bounding box synopsis and the CF-tree synopsis resulting in a FPB-tree and a FPC-tree. Table 1 shows the measured space and runtime overhead with respect to a FP-tree without a synopsis extension, confirming the expected complexity results.

## 5    Experiments

In this section, we show explorative scenarios of mining itemsets and association rules in geo-transactional databases using different spatial interestingness measures introduced above.

### 5.1    Dataset and Setup

For the experiments, we use a set of georeferenced and tagged photos that have been queried using the Flickr API. The set contains photos that have been taken from January 2006 to December 2009. We removed all photos having no tags as well as photos from the same user that occur at exactly the same position and have the same tags to limit the influence of photo series. The resulting data set has 55,650 photos, and the tag sets together with the associated georeference are used as geo-transactions.

### 5.2    Scenarios

We use our framework to mine the tag sets of the Flickr data set. For all mining tasks, we use a minimum support count $s = 60$, corresponding to a minimum support threshold $\delta = 0.1\%$. The total number of mined itemsets without spatial constraints is $|R| = 134,683$. Figure 1(a) shows a map of all geo-transactions in $\mathcal{O}$. We use a CF threshold $t = 1.0$ for all our mining tasks. A value of $t = 1.0$ corresponds to a bandwidth of around 110 km and provides sufficient smoothing to filter out small-scale characteristics, but allows to distinguish regional and global characteristics on a country scale.

**Area Filtering and Summarization.** To test whether our framework can find local and global itemsets, we performed a minimum, a maximum, and a combined area constraint mining task. Figure 1 shows the bounding boxes of spatial itemsets that fulfill the given area contraints. Area is measured in $degree^2$, with $1\ degree^2 \approx 12321\ km^2$. Figure 1(b) shows the bounding boxes for an area constraint $area < 40$. The corresponding list of frequent spatial itemsets is shown in part in Figure 2(a), which lists the top 12 out of all 334 results. All mined itemsets contain items with a strong local meaning such as city and place names ($newyorkcity$, $pebblebeach$, $manhatten$) or regions ($california$, $oregon$, $england$). The itemsets supporting the constraint $40 < area < 1000$ (Figure 1(c) and 2(b)) mostly adhere to larger regions ($pacific$, $coast$) and countries ($italy$, $germany$, $spain$, $usa$, $brazil$). The itemsets $\{england, london\}$ and $\{london, uk\}$ are not expected to have an area larger than 40. However, since the bounding box does not remove outliers those results arise from erroneous records. Using an area measure based on the confidence regions of a Gaussian distribution would be a promising choice to remove outliers. As expected, items of spatial itemsets that satisfy the constraint $area > 40000$ (Figure 1(d) and 2(c)) are comprised of global descriptive tags ($travel$, $trip$, $snow$, $landscape$) and camera manufactures (the latter being a common tag in Flickr).

Given the determined areas of the mined itemsets, we can describe the database by summarizing statistics. For example, there is no strong correlation between the support count of the itemsets and the area their bounding box covers, as shown in Figure 3(a). This indicates that local and global itemsets with large and small counts exist. However, a negative correlation between the area and the number of items, as well as between the support count and the number of items exists (as shown in Figure 3(b) and 3(c)), indicating that support count and area are strongly correlated given the number of items. Figure 3(d) shows the histogram of the area occupied by spatial itemsets, showing that a large number of itemsets exists that cover only a fairly small area. A second peak at $25000 < area < 35000$ indicates that itemsets with $area \approx 30000$ (area of almost half the earth) are common.

Applying area constraints allows to find local and global itemsets, which correspond to the meaning of items in the Flickr dataset. This allows to classify mined itemsets based on their scale very well. The summarizing statistics allow for a quick overview of the spatial itemset characteristics in a given database.

**Similar Itemset Distribution.** The spatial density distributions of three selected itemsets are depicted in Figure 4. They provide an overview of the spatial distribution characteristics of the itemsets. As can be seen in Figure 4(a), the itemset $\{travel\}$ and $\{travel, city\}$ have a number of peaks in America, Europe, and Asia, while $\{travel, asia\}$ only occurs in Asia. By clustering the itemsets based on their distributions we are able to discover groups of itemsets of similar geographic semantics. Here, $\{travel\}$ and $\{travel, city\}$ belong to a cluster of global semantics, while $\{travel, asia\}$ belongs to a cluster of regional semantics. We then might choose $travel \Rightarrow city$ as a rule expressing a semantic specialization (intra-cluster rule), while we choose $travel \Rightarrow asia$ as a rule expressing

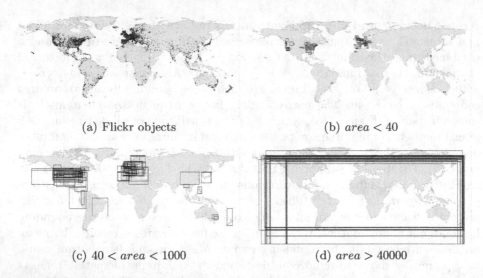

(a) Flickr objects

(b) *area* < 40

(c) 40 < *area* < 1000

(d) *area* > 40000

**Fig. 1.** Frequent spatial itemsets of Flickr dataset with minimum support count $s = 60$ and given area constraints

| s | area | itemset |
|---|---|---|
| 293 | 38.09 | california unitedstates |
| 205 | 00.06 | newyorkcity nyc |
| 198 | 06.37 | 2006 california |
| 197 | 00.00 | california pebblebeach |
| 197 | 00.00 | avalon catalina island |
| 196 | 00.00 | 2006 california pebble-beach september wedding |
| 188 | 35.00 | england unitedkingdom |
| 181 | 00.03 | manhattan newyork |
| 173 | 18.36 | california losangeles |
| 173 | 07.92 | coast oregon |
| 168 | 19.26 | newyork ny |
| 168 | 06.79 | coast ocean oregon |

(a) *area* ≤ 40

| s | area | itemset |
|---|---|---|
| 410 | 114.71 | italia italy |
| 317 | 827.85 | england london |
| 315 | 577.17 | espaa spain |
| 252 | 066.61 | deutschland germany |
| 206 | 743.63 | catalina island |
| 183 | 985.09 | coast pacific |
| 183 | 835.09 | london uk |
| 172 | 885.75 | coast ocean pacific |
| 167 | 986.05 | geotagged usa |
| 167 | 814.37 | brasil brazil |
| 159 | 884.72 | geotagged unitedstates |
| 158 | 063.82 | 2009 italy |

(b) 40 < *area* ≤ 1000

| s | area | itemset |
|---|---|---|
| 224 | 42316.39 | d80 nikon |
| 131 | 45327.77 | travel vacation |
| 125 | 40848.06 | travel trip |
| 96 | 45013.34 | ice snow |
| 92 | 45094.56 | nikon travel |
| 88 | 43082.56 | landscape nikon |
| 85 | 41947.31 | island water |
| 67 | 41551.23 | landscape travel |
| 66 | 43500.69 | nikon sea |
| 64 | 43674.28 | landscape sea |
| 63 | 40094.22 | trip vacation |
| 61 | 43978.02 | nikon snow |

(c) *area* > 40000

**Fig. 2.** Frequent spatial itemsets of Flickr dataset with minimum support count $s = 60$ and given area constraints ordered by support

a semantic variation (inter-cluster rule). The following table shows the itemset similarities on the basis of the $D_{sym}$ distance measure, see Eq. (1), representing a valid input to cluster itemsets based on their spatial distribution characteristics.

|  | $\{travel\}$ | $\{travel, city\}$ | $\{travel, asia\}$ |
|---|---|---|---|
| $\{travel\}$ | 0.0 | 0.126 | 1.595 |
| $\{travel, city\}$ |  | 0.0 | 1.154 |
| $\{travel, asia\}$ |  |  | 0.0 |

**Location-Conditional Support.** Next, we evaluate the task of mining itemsets that have a high location-conditional support. For this task, we mine spatial itemsets at three different locations, Berlin, San Francisco, and Capetown. At these locations, we look for itemset that occur with a location-conditional support $P(X|u)$ of at least 0.1%. The results are shown in Figure 5 and are ordered

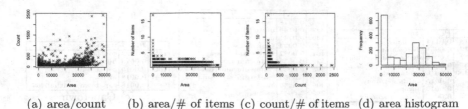

(a) area/count      (b) area/# of items  (c) count/# of items  (d) area histogram

**Fig. 3.** Summarizing statistics of area and support counts measures from frequent itemsets in the Flickr dataset

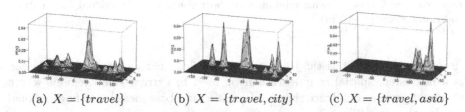

(a) $X = \{travel\}$          (b) $X = \{travel, city\}$          (c) $X = \{travel, asia\}$

**Fig. 4.** Spatial density distributions of three selected spatial itemsets mined from the Flickr dataset

by $P(X|u)$ in decreasing order. Figure 5 clearly shows the local characteristics of the mined itemsets. For all three locations, the country they are located in is the itemset with the highest probability (*germany, california, southafrica*), followed by itemsets that contain the city name (*berlin, sanfrancisco, capetown*) and having a strong local connection (*hiking*). This mining task is particularly useful if itemsets with a discriminative spatial meaning are to be discovered, for example, as input to recommendation models or mining association rules.

**Spatial Confidence.** In our last task we demonstrate that spatial confidence, which is the confidence that an association rule holds at a given location in space, is an effective measure to judge the validity of an association rule at a particular location $u$. We again use three sample locations, San Francisco, Berlin, and Bejing, and calculate the spatial confidence for the rules $\{travel\} \Rightarrow \{city\}$ and $\{travel\} \Rightarrow \{asia\}$. In the following, we list the spatial confidence for all combinations of the three locations and the two rules. The density estimates depicted in Figure 4 can be used to confirm the results.

$$conf(\{travel\}, u_{SanFran} \Rightarrow \{city\}) = 0.205$$
$$conf(\{travel\}, u_{Berlin} \Rightarrow \{city\}) = 0.101$$
$$conf(\{travel\}, u_{Bejing} \Rightarrow \{city\}) = 0.412$$
$$conf(\{travel\}, u_{SanFran} \Rightarrow \{asia\}) = 0.0$$
$$conf(\{travel\}, u_{Berlin} \Rightarrow \{asia\}) = 0.0$$
$$conf(\{travel\}, u_{Bejing} \Rightarrow \{asia\}) = 0.276$$

As can be seen for the consequent *city*, it is valid at all three locations, having a maximum for Bejing. However, for the consequent *asia*, which is an item with

**(a) Berlin**

| s | P(X\|u) | itemset |
|---|---|---|
| 618 | 0.0111 | germany |
| 372 | 0.0067 | deutschland |
| 252 | 0.0045 | deutschland germany |
| 237 | 0.0043 | berlin |
| 2400 | 0.0016 | geotagged |
| 83 | 0.0015 | berlin germany |
| 755 | 0.0013 | europe |
| 73 | 0.0011 | europe germany |

**(b) San Francisco**

| s | P(X\|u) | itemset |
|---|---|---|
| 1547 | 0.0278 | california |
| 832 | 0.0149 | unitedstates |
| 1447 | 0.0141 | usa |
| 523 | 0.0094 | sanfrancisco |
| 401 | 0.0072 | unitedstates usa |
| 326 | 0.0059 | california usa |
| 293 | 0.0053 | california unitedstates |
| 279 | 0.0050 | california sanfrancisco |
| 309 | 0.0050 | eyefi |
| 2400 | 0.0047 | geotagged |
| 2440 | 0.0045 | 2009 |
| 902 | 0.0043 | iphone |
| | | ... |

**(c) Capetown**

| s | P(X\|u) | itemset |
|---|---|---|
| 185 | 0.0033 | southafrica |
| 319 | 0.0019 | hiking |
| 105 | 0.0019 | capetown |
| 261 | 0.0018 | africa |
| 98 | 0.0018 | capetown southafrica |

**Fig. 5.** Frequent spatial itemsets of Flickr dataset mined at locations Berlin, San Francisco, and Capetown with minimum support count $s = 60$ ordered by location-conditional support $P(X|u)$

a strong local characteristic, the association rules clearly only hold for Bejing. As a next step, spatial confidence can be used to extract small regions where rules hold given a confidence threshold, making it a base measure to find regional association rules [17].

# 6    Conclusions and Ongoing Work

In this paper, we proposed a framework to mine spatial itemsets based on their spatial characteristics. We presented the conceptual basis for spatial itemset mining and introduced selected spatial interestingness measures. We outlined computational considerations to extend existing itemset mining approaches to aggregate synopses of spatial characteristics. We demonstrated the usefulness of spatial itemset mining by different exploration tasks using a real-world Flickr dataset. The exploration results show that spatial itemset mining allows to get a deep insight into spatial itemset databases. As such datasets will be more and more available, we believe that our approach to mining spatial data will prove useful in many application domains. Our ongoing work includes research for efficient synopsis data structures to capture interesting spatial characteristics and the evaluation of further exploration tasks using large-scale, real-world datasets.

# References

1. Rattenbury, T., Naaman, M.: Methods for extracting place semantics from Flickr tags. ACM Transactions on the Web 3(1), 1–30 (2009)
2. Sakaki, T.: Earthquake Shakes Twitter Users: Real-time Event Detection by Social Sensors. In: Proc. of WWW 2010, pp. 851–860 (2010)
3. Yin, Z., Cao, L., Han, J., Zhai, C., Huang, T.: Geographical Topic Discovery and Comparison. In: Proc. of WWW 2011, pp. 247–256 (2011)
4. Agrawal, R., Srikant, R.: Fast Algorithms for Mining Association Rules. In: Proc. of VLDB 1994, pp. 487–499 (1994)
5. Han, J., Koperski, K., Stefanovic, N.: GeoMiner: A system prototype for spatial data mining. In: Proc. of SIGMOD 1997, pp. 553–556 (1997)

6. Huang, Y., Shekhar, S., Xiong, H.: Discovering Colocation Patterns from Spatial Data Sets: A General Approach. IEEE Transactions on Knowledge and Data Engineering 16(12), 1472–1485 (2004)
7. Fukuda, T., Morimoto, Y., Morishita, S., Tokuyama, T.: Mining Optimized Numeric Association Attributes. In: Proc. of PODS 1996, pp. 182–191 (1996)
8. Srikant, R., Agrawal, R.: Mining Quantitative Association Tables Rules in Large Relational Tables. In: Proc. of SIGMOD 1996, pp. 1–12 (1996)
9. Wang, K., Tay, S.H.W., Liu, B.: Interestingness-Based Interval Merger for Numeric Association. In: Proc. of KDD 1998, pp. 121–127 (1998)
10. Yang, Y., Miller, J.: Association Rules over Interval Data. In: Proc. of SIGMOD 1997, pp. 452–461 (1997)
11. Aumann, Y., Lindell, Y.: A statistical theory for quantitative association rules. In: Proc. of KDD 1999, pp. 261–270 (1999)
12. Zhang, H., Padmanabhan, B., T.: On the discovery of significant statistical quantitative rules. In: Proc. of KDD 2004, pp. 374–383 (2004)
13. Huang, Y., Xiong, H., Shekhar, S., Pei, J.: Mining confident co-location rules without a support threshold. In: Proc. of SAC 2003, pp. 497–501 (2003)
14. Lin, Z., Lim, S.: Optimal candidate generation in spatial co-location mining. In: Proc. of SAC 2009, pp. 1441–1445 (2009)
15. Koperski, K., Han, J.: Discovery of Spatial Association Rules in Geographic Information Databases. In: Egenhofer, M., Herring, J.R. (eds.) SSD 1995. LNCS, vol. 951, pp. 47–66. Springer, Heidelberg (1995)
16. Sengstock, C., Gertz, M., Tran Van, C.: Spatial Interestingness Measures for Co-location Pattern Mining. In: Proc. of SSTDM (ICDM Workshop) 2012, pp. 821–826 (2012)
17. Ding, W., Eick, C.F., Wang, J., Yuan, X.: A Framework for Regional Association Rule Mining in Spatial Datasets. In: Proc. of ICDM 2006, pp. 851–856 (2006)
18. Hastie, T., Tibshirani, R., Friedman, J.: The Elements of Statistical Learning: Data Mining, Inference, and Prediction. Springer (2002)
19. Zhang, T., Ramakrishnan, R., Livny, M.: BIRCH: An Efficient Data Clustering Databases Method for Very Large Databases. In: Proc. of SIGMOD 1996, pp. 103–114 (1996)
20. Wasserman, L.: All of Statistics. Springer (2004)
21. Zhang, T., Ramakrishnan, R., Livny, M.: Fast Density Estimation Using CF-kernel for Very Large Databases. In: Proc. of KDD 1999, pp. 312–316 (1999)
22. Han, J., Pei, J., Yin, Y., Mao, R.: Mining Frequent Patterns without Candidate Generation: A Frequent-Pattern Tree Approach. Data Mining and Knowledge Discovery 8(1), 53–87 (2004)

# Mining Periodic Event Patterns
# from RDF Datasets

Anh Le and Michael Gertz

Database Systems Research Group, Heidelberg University, Germany
{anh.le.van.quoc,gertz}@informatik.uni-heidelberg.de

**Abstract.** Exposing and sharing data and information using linked data sources is becoming a major theme on the Web. Several approaches have been developed to model and efficiently query and match linked open data, primarily represented as RDF graphs from RDF facts and associated ontological frameworks. Interestingly, little work has yet been conducted to discover interesting patterns from such data.

In this paper, we present an approach that aims at discovering interesting periodic event patterns from RDF facts describing events, for example, music events or festivals. Our focus is on exploiting the temporal and geographic properties associated with such event descriptions as well as the concept hierarchies used to categorize the different components of event facts. Discovered patterns of periodic events can be used for prediction or detection of outliers in RDF datasets. We demonstrate the feasibility and utility of our framework using real event datasets extracted as RDF facts from the Website eventful.com.

## 1 Introduction

Over the past couple of years, there has been a significant increase in linked open data sources aimed at sharing, linking, and thus querying various types of distributed information more effectively[1]. Core of these initiatives are basically the representation of facts using RDF and the enrichment of fact descriptions using ontological frameworks. While there has been substantial research on various aspects of managing RDF (graph) data, including consolidating ontologies and querying linked RDF data using the query language SPARQL, interestingly there has been only little work on mining RDF data for interesting patterns. Given that with many RDF facts ontologies are associated, pattern mining poses some interesting challenges.

In this paper, we present a framework to discover interesting periodic event patterns from RDF data. Rather than looking at standard triples of the form subject, predicate, and object, event information represented in RDF also include time and location information, thus building quintuples, such as in YAGO2 [5]. The challenge in mining such type of data is that with each component ontologies are associated in which concepts are generalized and organized in hierarchies.

---

[1] http://linkeddata.org

B. Catania, G. Guerrini, and J. Pokorný (Eds.): ADBIS 2013, LNCS 8133, pp. 162–175, 2013.

Approaches to the discovery of periodic event patterns thus have to take such hierarchies, which are given as separate datasets, into account to derive patterns at different levels of granularity and abstraction, respectively. For example, in an event dataset there might be several facts about an event titled 'Wisdom of Yoga Workshop' happening at different times and locations, but from which no meaningful patterns can be derived. If one considers the type of event, specified, e.g., as the concept 'Yoga' in some hierarchy at a higher level of abstraction, interesting (periodic) event patterns could be derived from several such event instances. The challenge in finding such patterns is to have an efficient approach that generalizes facts using hierarchies not only for subject, predicates, and object specifications, but also for the temporal and geographic information included in respective RDF facts.

In the approach detailed in this paper, we focus in particular on periodic event patterns. That is, from a set of event instances, using concept hierarchies, we want to determine (a) periodicities associated with event types and (b) periodicities associated with co-occurring event types. While in previous work, a general approach to mining event patterns from RDF data has been introduced [1], in this paper, we develop a novel approach for mining periodic patterns. Such patterns can be used to make predictions about upcoming events, to detect outliers in a given dataset, or to explicitly add information about periodicity and co-occurrence to event descriptions. In order to guide the search for interesting periodic patterns, we allow the user to specify constraints related to the concepts, temporal, and geographic information associated with event descriptions. For example, form a large dataset, it might be trivial to derive that some music event is happing every Saturday (independent of the location), but if one is looking for patterns in music event ('music' being a generalization of music related events) that periodically occur in the same region or place, then such constraints can narrow down the search and allow for a more efficient discovery of patterns. We demonstrate the feasibility and utility of our approach using a large-scale event dataset extracted as RDF facts from the Website eventful.com. We show that in fact interesting periodic patterns can be extracted from such event data, which then can be used, e.g., to enrich the existing RDF dataset by further facts.

In the following section, we briefly discuss related work. In Section 3, we review the notations of events and event-templates. In Section 4, we introduce our specification language for periodic event patterns and then describe the method to mine such patterns in Section 5. After presenting some experimental results in Section 6, we conclude the paper in Section 7.

## 2   Related Work

Our work is closely related to approaches developed in (1) periodicity detection from time series data, (2) cyclic association rule mining from temporal databases, and (3) periodic pattern mining from spatio-temporal data, as outlined below.

There are several methods to detect unknown periods from time series or sequential data such as Fast Fourier Transform, auto-correlation [13], or

Lomb-Scargle periodogram [9]. These approaches aim at finding full-cycle periodicities in the sense that all points in a cycle have to be repeated. Thus, cases where only some points have a cyclic behavior are not detected [3]. For example, an event occurs every Saturday but it might or might not occur on other days of the week. Other approaches to the detection of periods focus on efficient detection algorithms [6], partial periodicity [4,10], noisy data [14], or incomplete observations [8]. Although our focus is on (1) types of data different from those studied in the above approaches and (2) on periodic patterns representing co-occurrences of multiple event topics, we still can employ any of the above methods to detect periods for single events and then use this information to find periodic patterns.

The problem of mining cyclic association rules has been introduced by Özden et al. [12] where an association rule $r$ is cyclic with respect to a period $p$ if $r$ holds in every $p$ time units. This approach aims at finding perfect periodicities and therefore it is impractical as patterns in real life are often imperfect [4]. Han et al. [3] consider partially periodic patterns where periodicities satisfying some confidence threshold are allowed. This approach can also be extended to mining periodic patterns at multiple levels of abstraction. However, the level of abstraction has to be given as input, whereas in our approach it is discovered. In general, all the above approaches are designed for time-related datasets, but not for datasets that also include spatial information or where explicit sequences of events do not exist.

Different from the above work, Mamoulis et al. [2,11] focus on moving object data to find periodic trajectories. For this, spatio-temporal data (movements of objects) are transformed into sequences of regions, e.g., districts or mobile communication cells. Periodic patterns are then cyclic sequences of visited regions. Rather than assuming predefined regions, Li et al. [7] propose a method to detect important regions, called reference spots, using a kernel method, from which then periodic behaviors are determined. To the best of our knowledge, there is no approach to determine periodic patterns from spatio-temporal data represented as sets of RDF facts describing the occurrence of events in terms of space and time. Furthermore, none of the approaches considers concept hierarchies that are associated with the description of event components. In [1], we studied the discovery of spatio-temporal patterns from RDF event data but did not address the more challenging problem of finding periodic event topics.

## 3    Basic Concepts and Notations

To model events and event templates based on RDF data, we adopt some notations and definitions from the framework presented in our previous work [1], which are briefly described in the following.

### 3.1    Concepts and Concept Hierarchies

An ontology is a knowledge representation that captures machine-interpretable definitions of basic concepts and relations among concepts in a specific domain.

An ontology together with a set of *individual instances* of *classes* constitutes a *knowledge base* where the classes are the semantic types of the individual instances. In our approach, we call the individual instances *entities* and the classes *categories*. For example, a particular music event might be categorized based on its genre. A category might be a sub-category of one or more other categories, that is, categories are organized in hierarchies. A hierarchy typically forms a *directed acyclic graph* (DAG) where a node corresponds to either a *category* or an *entity* An edge specifies a *belongs-to relationship* between two categories or an *is-a relationship* between an entity and a category. Figure 1 shows an example of a concept hierarchy, where entities are event titles, e.g., 'DevilDriver_in_Frankfurt', and categories are genres, e.g., 'Heavy_Metal' or 'Rock'.

Given a hierarchy $\mathcal{H}$ and a node $n \in \mathcal{H}$, we use two notations to describe hierarchical relationships among concepts: $n^\uparrow$ denotes the set of all *direct* generalizations of $n$, and $n^\Uparrow$ denotes the set of all generalizations of $n$, i.e., nodes having a paths from $n$, including $n$. For example, based on the hierarchy in Figure 1, we have Heavy_Metal$^\uparrow$={Rock} and Heavy_Metal$^\Uparrow$={Heavy_Metal, Rock, Music}.

To consider constraints on event patterns later on, we define the levels of a concept in a concept hierarchy as follows.

**Definition 1.** *(Concept Level) Given a hierarchy $\mathcal{H}$ and a node $n \in \mathcal{H}$, the level of $n$ in $\mathcal{H}$ is defined as follows:*
- *if $n$ is an entity, then $level(n) = 0$.*
- *if $n$ is a category, then $level(n) = \min\{level(c) \mid c \in \mathcal{H} \colon n \in c^\uparrow\} + 1$*

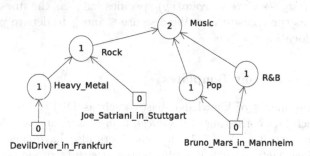

**Fig. 1.** Example of a concept hierarchy. Boxes represent entities (event titles), circles represent categories (genres). Numbers denote levels of concepts.

The level of a concept in a hierarchy is computed recursively. For example, the level of 'Heavy_Metal' is computed based on the level of 'DevilDriver_in_Frankfurt' (i.e., 0+1=1). The level of 'Rock' is computed based on the level of 'Heavy_Metal' and 'Joe_Satriani_in_Stuttgart' (i.e. min{1, 0}+1=1), and so on.

## 3.2   Time and Locations

To represent time associated with facts and event patterns, we assume different *timelines* such as $T_{day}$ for days or $T_{month}$ for months. Each timeline corresponds to a *temporal granularity*. A *timestamp* is an absolute time, e.g., '2012' anchored in the timeline $T_{year}$. A pair of a timeline and a timestamp, e.g., $\langle T_{year}, \text{'}2012\text{'}\rangle$, describes a *time point*. Different from [1] where relationships among time intervals are considered, here we only consider the start-time to identify occurrences of facts. A *temporal framework*, denoted $\mathcal{TF}$, then is a set of time points.

Timelines are organized in a hierarchy, e.g., $T_{day} \rightarrow T_{month} \rightarrow T_{year}$. An edge from a timeline to another timeline represents a *finer-than* (or contains) relationship. A mapping can be performed from a timestamp anchored in one timeline to a timestamp in a coarser timeline. Similar to entities, we define generalization operators $\uparrow$ and $\Uparrow$ to convert a time point to coarser granularities based on a temporal hierarchy. Given a time point $t$ in a timeline $T$, $t^{\uparrow}$ is a set of time points obtained from mapping $t$ to the next coarser timelines of $T$ and $t^{\Uparrow}$ is a set of time points obtained from mapping $t$ to all coarser timelines of $T$ (including $T$). For example, with a hierarchy $T_{day} \rightarrow T_{month} \rightarrow T_{year}$, $\langle T_{day}, \text{'May 1, 2012'}\rangle^{\Uparrow} = \{\langle T_{day}, \text{'May 1, 2012'}\rangle, \langle T_{month}, \text{'May 2012'}\rangle, \langle T_{year}, \text{'2012'}\rangle\}$.

Locations can be specified at different levels of granularity as well. A *spatial granularity*, e.g., country, represents a partition of the spatial domain in regions, called *spatial entities*. A pair of a granularity and a spatial entity, e.g., $\langle$Country, Germany$\rangle$, describes a location. A *spatial framework*, denoted $\mathcal{SF}$, is a set of locations. To define spatial constraints later, we assume that each spatial entity has a *spatial extent*. Such mappings might be explicitly provided in the knowledge base or they can be obtained using tools such as Yahoo Placemaker[2]. Similar to time, locations are organized in hierarchies. A spatial hierarchy is a DAG representing finer-than (or contains) relationships among spatial granularities. For example, ($city \rightarrow state \rightarrow country$) specifies 'city' as the finest granularity and 'country' as the coarsest one. We also use $\uparrow$ and $\Uparrow$ to denote generalization operators for locations.

## 3.3   Events and Event Templates

The basic components of knowledge bases such as DBPedia or Freebase are RDF facts, each fact consisting of a *subject*, a *predicate*, and an *object*. In some knowledge bases such as YAGO2 [5], some facts also have time and location component. Such facts are called *spatio-temporal events*, or simply *events*, conceptually represented as a quintuple $\langle$subject, predicate, object, time, location$\rangle$. The subject, predicate and object correspond to leaf nodes in some given concept hierarchies. We assume that the time and location components follow the temporal and spatial frameworks described in Section 3.2. For example, a quintuple $\langle$DevilDriver_in_Frankfurt, performed_by, Devil_Driver, $\langle T_{day}, \text{'August 9, 2013'}\rangle$, $\langle$City, 'Frankfurt'$\rangle$ $\rangle$ describes a music event in Frankfurt.

The five components of events can be generalized using the generalization operators to obtain so-called *event templates* (ETs). Subjects, objects and predicates are replaced by their (in)direct categories. The time and location components can be generalized to coarser granularities, too. An event might "produce"

---

[2] http://developer.yahoo.com/geo/placemaker

many different ETs and an ET might be derived from different events. If an ET $f$ can be derived from an event $e$, we say that $e$ is an *instance* of $f$. For example, a tuple $\langle$Heavy_Metal, performed_by, American_rock_band, $\langle T_{year}$, '2013'$\rangle$, $\langle$Country, 'Germany'$\rangle$ $\rangle$ is an ET of the event in the previous example.

We now define generalization operators ($\uparrow$ and $\Uparrow$) for ETs. Given an ET $f$, the set $f^{\uparrow}$ consists of all ETs obtained from $f$ by replacing one component, i.e., subject, predicate, object, time, or location, by its direct ancestors in the hierarchy. The set $f^{\Uparrow}$ then simply is the closure of $f$. The formal definition adopted from [1] is as follows.

**Definition 2.** *(ET Generalization) Given an event template $f = \langle s, r, o, t, l \rangle$ and a set of hierarchies for subjects, predicates, objects, time, and locations. The generalizations of $f$, denoted $f^{\uparrow}$ and $f^{\Uparrow}$, are determined as follows:*

- *$f^{\uparrow} := F_s \cup F_r \cup F_o \cup F_t \cup F_l$ with $\Gamma_x$ ($x \in \{s, r, o, t, l\}$) being a set of ETs obtained by replacing $x$ by an element in $x^{\uparrow}$.*
- *$f^{\Uparrow} := \{ \langle s^*, r^*, o^*, t^*, l^* \rangle \mid \forall x \in \{s, r, o, t, l\}\ x^* \in x^{\Uparrow}\}$, that is, a set of ETs obtained by replacing one or more components by their ancestors.*

For example, if we assume that for the hierarchy in Figure 1 only the subject can be generalized further, the two set of generalizations $f^{\uparrow}$ and $f^{\Uparrow}$ for an ET $f=\langle$'Heavy_metal',... $\rangle$ are $f^{\uparrow} =\{\langle$'Rock',... $\rangle\}$ and $f^{\Uparrow} =\{\langle$'Heavy_metal',... $\rangle$, $\langle$'Rock',... $\rangle$, $\langle$'Music',... $\rangle\}$.

## 4    Problem Formulation

In this section, we define *periodic patterns*, or *p-patterns* for short, which describe regularities of event topics (specified by event templates) over time. Before giving a formal definition of p-patterns, we introduce the notations of *time slots*, *support segments*, *constraints*, and *event instance vectors*.

### 4.1    Time Slots

Given a time interval of interest (e.g., 3 years) and a temporal granularity chosen as a time unit (e.g., a day), the time interval is divided into $n$ *time slots* where the duration of each slot is a time unit. For example, the time interval of 3 years from 1999-03-01 to 2002-03-31 shown in Figure 2 consists of 1106 days, and each day corresponds to a time slot. We assume that the time unit is chosen for the input dataset of events such that each event is anchored at only one slot based on its start-time.

Based on time slots, we introduce a notation of *segments* to model periodicities as follows. Given an integer $t > 0$, the $n$ slots are grouped into $\lfloor \frac{n}{t} \rfloor$ *segments*[3], such that each segment consists of $t$ consecutive slots and the $k$-th segment ($k \in \{0, 1, \ldots, \lfloor \frac{n}{t} \rfloor\}$), denoted $\mathcal{I}_k$, consists of $t$ slots from the slot $(t * k)$ to the slot $(t - 1 + t * k)$. The $i$-th slot ($i \in \{0, 1, \ldots, t-1\}$) of a segment $\mathcal{I}_k$, denoted

---

[3] If $n$ is not a multiple of $t$ then some last slots will be truncated.

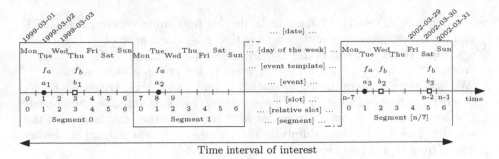

**Fig. 2.** Illustration of time slots and segments. The time interval of interest is divided into slots (days). The first segment (the first week) consists of the first 7 slots. The second segment (the second week) consists of the next 7 slots, and so on. The event template $f_a$ (whose instances are events $a_1$, $a_2$, $a_3$, etc.) is periodic but the event template $f_b$ (whose instances are events $b_1$, $b_2$, $b_3$, etc.) is not.

$\mathcal{I}_k[i]$, consists of all events belonging to the (absolute) slot $(i+t*k)$. For example, each segment in Figure 2 consists of 7 slots and each slot corresponds to a day of the week (Monday, Tuesday, etc.). With the above segmentation, an event template $f$ is said to be periodic if there exists $i \in \{0, 1, \ldots, t-1\}$ such that an instance of $f$ is found in the $i$-th slot of every segment (strictly periodic) or in most segments (relaxed periodicity). In Figure 2, the event template $f_a$ is periodic because its instances (events $a_1$, $a_2$, and $a_3$) are found at every second slot (Tuesday) of the segments. However, the event template $f_b$ is not periodic since its instances ($b_1$, $b_2$, and $b_3$) occur randomly. If there exists a periodic event template $f$ with a segmentation $t$, then $t$ is called a *period* for $f$.

In the following, we use $n$ to denote the number of slots in the time interval of interest and $t$ to denote the number of slots in each segment. We now introduce the notion of *support segments* in order to take into account a relaxed form of periodicity of event templates.

## 4.2   Support Segments

In practice, instances of an event template $f$ might be found at the $i$-th slot of most but not all segments. To relax the notion of periodicity, we define *support segments* and then introduce a *p-score* for an event template as follows.

**Definition 3 (Support Segments).** *Given an event template $f$, a segment $\mathcal{I}$ supports $f$ at the $i$-th slot if there exists an event $e$ in $\mathcal{I}[i]$ such that $e$ is an instance of $f$.*

For example, the first segment in Figure 2 supports an event template $f_a$ at the second slot and supports an event template $f_b$ at the fourth slot. To characterize the periodicity of an event template with respect to the $i$-th slot, we compute a p-score as the ratio of the number of the support segments to the total number of segments as follows.

**Definition 4 (p-score).** *Given an event template $f$ and a set of segments $\mathcal{IS}$, the p-score of $f$ with respect to the $i$-th slot is defined as*

$$p\text{-}score(f,i) := \frac{|\{\mathcal{I} \in \mathcal{IS} : \mathcal{I}\ supports\ f\ at\ the\ i\text{-}th\ slot\}|}{|\mathcal{IS}|} \in [0,1]$$

The p-score$(f,i)$ thus is simply the probability that instances of $f$ are found at the $i$-th slot of a segment. Since the function can be applied for any $i \in \{0, 1, \ldots, t-1\}$, we define a vector $V$, where $V[i]$=p-score$(f,i)$ $(i = 0 \ldots t-1)$ represents the periodic behavior of event templates. The vector $V$ is called an *event instance vector* (EIV).

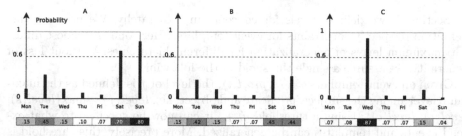

**Fig. 3.** Schematic representations of the EIVs for three sample event templates A, B, and C, which are shown by using bar charts and color gradient. With a threshold $\delta = 0.6$, A and C are periodic, but B is not.

As an example, assume an ET $A$ in Fig. 3, whose instances are events related to *'soccer'*, has the EIV [.15, .45, .15, .10, .07, .70, .80]. This EIV shows that it is likely to find events related to *'soccer'* on Saturday and Sunday every week.

Given a threshold $\delta \in [0,1]$, an ET $f$ with an EIV $V$ is called periodic if there exists an $i$ such that $V[i] \geq \delta$. For example, Figure 3 depicts three event templates, where A and C are periodic with a threshold $\delta = 0.6$, but B is not. Since our aim is to find periodic patterns consisting of multiple event templates that co-occur, we will extend the definitions and notations above to model periodicities of a set of event templates. Before doing so, we first focus on constraints that events specifying instances of a pattern have to satisfy.

### 4.3   Event Constraints

In practice, users might be interested in patterns that exhibit periodicities in a specific time interval and/or in a specific geographic region and simply refer to a specific concept. To allow for constraining the search space and also to obtain patterns that meet the users' requirements, we introduce three kinds of constraints for events.

- **Temporal Constraint**: A temporal constraint specifies the time interval of interest with a start-time and an end-time and a granularity of a slot.
- **Spatial Constraint**: The user might be interested in events occurring in a specific geographic region. Based on the spatial framework described in Section 3.2, the region of interest can be specified using a rectangle, a polygon, or simply based on instance names (e.g., 'Germany' or 'Munich').

- **Concept Constraint**: Based on the concept hierarchies for subjects, predicates and objects, the user might select some concepts that she is interested in. Then only events related to these concepts, i.e., events having paths to one of the concepts, are considered when mining for respective patterns.

In the rest of the paper, we use $\mathcal{C}$ to denote the set of event constraints (temporal, spatial and concept constraints) specified by the user.

## 4.4   Constraints on Event Templates

In Section 3, we defined levels for concepts in a hierarchy. We now use this definition to specify constraints for event templates. Here one can choose different maximum levels of generalization for different hierarchies. Without loss of generality, we assume a single threshold in the following.

Given an event template $f = \langle s, p, o, t, l \rangle$, the level of $f$ is defined as the maximum level among the subject, predicate and object, i.e., $level(f) = \max\{level(s), level(p), level(c)\}$. Using a threshold $g_{max}$, it can now be specified to what extend events and templates can be generalized. More precisely, this threshold is used to constraint the specificity of periodic event patterns to be discovered.

## 4.5   Periodic Patterns

We now extend Definition 3 to the support of a set of event templates and then define *periodic patterns*.

**Definition 5 (Support Segments).** *Let $\mathcal{C}$ be a set of event constraints and $F = \{f_1, f_2, ..., f_k\}$ be a set of event templates. A segment $\mathcal{I}$ supports $F$ at the $i$-th slot if there exists a set of $k$ events $E = \{e_1, e_2, ..., e_k\} \subseteq \mathcal{I}[i]$ such that: (1) $e_j$ ($j \in \{1, 2, ..., k\}$) is an instance of $f_j$ and (2) $E$ satisfies all constraints in $\mathcal{C}$.*

With this definition, the p-score of a set of event templates $F$ is defined similarly to the p-score of an event template (Definition 4). The vector $V$ with $V[i]$=p-score$(F,i)$ ($i = 0 \ldots t - 1$) is then called an event instance vector (EIV) of $F$. The formal definition of a periodic pattern is as follows.

**Definition 6 (Significant P-Pattern).** *Let $F$ be a set of event templates and $V$ be an event instance vector. A pair $\langle F, V \rangle$ is called a periodic pattern, or p-pattern, for short. Given a threshold $\delta \in [0, 1]$, a p-pattern with an EIV $V$ is called significant if there exists $i$ ($0 \le i \le t - 1$) such that $V[i] \ge \delta$.*

Given an RDF dataset $\mathcal{D}$ consisting of event instances in the form of $\langle$subject, predicate, object, time, location$\rangle$, a set of hierarchies $\mathcal{H}$ (including conceptual, temporal, and spatial hierarchies), a time unit $u$, a set of event constraints $\mathcal{C}$ (temporal, spatial and conceptual constraints), a generalization threshold $g_{max}$ for event templates and a p-score threshold $\delta$, we now focus on finding all *significant p-patterns* w.r.t. the threshold $\delta$ and the event constraints in $\mathcal{C}$ such that the level of each event template in the patterns is less than or equal to $g_{max}$.

# 5  Mining Periodic Patterns

We now describe our approach for finding all p-patterns that meet the requirements described in the previous section. The mining process consists of two steps: (1) identify periods for event templates and (2) level-wise search to find p-patterns.

## 5.1  Detecting Periods

Although the user can specify a period $t$ based on the kinds of periodicities of p-patterns (e.g., weekly or monthly) she is interested in, we also include a step for automatically detecting periods from a dataset of events to make our framework more general. This step is executed separately and helps the user to select meaningful periods for finding p-patterns in the next step. Obviously, if a pattern is periodic with a period $t$ then its ETs are also periodic with the period $t$. Therefore, this step identifies all possible periods for ETs as follows.

A straightforward way to detect periods for an ET $f$ is to construct a binary sequence $S$ from its instances (events), and then identify periods from that sequence. The binary sequence $S$ is a sequence of $n$ elements ($n$ is the number of time slots), where $S[i] = 1$ ($0 \leq i \leq n - 1$) if the time slot $i$ contains an event that is an instance of $f$, and $S[i] = 0$ otherwise. The sequence $S$ is processed further to detect periods by using any periodicity analysis method mentioned in Section 2. Since all periods of an ET can be detected by analyzing the periodicity of its generalizations, we apply the above process only for the most generalized ETs (according to the threshold $g_{max}$). Finally, we obtain a set of all possible periods and some of them can then be selected by the user for the next step.

## 5.2  Finding P-Patterns

Since each period is considered separately, this section describes a process to mine p-patterns under the assumption that the current period is $t$. Algorithm 1 shows the main procedure to mine p-patterns for a given period $t$ in two steps: (1) find size-1 patterns (consisting of one ET) and (2) level-wise search to find size-k ($k \geq 2$) patterns (consisting of $k$ ETs that are to be checked for co-occurrence).

First, a subset of events that satisfy the constraints in $C$ is created from the dataset $\mathcal{D}$ (Line 1). Lines 2-6 in Algorithm 1 show the process to discover size-1 p-patterns. This process is based on a function called *Search_PPatterns*. A size-1 candidate is created with an ET and an EIV initialized to zero. The function *Search_PPatterns* uses a Breadth-First-Search to find the most specialized generalizations of a candidate $P$ that are significant w.r.t. the threshold $\delta$. As shown in Algorithm 2, the true EIV of each candidate is computed to determine how significant the current candidate is. If the candidate is not significant, its generalizations will be added to a queue for further iterations. This process stops when no more candidate is in the queue.

The rest of Algorithm 1 (Lines 7 to 16) describes a while-loop to find p-patterns of size-2, size-3, etc. At an iteration step $k$ ($k \geq 1$), size-(k+1) p-patterns are discovered by using a size-k p-pattern $P_k$ and a size-1 p-pattern $P_1$. To ensure the size of new p-patterns is $(k+1)$, each ET in $P_k$ must not be a generalization of the ET in $P_1$ (Line 12). Then a size-$(k+1)$ candidate $(P_{k+1})$ is created and passed to the *Search_PPatterns* function to find size-$(k+1)$ p-patterns (Lines 13 and 14). Finally, a list of p-patterns of all sizes is returned.

---

**Algorithm 1.** P-Pattern-Miner

---

**Input:**
(a) $\mathcal{D}$: event dataset (quintuples)
(b) $\mathcal{H}$: set of hierarchies (for subjects, predicates, objects, time, locations)
(c) $t$: time period
(d) $\delta$: threshold for periodic score
(e) $\mathcal{C}$: set of constraints
(f) $g_{max}$: maximum level of generalization for event templates

**Output:** Set of all P-Patterns

1  $\mathcal{D}_c = \{e \in \mathcal{D} : e \text{ satisfies } \mathcal{C}\}$; /* only consider events satisfying constraints */
2  $L_1 = \{\}$;
3  **foreach** Event $e \in \mathcal{D}_c$ **do**
4  |    **foreach** Event Template $f \in e^{\uparrow}$ **do**
5  |    |    Create a size-1 candidate $P_1$ from $f$ and $t$;
6  |    |    $L_1 = L_1 \bigcup$ Search_PPatterns($P_1$, $\mathcal{D}_c$, $\mathcal{H}$, $\delta$, $g_{max}$);

7  $result = L_1$; k=1;
8  **while** $L_k \neq \{\}$ **do**
9  |    $L_{k+1} = \{\}$;
10 |    **foreach** $P_k = \langle F, V \rangle \in L_k$ **do**
11 |    |    **foreach** $P_1 = \langle \{g\}, V' \rangle \in L_1$ **do**
12 |    |    |    **if** $g$ is not a generalization of any $f \in F$ **then**
13 |    |    |    |    Create size-$(k+1)$ candidate $P_{k+1}$ from $F \cup \{g\}$ and $t$;
14 |    |    |    |    $L_{k+1} = L_{k+1} \bigcup$ Search_PPatterns($P_{k+1}$, $\mathcal{D}_c$, $\mathcal{H}$, $\delta$, $g_{max}$);

15 |    $result = result \bigcup L_{k+1}$;
16 |    $k = k + 1$;

17 **return** $result$;

---

## 6     Experimental Evaluation

We demonstrate the utility of our proposed framework using datasets crawled from the Website *eventful.com* for different topics from 2009 to 2013. The events of the datasets are transformed into RDF-style quintuples ⟨subject, predicate, object, time, location⟩. The objective of our evaluation is to show how to effectively apply the framework to mine p-patterns from RDF data and to demonstrate that interesting periodic event patterns can be discovered from such a dataset. In the following, we describe the setup of our experiments and present some interesting patterns extracted from different datasets. Our framework is implemented in Java and runs with 8GB heap size. All experiments were run on an Intel Core2Duo 3.16GHz with 4GB RAM, running Ubuntu 64bit.

**Algorithm 2.** Search_PPatterns(P, $\mathcal{D}_c$, $\mathcal{H}$, $\delta$, $g_{max}$)

/* *Breadth-first Search from P to find significant p-patterns satisfying threshold $\delta$* */

1  result={}; Queue=[P];
2  **while** *Queue is not empty* **do**
3  $\quad$ q=Queue.dequeue();
4  $\quad$ Compute the EIV $V$ of $q$ from $\mathcal{D}_c$;
5  $\quad$ **if** $\max_{i=0}^{t-1}\{V[i]\} \geq \delta$ **then**
6  $\quad$ $\quad$ $result = result \bigcup\{q\}$;
7  $\quad$ **else**
8  $\quad$ $\quad$ Compute the set $q^\uparrow$ from $q$ and $\mathcal{H}$;
9  $\quad$ $\quad$ **foreach** *Pattern* $r \in q^\uparrow$ **do**
10  $\quad$ $\quad$ $\quad$ **if** $level(r) \leq g\_max$ **then**
11  $\quad$ $\quad$ $\quad$ $\quad$ Queue.enqueue($r$);

12  **return** *result*;

## 6.1  Datasets and Experimental Setup

We map events from the website *eventful.com* to RDF data to demonstrate our framework. As raw data, each event consists of an event identifier, title, time, location, and a list of tags. Tags are keywords attached to events to categorize events. First, each such event specification is transform into a quintuple, where the predicate is *'happenedOnDate'*, the subject is the event identifier, the object is the start-time, the time and the location are the time and the location of the event. We employ concept constraints by selecting several topics based on event tags and for each topic, we create a subset of the data to mine patterns. A concept hierarchy for subjects is constructed from tag lists based on the method described in [1]. We set the threshold $g_{max}$ to 2 for this hierarchy.

**Table 1.** Statistics of exprimental results

| Dataset | Events related to | Events | threshold $\delta$ | Patterns | Exec.Time |
|---|---|---|---|---|---|
| W-Sport | 'sport' | 59,134 | 0.5 | 88 | 875s |
| W-Religion | 'religion' | 58,558 | 0.5 | 74 | 392s |
| W-Community | 'community' | 61,879 | 0.5 | 110 | 482s |
| M-Festival | 'festival' | 2,295 | 0.3 | 158 | 23s |
| M-Clubs&Associations | 'clubs & associations' | 6,682 | 0.3 | 310 | 72s |

In our experiments, we only consider weekly and monthly patterns. For weekly patterns, we select a day as time unit for slots. For monthly patterns, a month is divided into 3 sections of around 10 days (early: 1-10, middle: 11-20 and late: 21-end), and each such section is considered a time unit. Table 1 shows the list of the datasets used for our experiments where the first three datasets ('W-...') are used to mine weekly patterns and the last two ('M-...') are used to mine monthly patterns.

## 6.2 Experimental Results

For each dataset, we ran the algorithm presented in Section 5 with different settings of the threshold $\delta$, ranging from 0 to 1. Obviously, the lower the value, the more p-patterns are found. Table 1 shows the number of patterns (of all sizes) and execution times for $\delta = 0.5$ (to mine weekly patterns) and $\delta = 0.3$ (to mine monthly patterns).

Due to space constraints, we only show five of the discovered interesting patterns for each dataset in Table 2. Since we consider hierarchies for subjects and locations, we present ETs of patterns in the form of *subject.location*.

**Table 2.** Sample patterns with their p-scores extracted from the experimental datasets. Predicates, objects and time are omitted, * denotes all locations.

| Event templates | Mon | Tue | Wed | Thu | Fri | Sat | Sun |
|---|---|---|---|---|---|---|---|
| Some patterns extracted from W-Sport | | | | | | | |
| {training.*} | 0.16 | 0.1 | 0.13 | 0.18 | 0.13 | **0.54** | 0.05 |
| {family.California} | 0.12 | 0.31 | 0.14 | 0.31 | 0.05 | 0.23 | **0.58** |
| {outdoor & recreation.Idaho} | 0.1 | 0.45 | 0.02 | **0.54** | 0.02 | 0.16 | 0.03 |
| {club & association.*; learning.*} | 0 | 0.01 | 0.2 | 0 | 0.2 | **0.62** | 0 |
| {health.*; sport.*} | **0.55** | 0.25 | 0.15 | 0.4 | 0.14 | 0.25 | 0.05 |
| Some patterns extracted from W-Religion | | | | | | | |
| {pray.United States} | 0.33 | 0 | 0.39 | 0 | 0.02 | 0 | **0.59** |
| {fundrais.*} | 0.08 | 0.13 | **0.53** | 0.33 | 0.27 | 0.38 | 0.4 |
| {taichi.United States} | 0.32 | **0.52** | 0.01 | 0.03 | 0 | 0.35 | 0.32 |
| {music.*; religion & spirituality.*} | 0.32 | 0.18 | 0.14 | 0.39 | 0.32 | 0.37 | **0.97** |
| {yoga.*; religion & spirituality.*} | 0.38 | 0.41 | 0.35 | **0.55** | 0.32 | 0.43 | 0.34 |
| Some patterns extracted from W-Community | | | | | | | |
| {music.New York} | 0 | 0.1 | 0.12 | 0.15 | 0.09 | 0.12 | **0.74** |
| {party.*} | 0.42 | 0.1 | 0.1 | 0.12 | **0.58** | 0.33 | 0.2 |
| {concert.*} | **0.58** | 0.28 | 0.12 | 0.33 | **0.67** | 0.15 | 0.16 |
| {book.*; fun.*} | 0.32 | 0.38 | 0.28 | **0.5** | 0.03 | 0.2 | 0.01 |
| {religion & spirituality.*; community.*} | 0.33 | 0.47 | **0.61** | 0.37 | 0.32 | 0.2 | **0.55** |

| Event templates | Early | Mid | Late |
|---|---|---|---|
| Some patterns extracted from M-Festival | | | |
| {livemusic.*} | **0.34** | 0.26 | 0.03 |
| {shopping.*} | 0.1 | 0.12 | **0.66** |
| {food.Michigan} | **0.68** | 0.05 | 0.05 |
| {food.*; music.*} | **0.49** | 0.08 | 0.18 |
| {outdoor & recreation.*; festival & parad.*; art.*} | 0 | **0.36** | 0.19 |
| Some patterns extracted from M-Club & association | | | |
| {business.Texas} | 0.14 | **0.45** | 0.03 |
| {health.*} | 0.14 | 0.29 | **0.43** |
| {sport.*} | **0.72** | 0.28 | 0.24 |
| {fundrais.*; club & association.*} | **0.32** | 0.2 | 0.06 |
| {family.*; club & association.*} | 0.08 | **0.53** | 0.24 |

As shown in Table 2, some interesting patterns are found. For example, it is likely to find events in the US related to *'pray'* on a Sunday every week. An event related to *'music'* and an event related to *'religion'* usually co-occur on Sunday. There is a high probability that an event related to *'concert'* occurs on Monday and Friday. Also several monthly patterns are found, such as events related to *'live music'* and *'festival'*, which usually occur early each months, whereas events related to *'shopping'* usually occur late each months. Such patterns can be used, for instance, to predict upcoming events or to detect outliers in a given dataset. In addition, they can be used to enrich the existing RDF dataset by adding information about periodicity and co-occurrence to event descriptions.

# 7   Conclusions and Ongoing Work

In this paper, we presented a framework to mine periodic patterns from RDF datasets describing events. The patterns represent topics of events that co-occur and repeat over time. Event topics are modeled based on event templates that are derived from events by generalizing subjects, predicates, objects, time and locations. As a separate input to our approach, hierarchies for concepts, time and locations are employed to obtain event templates and their generalizations. We demonstrated our framework using real datasets and found some interesting patterns.

As shown in Section 4.4, concept constraints are based on the levels of subjects, predicates and objects. We are currently extending our framework to support more expressive types of such constraints, e.g., stating that ETs of a pattern must be similar in terms of concepts, time or locations. We are also investigating how to apply the framework to predict future events and to detect outliers in RDF-based event datasets.

# References

1. Anh, L.V.Q., Gertz, M.: Mining spatio-temporal patterns in the presence of concept hierarchies. In: ICDM Workshops, pp. 765–772 (2012)
2. Cao, H., Mamoulis, N., Cheung, D.: Discovery of Periodic Patterns in Spatiotemporal Sequences. TKDE 19(4), 453–467 (2007)
3. Han, J., Gong, W., Yin, Y.: Mining Segment-Wise Periodic Patterns in Time-Related Databases. In: KDD 1998, pp. 214–218 (1998)
4. Han, J., Yin, Y.: Efficient mining of partial periodic patterns in time series database. In: ICDE, pp. 106–115 (1999)
5. Hoffart, J., L-Kelham, E., Suchanek, F.M., de Melo, G., Berberich, K., Weikum, G.: YAGO2: Exploring and Querying World Knowledge in Time, Space, Context, and Many Languages. In: WWW, pp. 229–232 (2011)
6. Indyk, P., Koudas, N., Muthukrishnan, S.: Identifying representative trends in massive time series data sets using sketches. In: VLDB, pp. 363–372 (2000)
7. Li, Z., Ding, B., Han, J., Kays, R., Nye, P.: Mining periodic behaviors for moving objects. In: KDD 2010, pp. 1099–1108 (2010)
8. Li, Z., Wang, J., Han, J.: Mining event periodicity from incomplete observations. In: KDD 2012, pp. 444–452 (2012)
9. Lomb, N.R.: Least-squares frequency analysis of unequally spaced data. Astrophysics and Space Science 39, 447–462 (1976)
10. Ma, S., Hellerstein, J.L.: Mining partially periodic event patterns with unknown periods. In: ICDE, pp. 205–214 (2001)
11. Mamoulis, N., Cao, H., Kollios, G., Hadjieleftheriou, M., Tao, Y., Cheung, D.W.: Mining, indexing, and querying historical spatiotemporal data. In: KDD 2004, pp. 236–245 (2004)
12. Ozden, B., Ramaswamy, S., Silberschatz, A.: Cyclic Association Rules. In: ICDE, pp. 412–421 (1998)
13. Vlachos, M., Yu, P., Castelli, V.: On Periodicity Detection and Structural Periodic Similarity. In: SDM, pp. 449–460 (2005)
14. Yang, J., Wang, W., Yu, P.S.: Mining asynchronous periodic patterns in time series data. In: KDD 2000, pp. 275–279 (2000)

# Honey, I Shrunk the Cube

Matteo Golfarelli and Stefano Rizzi

DISI, University of Bologna, Italy
{matteo.golfarelli,stefano.rizzi}@unibo.it

**Abstract.** Information flooding may occur during an OLAP session when the user drills down her cube up to a very fine-grained level, because the huge number of facts returned makes it very hard to analyze them using a pivot table. To overcome this problem we propose a novel OLAP operation, called shrink, aimed at balancing data precision with data size in cube visualization via pivot tables. The shrink operation fuses slices of similar data and replaces them with a single representative slice, respecting the constraints posed by dimension hierarchies, until the result is smaller than a given threshold. We present a greedy agglomerative clustering algorithm that at each step fuses the two slices yielding the minimum increase in the total approximation, and discuss some experimental results that show its efficiency and effectiveness.

## 1 Introduction

OLAP is the main paradigm for querying multidimensional databases and data marts. A typical OLAP query asks for returning the values of one or more numerical measures, grouped by a given set of analysis attributes, possibly with reference to a subset of attribute values. The data returned by an OLAP query takes the form of a multidimensional cube, and can be represented either graphically or textually. While graphical representations range from simple line, bar, and pie charts to more sophisticated dashboards involving gauges and geographical maps, the most commonly used textual representation is the so-called *pivot table*. A pivot table usually consists of row, column, and data fields that allow aggregating measures using operators such as sum and average (see Figure 1).

To simplify the querying process for non-ICT users and to support more effectively the decision-making process, OLAP analyses do not normally come in the form of stand-alone queries, but rather of *session*. An OLAP session is a sequence of OLAP queries, each obtained by transforming the previous one through the application of an *OLAP operation*, so as to generate a sort of path that explores relevant areas of the cube. OLAP operations commonly supported by specialized front-end tools are *roll-up*, that further aggregates the data returned by the previous query along dimension hierarchies, *drill-down*, that disaggregates them, and *slice-and-dice*, that selects a subset of values based on some predicate [1].

The effectiveness of OLAP analyses depends on several factors, such as the ability of the front-end tool in intuitively displaying metadata to users and

B. Catania, G. Guerrini, and J. Pokorný (Eds.): ADBIS 2013, LNCS 8133, pp. 176–189, 2013.

|  | Year | | |
|---|---|---|---|
|  | 2010 | 2011 | 2012 |
| Miami | 47 | 45 | 50 |
| Orlando | 44 | 43 | 52 |
| Tampa | 39 | 50 | 41 |
| Washington | 47 | 45 | 51 |
| Richmond | 43 | 46 | 49 |
| Arlington | – | 47 | 52 |

**Fig. 1.** A simple pivot table showing data per City and Year

the performance of the underlying multidimensional engine. In particular, when pivot tables are used, one of factors is the achievement of a satisfactory compromise between the precision and the size of the data being visualized. Indeed, as argued in [2], more detail gives more information, but at the risk of missing the overall picture, while focusing on general trends of data may prevent users from observing specific small-scale phenomena. This is also strictly related to the "information flooding" problem, often occurring when a non-expert user is provided with the full expressiveness of OLAP, that may happen because the user drilled down her cube up to a very fine-grained level, where data are very detailed and a huge number of measure values are to be returned [2]. In this case, it is very hard for the user to browse and analyze the resulting data using a pivot table, especially if the device used has limited visualization and data-transmission capabilities (which is becoming more and more common, considering the wide diffusion of smartphones and tablets as supports for individual productivity).

A possible solution to the above-mentioned problem, often adopted in commercial tools, is to put some constraints to the accessible cube exploration paths, e.g., by disallowing drill-downs to the maximum level of data detail (so-called *semi-static reporting*). However, this often leaves users puzzled and unsatisfied. Other approaches devised in the literature that may be used to cope with information flooding are:

- *Query personalization*: when querying, users are allowed to express their *preferences*, e.g., by stating that data for Italian cities are more relevant than those for other cities, or that measure values below a given threshold are the most interesting [3].
- *Intensional query answering*, where the data returned by a query are summarized with a concise description of the properties they share [2].
- *Approximate query answering*, aimed at increasing querying efficiency by quickly returning a concise answer at the price of some imprecision in the returned values [4].
- *OLAM—On-Line Analytical Mining*: the OLAP paradigm is coupled with data mining techniques to create an approach where multidimensional data can be mined "on-the-fly" to extract concise patterns for user's evaluation,

but at the price of an increased computational complexity and an overhead for analyzing the generated patterns [5].

The approach described in this paper can be seen as a form of OLAM based on hierarchical clustering. Starting from the observation that approximation is a key to balance data precision with data size in cube visualization via pivot tables, we propose a novel OLAP operation called *shrink* that can be applied to the cube resulting from an OLAP query to decrease its size while controlling the loss in precision. The shrink operation is ruled by a parameter expressing the maximum size allowed for the resulting data. The idea is to fuse slices of similar data and replace them with a single representative slice (computed as their average), respecting the constraints posed by dimension hierarchies. To this end we present a greedy agglomerative clustering algorithm that at each step fuses the two slices yielding the minimum increase in the total approximation.

The paper outline is as follows. Section 2 introduces a formalization and our working example. Section 3 describes the shrink approach, while Section 4 shows some experimental results. After discussing the related literature in Section 5, in Section 6 we draw the conclusions.

## 2    Background on Cubes

In this section we introduce a basic formal setting to manipulate multidimensional data. For simplicity we will consider hierarchies without branches, i.e., consisting of chains of attributes, and focus on schemata that include a single measure.

**Definition 1 (Multidimensional Schema).** *A multidimensional schema (or, briefly, a* schema*)* $\mathcal{M}$ *is a couple of*

- *a finite set of disjoint hierarchies,* $\{h_1, \ldots, h_n\}$, *each characterized by a set* $A_i$ *of attributes and a roll-up total order* $\prec_{h_i}$ *of* $A_i$. *Each hierarchy attribute* $a$ *is defined over a categorical domain* $Dom(a)$.
- *a family of roll-up functions that, for each pair of attributes* $a_k, a_j \in A_i$ *such that* $a_k \prec_{h_i} a_j$, *roll-up each value in* $Dom(a_k)$ *to one value in* $Dom(a_j)$.

To simplify the notation, we will use letter $a$ for the attributes of $h_1$, letter $b$ for the attributes of $h_2$, and so on; besides, we will order the indexes of attributes in each hierarchy according to their roll-up order: $a_1 \prec_{h_1} a_2 \prec_{h_1} \ldots$.

A group-by includes one level for each hierarchy, and defines a possible way to aggregate data.

**Definition 2 (Group-by).** *A group-by of schema* $\mathcal{M}$ *is an element* $G \in A_1 \times \ldots \times A_n$. *A coordinate of* $G = \langle a, b, \ldots \rangle$ *is an element* $g \in Dom(a) \times Dom(b) \times \ldots$.

*Example 1.* IPUMS is a public database storing census microdata for social and economic research [6]. As a working example we will use a simplified form of its CENSUS multidimensional schema based on two hierarchies, namely RESIDENCE and TIME. Within RESIDENCE it is City $\prec_{\text{RESIDENCE}}$ State, and Miami $\in$

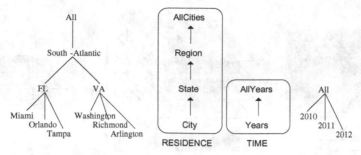

**Fig. 2.** Roll-up orders and functions for two hierarchies in the CENSUS schema

$Dom(City)$ rolls-up to FL $\in Dom(State)$ (roll-up orders and functions are shown in Figure 2). Some examples of group-by are $G_1 = \langle City, Year \rangle$ and $G_2 = \langle State, AllYears \rangle$. A coordinate of $G_1$ is $\langle Miami, 2012 \rangle$.

An instance of a schema is a set of facts; each fact is characterized by a group-by $G$ that defines its aggregation level, by a coordinate of $G$, and by a numerical value $m$. Our shrink operation will be applied to a cube, i.e., to the subset of facts resulting from any OLAP query launched; this can be formalized as follows:

**Definition 3 (Cube).** *A cube at group-by $G$ is a partial function $C$ that maps a coordinate of $G$ to a numerical value (measure). Each couple $\langle g, m \rangle$ such that $C(g) = m$ is called a* fact *of $C$.*

The reason why function $C$ is partial is that cubes are normally *sparse*, i.e., some facts are missing (their measure value is null). An example of missing fact is the one for the Arlington city and year 2010 in Figure 1.

*Example 2.* Two examples of facts of CENSUS are $\langle\langle Miami, 2012 \rangle, 50 \rangle$ and $\langle\langle Orlando, 2011 \rangle, 43 \rangle$. The measure in this case quantifies the average income of citizens. A possible cube at $G_1$ is depicted in Figure 1.

## 3   The Shrink Approach

For explanation simplicity we will start by assuming that the shrink operation is applied along hierarchy $h_1$ to a cube $C$ at the finest group-by, $G = \langle a_1, b_1, \ldots \rangle$.

First of all we observe that, given attribute value $v \in Dom(a_1)$, cube $C$ can be equivalently rewritten as a set of value-slice couples:

$$C = \{\langle v, C^v \rangle, v \in Dom(a_1)\}$$

where $C^v$ is the *slice* of $C$ corresponding to $a_1 = v$ (in the common OLAP sense).

When the shrink operation is applied to $C$, the slices of $C$ are (completely and disjointly) partitioned into a number of clusters, and all the slices in each cluster are fused into a single, approximate slice, which we call *f-slice*, by averaging their non-null measure values. This means that an f-slice in the shrunk cube

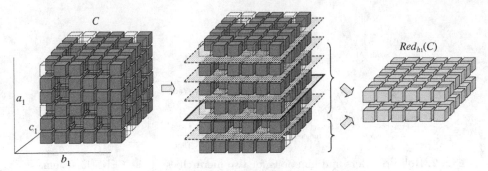

**Fig. 3.** The shrinking intuition

may not refer to a single value of the domain of $a_1$, but rather to a set of values —or even to a set of values of a more aggregate attribute $a_2, a_3, \ldots$ in $h_1$. This process is exemplified in Figure 3, where $C$ is first graphically decomposed into its slices over $a_1$; these six slices are then partitioned into two clusters including four and two slices, respectively. Finally, the fusion step creates the two f-slices that constitute the shrunk cube.

More precisely, let $a_j$, $j \geq 1$, be an attribute of $h_1$, and $V \subseteq Dom(a_j)$ be a set of values of $a_j$. We denote with $Desc_1(V)$ the set of all values of $a_1$ that roll-up to a value in $V$ (conventionally, if $j = 1$ it is $Desc_1(V) = V$ for all $V \subseteq Dom(a_1)$). The shrink operation takes cube $C$ in input and returns a *reduction* $Red_{h_1}(C)$ of $C$, i.e., a set of couples, each formed by a set $V$ of values of any $a_j$, $j \geq 1$ and by the f-slice $F^V$ that results from fusing the slices in $Desc_1(V)$:

$$Red_{h_1}(C) = \{\langle V, F^V \rangle, V \subseteq Dom(a_j)\}$$

The measure value of each fact in $F^V$ is computed as the average of the non-null measure values of the corresponding facts in the slices belonging to $Desc_1(V)$[1]. F-slice $F^V$ is said to *have level* $j$; f-slices with different levels can be mixed in a reduction. The *size* of a reduction is the number of f-slices it includes: $size(Red_{h_1}(C)) = |Red_{h_1}(C)|$.

Noticeably, to preserve the semantics of hierarchies in the reduction, the clustering of the slices for fusion must meet some further constraints besides disjointness and completeness:

1. Two slices corresponding to values $v$ and $v'$ of $a_1$ can be fused in a single f-slice with level $j$ only if both $v$ and $v'$ roll-up to the same value of $a_{j+1}$ (and therefore to the same value of all the subsequent attributes in $h_1$).
2. If the slices corresponding to all the values of $a_1$ that roll-up to a single value $\overline{v}$ of $a_j$ ($j > 1$) are *all* fused together, then the corresponding f-slice has level $j$ and is coupled in the reduction with a set $V$ such that $\overline{v} \in V$ .

*Example 3.* Figure 4 shows two possible reductions, with size 3 and 1 respectively, of the cube shown in Figure 1: in (a) the first two rows have been fused

---

[1] Of course, if $Desc_1(V) = \{v\}$, the measure values in $F^V$ are those in $C^v$.

|        | Year |      |      |
|--------|------|------|------|
|        | 2010 | 2011 | 2012 |
| Miami, Orlando | 45.5 | 44 | 51 |
| Tampa | 39 | 50 | 41 |
| Virginia | 45 | 46 | 50.6 |

(a)

|        | Year |      |      |
|--------|------|------|------|
|        | 2010 | 2011 | 2012 |
| South-Atlantic | 44 | 46 | 49.2 |

(b)

**Fig. 4.** Two reductions of the same cube

into a single f-slice (with level 1) referring to {Miami,Orlando} and the last three rows have been replaced by a single f-slice (with level 2) referring to the Virginia state, while in (b) all six rows have been replaced by a single f-slice (with level 3) referring to the South-Atlantic region. Note that, in the f-slice for Virginia shown in Figure 4.a, the income measure has been averaged on three facts for years 2011 and 2012, and on two facts for year 2010 (since the income value for Arlington in 2010 is null in Figure 1). Some examples of slice clustering that would violate the hierarchy semantics are as follows:

$$\{\mathsf{Miami, Orlando, Washington}\}, \{\mathsf{Tampa, Richmond, Arlington}\} \qquad (1)$$
$$\{\mathsf{FL, Washington}\}, \{\mathsf{Richmond}\}, \{\mathsf{Arlington}\} \qquad (2)$$
$$\{\mathsf{FL}\}, \{\mathsf{Tampa}\}, \{\mathsf{Washington}\} \qquad (3)$$

(in (1) two slices corresponding to cities of different states are put together; in (2) values of attributes with different levels are mixed; in (3) the clustering is neither complete nor disjoint).

This approach can easily be generalized to be applied to any cube at any group-by $G = \langle a_k, \ldots \rangle$ resulting from an OLAP query; of course, all f-slices in the reduction will have levels not lower than $k$ in this case.

### 3.1   Measuring the Approximation Error

Fusing some slices into a single f-slice of the reduction when applying the shrink operation gives raise to an approximation. Like in [7], to measure this approximation we use the *sum squared error*.

Given cube $C$ at group-by $G = \langle a_k, b, \ldots \rangle$, let $V \subseteq Dom(a_j)$, $k \le j$. We denote with $Desc_k(V)$ the set of all values of $a_k$ that roll-up to a value in $V$, and with $F^V$ the corresponding f-slice. Now let $\bar{g} \in Dom(b) \times Dom(c) \times \ldots$ be an incomplete coordinate of $G$ (no value for attribute $a_k$ is given). The *sum squared error* (SSE) associated to $F^V$ is

$$SSE(F^V) = \sum_{\bar{g} \in Dom(b) \times Dom(c) \times \ldots} \ \sum_{v \in Desc_k(V)} (C^v(\bar{g}) - F^V(\bar{g}))^2 \qquad (4)$$

(conventionally, $C^v(\overline{g}) - F^V(\overline{g}) = 0$ if $C^v(\overline{g})$ is null). Given reduction $Red_{h_1}(C)$, the SSE associated to $Red_{h_1}(C)$ is

$$SSE(Red_{h_1}(C)) = \sum_{\langle V, F^V \rangle \in Red_{h_1}(C)} SSE(F^V) \tag{5}$$

*Example 4.* The SSEs associated to the f-slices in Figure 4.a are $(2.25 + 2.25) + (1 + 1) + (1 + 1) = 8.5$, 0, and 14.68 respectively; the overall SSE associated to the reduction is 23.2. The SSE associated to the reduction in Figure 4.b is 158.8.

## 3.2    A Heuristic Algorithm for the Shrink Operation

Given a cube $C$, a combinatorial number of possible reductions can be operated on it, one for each way of clustering the slices of $C$ by preserving the hierarchy semantics. Of course, the more the reduction process is pushed further, the lower the number of resulting f-slices; hence, the lower the size of the data returned to the user but the higher the approximation introduced. So, it is apparent that the reduction process should be driven by a parameter $size_{max}$ expressing the trade-off between size and precision, in particular the maximum tolerable number of f-slices in the reduction (determined for instance by the size of the display and/or by the network bandwidth of the device). In the light of this, the problem of finding a reduction of $C$ along $h_1$ can be so formulated:

*Problem 1 (Reduction Problem).* Find the reduction that yields the minimum SSE among those whose size is not larger than $size_{max}$.

The reduction problem has exponential complexity, which is hardly compatible with the inherent interactivity of OLAP sessions. Indeed, the presence of hierarchy-related constraints reduces the problem search space, so the worst case for the reduction problem is the one where no such constraints are present (i.e., all values of $a_k$ roll-up to the same value of $a_{k+1}$). In this case, the size of the search space is given by the $|Dom(a_k)|$-th Bell number[2], i.e., the number of different partitions of a set with $|Dom(a_k)|$ elements [8]. So, there is a need for a heuristic approach that satisfies real-time computational feasibility while preserving the quality of the solutions obtained. In this direction, we observe that a reduction is determined starting from a clustering of the slices in $C$, where each cluster determines an f-slice and the size of the reduction is given by the number of clusters. Then, we show how the reduction problem can be solved in a sub-optimal way by applying an agglomerative algorithm for hierarchical clustering with constraints to the set of slices in $C$.

*Hierarchical clustering* aims at building a hierarchy of clusters [9]; this can be done following either a top-down or a bottom-up approach. The algorithms based on a bottom-up approach are called *agglomerative*: each element (each slice of $C$, in our case) initially stands in its own cluster, then pairs of clusters are progressively merged. The decision of which pair of clusters will be merged is

---

[2] The Bell number is defined as follows: $B_{n+1} = \sum_{k=0}^{n} \binom{n}{k} B_k$, $B_0 = B_1 = 1$.

## Algorithm 1. The Shrinking Algorithm

**Require:** $C$ at group-by $G = \langle a_k, b, \ldots \rangle$, $size_{max}$
**Ensure:** $Red_{h_1}(C)$

1: $Red_{h_1}(C) \leftarrow \emptyset$                                   ▷ Initialize reduction...
2: **for all** $v \in Dom(a_k)$ **do** $Red_{h_1}(C) \leftarrow Red_{h_1}(C) \cup \{\langle\{v\}, C^{\{v\}}\rangle\}$      ▷ ...one slice per f-slice
3: **while** $size(Red_{h_1}(C)) > size_{max}$ **do**             ▷ Check constraint on maximum size
4:     find $F^{V'}, F^{V''} \in Red_{h_1}(C)$ s.t. $F^{V'}$ and $F^{V''}$ have the same level $j$
5:        **and** $SSE(F^{V' \cup V''}) - SSE(F^{V'}) - SSE(F^{V''})$ is minimal
6:        **and** all values in $V' \cup V''$ roll-up to the same value of $a_{j+1}$     ▷ Hierarchy constraints
7:     $Red_{h_1}(C) \leftarrow Red_{h_1}(C) \setminus \{F^{V'}, F^{V''}\}$            ▷ Merge into an f-slice...
8:     **if** $\exists \overline{v} \in Dom(a_{j+1})$ s.t. $Desc_j(\overline{v}) = V' \cup V''$ **then**
9:        $Red_{h_1}(C) \leftarrow Red_{h_1}(C) \cup \{F^{\{\overline{v}\}}\}$            ▷ ...either at level $j+1$
10:     **else**
11:        $Red_{h_1}(C) \leftarrow Red_{h_1}(C) \cup \{F^{V' \cup V''}\}$            ▷ ...or at level $j$
12: **return** $Red_{h_1}(C)$

usually taken in a greedy (i.e., locally optimal) manner. In our context, merging two clusters means merging two f-slices (i.e., fusing all the slices belonging to each of the two f-slices into a single f-slice). As a merging criterion we adopted the *Ward's minimum variance method* [9], i.e., at each step we merge the pair of f-slices that leads to minimum increase $\Delta SSE$ in the total SSE of the corresponding reduction. Of course, two f-slices can be merged only if the resulting reduction preserves the hierarchy semantics as explained above. Finally, the agglomerative process is stopped when the next merge would violate the constraint expressed by $size_{max}$. The overall process is outlined in Algorithm 1; remarkably, since the SSE grows monotonically at each merge, by simply changing line 3 the same algorithm can be used to solve a symmetrical formulation of the reduction problem asking for the reduction that has minimum size among those whose SSE is below a threshold.

Interestingly, it can be proven that the SSE of a reduction can be incrementally computed, i.e., the SSE of an f-slice $F^{V' \cup V''}$ obtained by merging two f-slices $F^{V'}$ and $F^{V''}$ (line 5) can be computed from the SSEs of the f-slices to be merged as follows:

$$SSE(F^{V' \cup V''}) = SSE(F^{V'}) + SSE(F^{V''}) +$$

$$\times \sum_{\overline{g} \in Dom(b) \times Dom(c) \times \ldots} \frac{H'_{\overline{g}} \cdot H''_{\overline{g}}}{H'_{\overline{g}} + H''_{\overline{g}}} \cdot (F^{V'}(\overline{g}) - F^{V''}(\overline{g}))^2 \qquad (6)$$

where $H'_{\overline{g}} = |\{v \in Desc_k(V') \text{ s.t. } C^v(\overline{g}) \text{ is not null}\}|$ (similarly for $H''_{\overline{g}}$).

*Example 5.* Consider again the cube in Figure 1. In the following we show in detail how Algorithm 1 computes a reduction with $size_{max} = 3$ (Figure 5).

1. In the initialization step (line 2), six singleton f-slices are created, one for each slice of $C$. Since this first reduction has size 6, it violates the $size_{max}$ constraint and the *while* cycle is entered.
2. The most promising merge is the one between the Arlington and the Washington slices, that yields $\Delta SSE = 2.5$ (Figure 5.a, right).

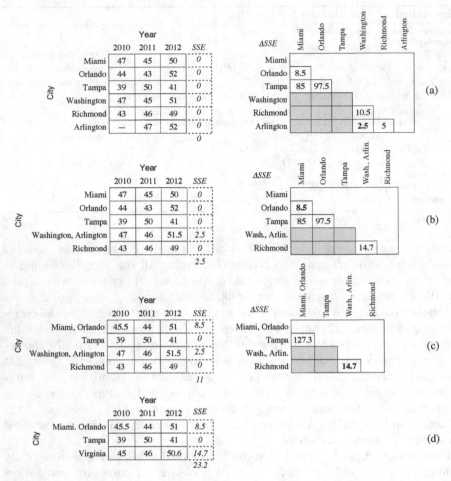

**Fig. 5.** Applying the agglomerative clustering algorithm

3. In the resulting reduction (Figure 5.b, left), that still violates the $size_{max}$ constraint, the most promising merge is the one between the Miami and the Orlando slices (Figure 5.b, right).
4. At this step, the Richmond slice is fused with the Washington-Arlington f-slice (Figure 5.c, right). Since the resulting f-slice covers all the Virginia state, its level is changed to 2 (Figure 5.d).
5. The resulting reduction meets the $size_{max}$ constraint, so the algorithm stops.

The worst-case complexity of Algorithm 1 in terms of operations for computing $\Delta SSE$ refers to the case where no hierarchy-related constraints are present and $size_{max} = 2$, in which case it is $O(\delta^3 \cdot |C^v|)$ where $\delta = |Dom(a_k)|$ and $|C^v|$ is the size of each slice. Specifically, each $\Delta SSE$ computation requires $|C^v|$ operations according to Equation 6 (because it is $|Dom(b)| \cdot |Dom(c)| \cdot \ldots = |C^v|$); if $size_{max} = 2$ the while loop (line 3 of Algorithm 1) is executed $\delta - 2$ times, and

**Table 1.** The four cubes used for testing

| Cube | # facts | Sparsity | # facts per slice |
|------|---------|----------|-------------------|
| $C_0$ | 34,008,000 | 0.6% | 21,800 |
| $C_1$ | 13,603,200 | 11.4% | 8,720 |
| $C_2$ | 1,622,400 | 4.5% | 1,040 |
| $C_3$ | 28,080 | 22.2% | 18 |

it requires $\frac{(\delta-i+1)(\delta-i)}{2} \Delta SSE$ computations (line 5) at the $i$-th time, yielding a total of $\frac{\delta(\delta^2-2)}{6} - 1$ computations.

## 4  Experimental Results

This section collects the main results of the tests we carried out to evaluate the behavior of the shrink operation. The shrink algorithm is implemented in C++ and has run on a Pentium i5 quad-core (2.67GHz, 4 GB RAM, Windows 7-64 bit). The experiments have been carried out on four 4-dimensional cubes extracted from the IPUMS database, whose properties are summarized in Table 1. The cubes from $C_1$ to $C_3$ are obtained by grouping the facts in $C_0$ at progressively coarser granularities.

For each cube, we applied the shrink operation to the City attribute of the RESIDENCE hierarchy. Figure 6 shows the SSE and the corresponding number of non-null facts in the reduction for decreasing values of $size_{max}$ from 1560 (no shrinking, all city slices are kept distinct) to 1 (all city slices are fused into a single f-slice). As expected, the SSE shows an exponential behavior that is more apparent for cubes at coarser group-by's. The decrease in the number of non-null facts is also related to the features of the cube to which shrinking is applied, namely, their sparsity and their distribution of measure values. More details on these issues follow:

- When a non-null fact is fused with a null one, the contribution to the SSE is 0 (see Section 3.1). Thus, the SSE increases slowly in cubes with a fine group-by due to their high sparsity.
- The SSE increases slowly for high values of $size_{max}$ because Algorithm 1 merges the f-slices with lowest SSE first and because sparsity decreases as the shrinking process is pushed further.
- While the decrease in the total number of facts of a reduction strictly depends on the size of the slices to be fused (when two slices are fused, the total number of facts is always reduced by the total number of facts in a slice), the decrease of non-null facts varies according to the sparsity.
- Sparsity is not the only factor that determines the SSE values and trend. Facts at a specific group-by can be characterized by measure values with particularly high variance. For instance, this is the case for $C_2$, which is sparser than $C_3$ and has a finer group-by, but initially leads to higher SSEs.

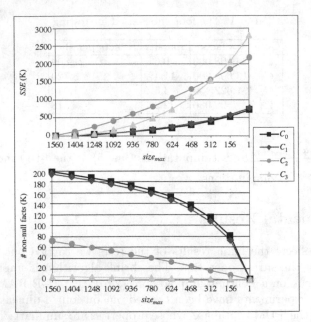

**Fig. 6.** SSE (top) and number of non-null facts in the reduction (bottom) for decreasing values of $size_{max}$

The overall execution time for reducing our largest cube, $C_0$, to two f-slices ($size_{max} = 2$) is 3.08 secs, with an average duration of each reduction step of about 2 milliseconds. The time is not significantly smaller for the other cubes, because its major component is the search of the minimum SSE, which only depends on the number of f-slices to be compared (due to the enforcement of hierarchy-related constraints, only a small number of SSE computations are required, so the size of each slice is not very influential).

To better evaluate the trade-off between sub-optimality and time-saving introduced by our heuristics, we also implemented an optimal algorithm based on a branch-and-bound approach that carries out an exhaustive enumeration of the feasible solutions. The enumeration technique we adopted was initially proposed in [8] for enumerating in a systematic manner all the partitions of the leafs of a generalization hierarchy. A node in the search space, corresponding to a possible partition, is cut when its SSE is greater or equal to the one of the best feasible solution found so far; the initial upper-bound was set to our heuristic solution. Among the different criteria for choosing the next node to be expanded (i.e., the two f-slices to be merged), the most effective one turned out to be a best-first search choosing to merge the two f-slices for which the ratio of the SSE obtained and the relative reduction in cube size is minimum. Unfortunately, the huge size of the search space allowed us to solve to optimality only very small problems ($|Dom(a_k)| \leq 23$ and $size_{max} = 0.3 \cdot |Dom(a_k)|$). In all the experiments, the gap between the optimal and heuristic solution in terms of SSE value turned out to be less than 10%. The execution time of the optimal algorithm was about

30 mins when $|Dom(a_k)| = 23$; when $|Dom(a_k)| = 55$, no optimal solution was found after about 1 day.

## 5  Related Literature

Many studies in the literature can be related to the information flooding problem.

*Query personalization* allows a reduction in the output size by providing a mechanism for selecting only the information relevant to the users. In [3] the authors propose an algebra for defining preferences in the OLAP context. This approach is powerful from an expressiveness point of view, but differently from the shrink operation, it requires users to manually specify their preference criteria. In [10] a technique is proposed for obtaining a personalized visualization of OLAP query results in presence of constraints on the maximum number of returned cells for each dimension of analysis; user preferences are manually defined in terms of a total order of the dimensional values. Also most other works on OLAP content personalization, such as [11], share with the two mentioned above the need for some manual intervention to define preferences. In a few cases, such as [12], preferences are automatically derived by analyzing the log of the user queries, which makes the results of current queries dependent on the past queries —while the results of shrinking are always independent of the past.

According to [13], an *intensional answer* to a query, instead of returning a precise and complete set of tuples, summarizes them with a concise description of the properties these tuples share. Intensional query answering has been applied in many area of computer science (e.g., object-oriented databases [14], deductive database [15], and question answering systems [16,17]) but, to the best of our knowledge, the only work related to the OLAP area is [2], that proposes a framework for computing an intensional answer to an OLAP query by leveraging the previous queries in the current session. The idea is to use an intensional answer to concisely characterize the cube regions whose data do not change along the sequence, coupled with an extensional answer to describe in detail only the cube regions whose data significantly differ from the expectation. Like for [12], even here query results strongly depend on the user history.

Another research topic that is related to our is *approximate query answering*, whose main goal is to increase query efficiency by returning a reduced result while minimizing the approximation introduced —which is clearly similar to our goal. Some approximate query answering approaches were specifically devised for OLAP. For instance, in [18] the authors propose a set of techniques that, given a fixed amount of space, return the cells that maximize the accuracy for all possible group-by queries on a set of columns. While the shrink operation uses approximation to reduce the size of query results, in [18] using sampling to quickly compute measure values for each group introduces an approximation but does not change the size of results. In [19] the focus is different: tuples are sent to a data warehouse as separate streams that must be integrated, and the approximation of queries when a small quantity of memory is available is studied. OLAP query approximation was also studied in the context of data exchange [20]. In particular, in [19] a data warehouse is fed by several sources and the authors study how to sample each source to guarantee that the union of the samples correctly approximates the union of the sources. [21] considers a peer-to-peer data warehouse and studies how to approximately answer OLAP queries by mapping the information available on a source peer on the schema of a target peer.

Finally, an approach similar to ours in the area of temporal databases is *parsimonious temporal aggregation* (PTA), a novel temporal aggregation operator that merges tuples related to the same subject and with consecutive time intervals to computes aggregation summaries that reflect the most significant changes in the data over time [22]. Though our technique shares the same basic principle, it bears more complexity because (a) the constraints deriving from temporal consecutiveness strongly reduce the PTA search space; (b) preserving hierarchy semantics introduces additional complexity; (c) PTA computes the approximation level on a single numerical attribute while we work with complex multidimensional slices.

## 6    Conclusions and Future Works

In this paper we have proposed a novel OLAP operation, called shrink, to cope with the information flooding problem by enabling users to balance the size of query results with their approximation. We described a heuristic implementation of the shrink operation and discussed its efficiency and effectiveness. Remarkably, since the proposed algorithm works by progressively merging couples of f-slices, when using the shrink operation during an OLAP session the shrunk results obtained for a given size threshold can always be reused to compute more efficiently the results for a lower size threshold.

To enhance the shrink approach, we are currently working on two relevant issues. Firstly, to improve effectiveness, we will extend the formulation of the operation to work on several hierarchies simultaneously. This will obviously make the problem much harder from the computational point of view, thus bringing to the forefront efficiency issues. To cope with this, we will study smarter heuristics to obtain good solutions through a greedy approach on the one hand, specific optimization techniques to obtain optimal solution through dynamic programming approaches on the other.

## References

1. Golfarelli, M., Rizzi, S.: Data Warehouse design: Modern principles and methodologies. McGraw-Hill (2009)
2. Marcel, P., Missaoui, R., Rizzi, S.: Towards intensional answers to OLAP queries for analytical sessions. In: Proc. DOLAP, Maui, USA, pp. 49–56 (2012)
3. Golfarelli, M., Rizzi, S., Biondi, P.: myOLAP: An approach to express and evaluate OLAP preferences. IEEE Trans. Knowl. Data Eng. 23(7), 1050–1064 (2011)

4. Vitter, J.S., Wang, M.: Approximate computation of multidimensional aggregates of sparse data using wavelets. In: Proc. SIGMOD, Philadelphia, USA, pp. 193–204 (1999)
5. Han, J.: OLAP mining: Integration of OLAP with data mining. In: Proc. Working Conf. on Database Semantics, Leysin, Switzerland, pp. 3–20 (1997)
6. Minnesota Population Center: Integrated public use microdata series (2008), http://www.ipums.org
7. Gordevicius, J., Gamper, J., Böhlen, M.H.: Parsimonious temporal aggregation. VLDB J. 21(3), 309–332 (2012)
8. Li, T., Li, N.: Towards optimal k-anonymization. DKE 65(1), 22–39 (2008)
9. Tan, P.N., Steinbach, M., Kumar, V.: Introduction to Data Mining. Pearson International (2006)
10. Bellatreche, L., Giacometti, A., Marcel, P., Mouloudi, H., Laurent, D.: A personalization framework for OLAP queries. In: Proc. DOLAP, Bremen, Germany, pp. 9–18 (2005)
11. Jerbi, H., Ravat, F., Teste, O., Zurfluh, G.: A framework for OLAP content personalization. In: Catania, B., Ivanović, M., Thalheim, B. (eds.) ADBIS 2010. LNCS, vol. 6295, pp. 262–277. Springer, Heidelberg (2010)
12. Aligon, J., Golfarelli, M., Marcel, P., Rizzi, S., Turricchia, E.: Mining preferences from OLAP query logs for proactive personalization. In: Eder, J., Bielikova, M., Tjoa, A.M. (eds.) ADBIS 2011. LNCS, vol. 6909, pp. 84–97. Springer, Heidelberg (2011)
13. Motro, A.: Intensional answers to database queries. IEEE Trans. Knowl. Data Eng. 6(3), 444–454 (1994)
14. Yoon, S.C., Song, I.Y., Park, E.K.: Intelligent query answering in deductive and object-oriented databases. In: Proc. CIKM, Gaithersburg, USA, pp. 244–251 (1994)
15. Flach, P.: From extensional to intensional knowledge: Inductive logic programming techniques and their application to deductive databases. Technical report, University of Bristol, Bristol, UK (1998)
16. Benamara, F.: Generating intensional answers in intelligent question answering systems. In: Proc. Int. Conf. Natural Language Generation, Brockenhurst, UK, pp. 11–20 (2004)
17. Cimiano, P., Rudolph, S., Hartfiel, H.: Computing intensional answers to questions – an inductive logic programming approach. DKE 69(3), 261–278 (2010)
18. Acharya, S., Gibbons, P.B., Poosala, V.: Congressional samples for approximate answering of group-by queries. In: Proc. SIGMOD Conference, Dallas, USA, pp. 487–498 (2000)
19. de Rougemont, M., Cao, P.T.: Approximate answers to OLAP queries on streaming data warehouses. In: Proc. DOLAP, Maui, USA, pp. 121–128 (2012)
20. Fagin, R., Kolaitis, P.G., Miller, R.J., Popa, L.: Data exchange: Semantics and query answering. In: Calvanese, D., Lenzerini, M., Motwani, R. (eds.) ICDT 2003. LNCS, vol. 2572, pp. 207–224. Springer, Heidelberg (2002)
21. Golfarelli, M., Mandreoli, F., Penzo, W., Rizzi, S., Turricchia, E.: OLAP query reformulation in peer-to-peer data warehousing. Inf. Syst. 37(5), 393–411 (2012)
22. Gordevicius, J., Gamper, J., Böhlen, M.H.: Parsimonious temporal aggregation. VLDB J. 21(3), 309–332 (2012)

# OLAP in Multifunction Multidimensional Databases

Ali Hassan[1], Frank Ravat[1], Olivier Teste[2],
Ronan Tournier[1], and Gilles Zurfluh[1]

[1] Université Toulouse 1 Capitole, IRIT (UMR 5505),
2 Rue du Doyen Gabriel Marty, 31042 Toulouse cedex 9, France
[2] Université Toulouse 2 / IUT Blagnac, IRIT (UMR 5505),
1 Place Georges Brassens, BP 60073, 31703 Blagnac cedex, France
{hassan,ravat,teste,tournier,zurfluh}@irit.fr

**Abstract.** Most models proposed for modeling multidimensional data warehouses consider a same function to determine how measure values are aggregated. We provide a more flexible conceptual model allowing associating each measure with several aggregation functions according to dimensions, hierarchies, and levels of granularity. This article studies the impacts of this model on the multidimensional table (MT) and the OLAP algebra [11]. It shows how the MT can handle several aggregation functions. It also introduces the changes of the internal mechanism of OLAP operators to take into account several aggregation functions especially if these functions are non-commutative.

**Keywords:** Multidimensional database, OLAP analysis, OLAP operators, aggregation function, multidimensional table.

## 1 Introduction

Within multidimensional models, analysis indicators are analyzed according to several dimensions. Decision-makers can use OLAP operators [11] to study measures according to different levels of granularity. In this way, data is regrouped according to selected levels and aggregated using aggregation functions. Classical multidimensional databases provide the use of only the same aggregation function to aggregate a measure over all the multidimensional space. This capacity is not sufficient to face situations which require several aggregation functions to aggregate a same measure. For example, the monthly average temperatures are obtained from the calculus of the average of daily temperatures. However, it is possible to calculate the average department (a subdivision of regions in the French administrative geographical system) temperatures according to two ways. The first, i.e. **simple** way, uses the same aggregation that is used by television weather forecasts by choosing the main city ('prefecture') that is considered as representative of the considered department. The second, i.e. **scientific** way, takes into account all the temperatures of all the cities of the department.

**Related Work.** Most of the existing works consider that a measure is associated to one aggregation function that will be used for all the different modeled

B. Catania, G. Guerrini, and J. Pokorný (Eds.): ADBIS 2013, LNCS 8133, pp. 190–203, 2013.

aggregation levels. [6, 13] do not specify aggregation functions at the measure level; however, they leave the possibility to use several aggregation functions during OLAP data explorations. This provides great flexibility, but allows the user to do errors by using inappropriate aggregation functions. The authors of [9] suggest linking each measure to a set of functions that contain only those which are valid for the measure. However, each function is used uniformly for all dimensions and all hierarchical levels that compose the multidimensional space. Recent works [1] allow the use of a different aggregation function for each dimension but without allowing the function to be changed according to the selected hierarchical level. This problem was solved by recent aggregation models [10, 2]. These works allow associating each measure with an aggregation function for each dimension, each hierarchy or each aggregation level. However, they do not consider non-commutative aggregation functions.

Regarding commercial tools, Business Objects uses a single aggregation function for each measure. By contrast, Microsoft Analysis Services offers the possibility of applying a 'custom rollup' in a hierarchy [7]. However aggregation functions are related to neither a specific dimension nor an aggregation level. They are related to a member (an instance) of an aggregation level in a hierarchy (i.e. a line in the dimension table). Therefore, applying this 'custom rollup' to a single aggregation level requires repeating it for all the instances of that level. This causes storage and performance problems [7].

**Contribution.** In order to overcome these limits, we have developed a conceptual model for representing multidimensional data [8]. This model associates a measure with different aggregation functions according to dimensions, hierarchies and aggregation levels called parameters. Moreover, the model controls the validity of the aggregated values by defining an order of execution for non-commutative functions. The model considers also the case where an aggregated measure cannot be calculated using aggregation constraints. These constraints define the starting level from which the aggregation can be calculated.

The application of this model has several impacts on the resulting multidimensional table (MT) as well as on the OLAP manipulation operators [11]. In this article, we present modifications that have to be applied to adapt the MT to present several aggregation functions and the changes on the OLAP operators to be able to deal with cases of multiple and non-commutative functions.

This paper is organized as follows: Section 2 presents our conceptual model for integrating several aggregation functions for a same measure. Section 3 discusses the changes that we make on the OLAP operators and MT to adapt to our model. Finally, Section 4 details our experiments.

## 2   Multifunction Conceptual Data Model

Let $\mathcal{N} = \{n_1, n_2, ...\}$ be a finite set of non redundant names and $\mathcal{F} = \{f_1, f_2, ...\}$ a finite set of aggregation functions.

**Definition 1.** A *fact*, noted $F_i$, is defined by $(n^{Fi}, M^i)$.

- $n^{Fi} \in \mathcal{N}$ is the name that identifies the fact,
- $M^i = \{m_1, ..., m_{pi}\}$ is a set of *measures*.

**Definition 2.** A *dimension*, noted $D_i$, is defined by $(n^{Di}, A^i, H^i)$.

- $n^{Di} \in \mathcal{N}$ is the name that identifies the dimension,
- $A^i = \{a_1^i, ..., a_{ri}^i\} \bigcup \{Id^i, All^i\}$ is the set of *dimension attributes*,
- $H^i = \{H_1^i, ..., H_{si}^i\}$ is the set of *hierarchies*.

Hierarchies organize the attributes of a dimension, called parameters, from the finest granularity (root parameter noted $Id^i$) up to the most general granularity (the extremity parameter noted $All^i$).

**Definition 3.** A *hierarchy*, noted $H_j$ (abusive notation of $H_j^i, \forall\, j \in [1..s_i]$), is defined by $(n^{Hj}, P^j, \prec^{Hj}, \text{Weak}^{Hj})$.

- $n^{Hj} \in \mathcal{N}$ is the name that identifies the hierarchy,
- $P^j = \{p_1^j, ..., p_{qj}^j\}$ is the set of attributes of the dimension called *parameters*, $P^j \subseteq A^i$,
- $\prec^{Hj} = \{(p_x^j, p_y^j) \mid p_x^j \in P^j \wedge p_y^j \in P^j\}$ is a binary relation, antisymmetric and transitive,
- $\text{Weak}^{Hj} : P^j \to 2^{A^i \backslash P^j}$ is an application that associates to each parameter a set of attributes of the dimension, called *weak attributes*.

Assuming $M = \bigcup_{i=1}^n M^i$, $H = \bigcup_{i=1}^m H^i$, $P^i = \bigcup_{j=1}^{s_i} P^j$ and $P = \bigcup_{i=1}^m P^i$.

**Definition 4.** A *multidimensional schema*, noted S, is defined by (F, D, Star, Aggregate). This schema is an extension of our previous works [8] where we add a new type of aggregation (multiple hierarchical) and we revisit the execution order mechanism in order to allow our model to be more expressive.

- $F = \{F_1, ..., F_n\}$ is a set of facts, if $\mid F \mid = 1$ then the schema is called a star schema, while if $\mid F \mid > 1$ then the schema is called a constellation schema
- $D = \{D_1, ..., D_m\}$ is a finite set of dimensions
- Star : $F \to 2^D$ is a function that associates each fact to a set of dimensions according to which it can be analyzed
- Aggregate : $M \to 2^{N^* \times \mathcal{F} \times 2^D \times 2^H \times 2^P \times N^-}$ associates each measure to a set of aggregation functions [8]. It allows defining 4 types of aggregation functions:
  - General (if $2^D = \varnothing$, $2^H = \varnothing$ and $2^P = \varnothing$): aggregates measure values with any parameter. This function represents the aggregation function of the classical model,
  - Multiple dimensional (if $2^H = \varnothing$ and $2^P = \varnothing$): aggregates the measure on the whole considered dimension,
  - Multiple hierarchical (if $2^P = \varnothing$): aggregates the measure on all the considered hierarchy,
  - Differentiated (if $2^D \neq \varnothing$, $2^H \neq \varnothing$ and $2^P \neq \varnothing$): aggregates the measure between the considered parameter and the one directly above.

When considering the case of non-commutative functions, $N^*$ associates to each function an order number that represents the execution priority. The function with the lowest order has the highest priority. Commutative functions have the same order. $N^-$ is used to constrain aggregations by indicating a specific aggregation level from which the considered aggregation must be calculated. A non constrained aggregation will be associated to 0 while a constrained aggregation will be associated to a negative value to force the calculation from a chosen level lower than the considered level.

Using the 'Star' function we obtain a structural schema that visualizes structural elements (facts, dimensions and hierarchies) by hiding the aggregation mechanism, Fig. 1 (a). The graphical formalism used is inspired by [4, 11, 12]. The schema in Fig. 1 (a) corresponds to the analysis of average temperatures. The fact 'Temperature' is associated to three dimensions: 'Geography', 'Dates' and 'Time'. The 'Geography' dimension is composed of two hierarchies that correspond to different means for aggregating data: 'Simple' and 'Scientific'. The 'Time' dimension has only one hierarchy that orders the hourly granularities at which temperatures are recorded during the day. The 'Dates' dimension has several hierarchies that organize the granularity levelss of days.

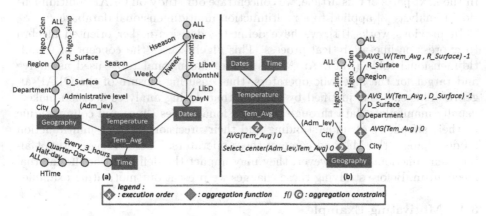

**Fig. 1.** Graphical representation of the multidimensional schemas

For each measure, an aggregation schema can be obtained using the 'Aggregate' function. This schema details the aggregation mechanism that applies to the measure by simplifying the structural elements as much as possible. Fig. 1 (b) shows the aggregation schema of the average temperatures measure 'Tem_Avg'. It has a general aggregation function 'Avg' and a multiple hierarchical function 'Select_centre' on the hierarchy 'Hgeo_simp'. It has also differentiated functions 'Avg' and 'Avg_W' on the hierarchy 'Hgeo_Scien'. 'Avg_W' aggregates the values from the level directly below the considered one (constraint -1).

The function Select_centre(I, M) takes two numeric inputs. It returns the value M that corresponds to Max(I). For example, if one applies Select_center (administrative_level, Tem_Avg) at the region level, it returns the temperatures of the region prefecture (the city that has the highest administrative level).

The function Avg_W(X, Y) takes two numeric inputs. It returns the average of the X values weighed by Y. In other words, a weighted average Avg_W(X, Y) = $\frac{\sum(X \times Y)}{\sum Y}$.

Within the aggregation schema the hierarchies are presented in a disjoint way (each hierarchy is represented by a complete path from the root parameter to the extremity parameter ALL) contrarily to the structural schema where they are represented in a compact way (the common parts of the hierarchies are fused, forming a tree-like structure of parameters). Aggregation functions are modeled by diamonds. Each diamond shows the execution order and the aggregation constraint. Aggregations with constraints set at -1 are calculated using the level directly below; for example, the average temperature 'Tem_Avg' for each region must be calculated from department average temperatures. In the hypothesis where we would have chosen to calculate this average temperature by region using the city temperatures, the constraint would have been set to -2.

## 3   OLAP Manipulation

In the next parts of this article, we concentrate our study on OLAP multidimensional analysis [3] applied to a multifunction multidimensional database.

In previous works [11], we have defined a decision-maker oriented algebra to express on-line analytical process. This algebra uses the concept of a multidimensional table (MT). An MT is a two dimensional table used as source and target for the algebraic operators, thus ensuring closure of the OLAP algebra. This choice is justified by the fact that during analyses decision-makers usually manipulate data through bi-dimensional tables (lines and columns) due to their simplicity of understanding and their precision [6]. Our multifunction model changes the MT and some of these operators. These changes do not alter their functionality; however, they may impact the definition or the internal mechanism. Before studying these changes we present our motivating example.

### 3.1   Motivating Example

Table 1 contains a simplified data sample consistent with our example above. In this example, temperature is recorded twice a day in the department 'Rhône' while being recorded once in the department 'Isère'.

**Table 1.** Average temperatures of departments by day and time

| Region | Department | D_Surface | Date | Time | Tem_Avg |
|--------|-----------|-----------|------|------|---------|
| Rhône-Alpes | Rhône | 3249 | 1/1/2012 | 00:00 | -1 |
| Rhône-Alpes | Rhône | 3249 | 1/1/2012 | 12:00 | 1 |
| Rhône-Alpes | Isère | 7431 | 1/1/2012 | 12:00 | 2 |
| Rhône-Alpes | Rhône | 3249 | 2/1/2012 | 00:00 | 0 |
| Rhône-Alpes | Rhône | 3249 | 2/1/2012 | 12:00 | 2 |
| Rhône-Alpes | Isère | 7431 | 2/1/2012 | 12:00 | 2 |

If the user wishes to analyze the average temperatures of regions by month, then two aggregation functions must be used:

- 'Avg_W(Tem_Avg, D_Surface)' to calculate the average temperatures of regions from the temperatures of departments weighted by the surfaces of the departments 'D_Surface',
- 'Avg(Tem_Avg)' to calculate the average monthly temperatures from the daily temperatures.

If the function 'Avg_W(Tem_Avg, D_Surface)' is applied first, we obtain the average temperatures of regions by day and time (Table 2). If we apply the function 'Avg(Tem_Avg)' after we obtain the required average temperatures of regions by month (Table 3).

**Table 2.** Average temperatures of regions by day and time

| Region | Date | Time | Tem_Avg |
|--------|------|------|---------|
| Rhône-Alpes | 1/1/2012 | 00:00 | -1 |
| Rhône-Alpes | 1/1/2012 | 12:00 | 1.7 |
| Rhône-Alpes | 2/1/2012 | 00:00 | 0 |
| Rhône-Alpes | 2/1/2012 | 12:00 | 2 |

**Table 3.** Average temperatures of regions by month

| Region | Month | Tem_Avg |
|--------|-------|---------|
| Rhône-Alpes | 2012-1 | **0.67** |

But if the function 'Avg(Tem_Avg)' is applied first, we obtain the average temperatures of departments by month (Table 4). If we apply the function 'Avg_W(Tem_Avg, D_Surface)' afterwards, we obtain the average temperatures of regions by month (Table 5).

**Table 4.** Average temperatures of departments by month

| Region | Department | D_Surface | Month | Tem_Avg |
|--------|------------|-----------|-------|---------|
| Rhône-Alpes | Rhône | 3249 | 2012-1 | 0.5 |
| Rhône-Alpes | Isère | 7431 | 2012-1 | 2 |

**Table 5.** Average temperatures of regions by month

| Region | Month | Tem_Avg |
|--------|-------|---------|
| Rhône-Alpes | 2012-1 | **1.54** |

The difference between the results obtained in Table 3 and Table 5 proves the need to use the execution order. The key role of the execution order in our model is to force the execution of aggregation functions in a specific order so as not to have an erroneous result due to non-commutativity. The choice of a valid execution order depends on the requirements of the user. It may differ from one case to another, even if the functions are the same in both cases. Our model allows setting the order which gives a valid result for the user. In this example, the result that correspond to the user's requirements are in Table 3. This justifies why we give a value of 1 for the execution order of the function 'Avg_W' while we give 2 for the execution order of the function 'Avg' Fig. 1 (c).

We conclude from the previous example that the application of an aggregation function with an execution order lower after an aggregation function with an execution upper give an erroneous result.

For RollUp[1] and Rotate[2] operators, the aggregation level has to be changed from the current level to another higher level.

- For the RollUp operator: from the current level to the targeted level
- For the Rotate operator: from the current level to the ALL level of the dimension that will be removed from the MT

In order to do that, it is necessary to apply new aggregation functions after having applied the previous functions that built the current MT (before applying the RollUp and Rotate operators). But, the execution orders of the previous functions can be higher than those of the new functions, i.e. it is not possible to apply the new functions after having applied the previous functions. This requires setting up costly internal mechanisms for these RollUp and Rotate operators. We study these mechanisms in the following section.

### 3.2 Extended OLAP

Within this section, we study the RollUp operator as it is not only one of the most emblematic OLAP manipulation operator and but also one of the most used. First the classical RollUp operator is presented, then changes required for this operator to work within our multifunction model are described.

**Classical RollUp.** In order to study this operator, let us have an MT that analyses average temperatures of departments by months (Fig. 2(a)). From this MT, a RollUp operator will be used to analyze the average temperatures by regions instead of departments (Fig. 2(b)).

To study the classical RollUp operator, we take into account the classical multidimensional model where there is only one function for aggregating the

---

[1] RollUp($T_{SRC}$, D, $p_i$) [11]: modifies the granularity level on the current hierarchy of the dimension (D) towards an upper level or less detailed level ($p_i$) by removing one (or several) parameters in line or column of an MT ($T_{SRC}$).

[2] Rotate($T_{SRC}$, $D_{old}$, $D_{new}$ [,$H_{new}$]) [11]: allows changing an analysis axis ($D_{old}$) by another one ($D_{new}$) in an MT ($T_{SRC}$). It is possible to specify the hierarchy ($H_{new}$) of the new dimension to use it in the resulting MT.

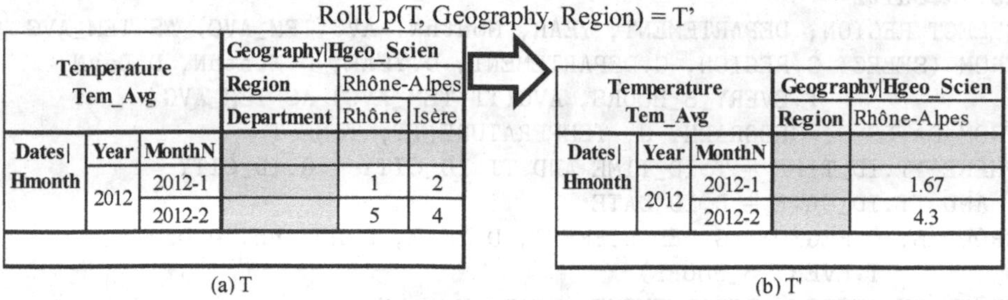

RollUp(T, Geography, Region) = T'

| Temperature Tem_Avg | | Geography\|Hgeo_Scien | |
|---|---|---|---|
| | | Region | Rhône-Alpes |
| | | Department | Rhône | Isère |
| Dates\| Hmonth | Year | MonthN | | |
| | 2012 | 2012-1 | 1 | 2 |
| | | 2012-2 | 5 | 4 |

| Temperature Tem_Avg | | Geography\|Hgeo_Scien | |
|---|---|---|---|
| | | Region | Rhône-Alpes |
| Dates\| Hmonth | Year | MonthN | |
| | 2012 | 2012-1 | 1.67 |
| | | 2012-2 | 4.3 |

(a) T                                                     (b) T'

**Fig. 2.** RollUp

measure. The SQL query that performs the analysis of the average temperatures by departments and by months (Fig. 2(a)) is the following:

```
R1:Result1 =
SELECT G.REGION, G.DEPARTMENT, D.YEAR, D.MonthN,
       AVG(TT.TEM_AVG) AS TEM_AVG, SUM(TT.TEM_AVG) AS sum_Tem_Avg,
       COUNT(TT.TEM_AVG) AS count_Tem_Avg
FROM DATES D, GEOGRAPHY G, TEMPERATURE TT
WHERE TT.ID_CITY = G.ID_CITY AND TT.ID_DATE = D.ID_DATE
GROUP BY G.REGION, G.DEPARTMENT, D.YEAR, D.MonthN;
```

To execute the desired drilling operation (RollUp), one may profit from the results of the previous MT, if the aggregation function is distributive or algebraic (in this later case, intermediate values have to be stored) [5]. In our example, the aggregation function (Avg) is algebraic. Intermediate values required are the sum of the temperatures of cities (sum_Tem_Avg) for each department and the number of occurrences (count_Tem_Avg). Thus, the R2 query that performs the RollUp and corresponds to the MT in Fig. 2(b), benefits from the intermediate values of the results (Result1) of the previous query R1. In other words, in order to perform a RollUp operation, it is not necessary to load the original base measure values, as the MT values can be used as an intermediate values.

```
R2:
SELECT REGION, YEAR, MonthN,
       SUM(sum_Tem_Avg)/SUM(count_Tem_Avg) AS TEM_AVG
FROM Result1
GROUP BY REGION, YEAR, MonthN;
```

**Extended RollUp.** The changes of the RollUp are due to the difference between the execution orders of the aggregation functions of a measure. In this multifunction model, as there are several functions that aggregate the measure in the multidimensional space, the SQL query that performs the analysis of the average temperatures by departments (on the hierarchy 'Hgeo_scien') by month (Fig. 2 (a)) becomes more complex:

```
R3: Result2 =
SELECT REGION, DEPARTEMENT, YEAR, MonthN, AVG(TEM_AVG) AS TEM_AVG
FROM (SELECT G.REGION, G.DEPARTEMENT, D.YEAR, D.MonthN, D.DayN,
                T.EVERY_3_HOURS, AVG(TT.TEM_AVG) AS TEM_AVG
FROM DATES D, GEOGRAPHY G, TEMPERATURE TT, TIME T
WHERE TT.ID_TIME = T.ID_TIME AND TT.ID_CITY = G.ID_CITY
   AND TT.ID_DATE = D.ID_DATE
GROUP BY G.REGION, G.DEPARTEMENT, D.YEAR, D.MonthN, D.DayN,
                T.EVERY_3_HOURS)
GROUP BY REGION, DEPARTEMENT, YEAR, MonthN;
```

To perform a RollUp in a multifunction model, we can distinguish two cases according to aggregation functions that correspond to the requeste RollUp:

*Case 1.* All the execution orders of the aggregation functions that aggregate the measure between the current parameters and the requested parameters are higher or equal to the execution orders of the aggregation functions that aggregate the measure between the base parameters and the current parameters (including the ALL levels of the non-shown dimensions in the MT).

For example, if we want to perform a RollUp to analyze the average temperatures of departments by year, the function that aggregates the average temperatures between the level 'MonthN' and 'Year' is the general function 'Avg(Tem_Avg)'. This function has an execution order of 2 that is greater or equal to the execution orders of aggregation functions that aggregate the average temperatures between the base levels and the levels 'Department', 'MonthN' and 'ALL' of the 'Time' dimension. In this case, in the same way as the classic RollUp, we can benefit from the values already in the MT, as in query R4.

```
R4:
SELECT REGION, DEPARTEMENT, YEAR,
        SUM(sum_Tem_Avg)/SUM(count_Tem_Avg)AS TEM_AVG
FROM Result2
GROUP BY REGION, DEPARTEMENT, YEAR;
```

*Case 2.* The execution order of an aggregation function that aggregates the measure between a current parameter and a requested parameter is less than an execution order of an aggregation function that aggregates the measure between a base parameter and a current parameter.

For example, if we want to perform a RollUp to analyze the average temperatures of regions by months from the average temperatures of departments by months, the function that aggregates the average temperatures between the levels 'Department' and 'Region' on the hierarchy 'Hgeo_Scien' is the function 'Avg_W(Tem_Avg, D_Surface)'. This function has an execution order of 1 that is less than the execution order of the general function that aggregates the average temperatures between the base level and the 'MonthN' level. This means that it is necessary to calculate the average temperature by region before calculating the average temperatures by months. In this case, it is not possible to benefit from the values of the measure displayed within the MT. Base values of the measure have to be used to get the aggregated values as in query R5[3].

```
R5:
SELECT REGION, YEAR, MonthN, AVG(TEM_AVG) AS TEM_AVG
FROM (SELECT REGION, YEAR, MonthN, DayN, EVERY_3_HOURS,
          AVG_W(DATA_WEIGHTED(TEM_AVG, D_SURFACE)) AS TEM_AVG
FROM (SELECT G.REGION, G.DEPARTEMENT, D.YEAR, D.MonthN, D.DayN,
          T.EVERY_3_HOURS, G.D_SURFACE, AVG(TT.TEM_AVG) AS TEM_AVG
FROM DATES D, GEOGRAPHY G, TEMPERATURE TT, TIME T
WHERE TT.ID_TIME = T.ID_TIME AND TT.ID_CITY = G.ID_CITY
   AND TT.ID_DATE = D.ID_DATE
GROUP BY G.REGION, G.DEPARTEMENT, D.YEAR, D.MonthN, D.DayN,
          T.EVERY_3_HOURS, G.D_SURFACE)
GROUP BY REGION, YEAR, MonthN, DayN, EVERY_3_HOURS)
GROUP BY REGION, YEAR, MonthN;
```

### 3.3 Extended Multidimensional Table

Within the scope of this article, we extend the concept of the MT in order to support the multifunction principles of our multidimensional model, especially by integrating the associated aggregation functions within its definition. An extended MT is thus defined as follows:

$$TM = (F, <(D_L, h_L, <p_{L1}, p_{L2}, ...>), (D_C, h_C, <p_{C1}, p_{C2}, ...>)>,$$
$$<\{Aggregate(m_1)\}, \{Aggregate(m_2)\}, ...>, Pred)$$

- F: analyzed fact,
- $D_L$, $D_C$: dimensions displayed in lines and columns respectively,
- $h_L$, $h_C$: hierarchies, used to respectively navigate in lines or columns,
- $p_{L1}$, $p_{L2}$..., $p_{C1}$, $p_{C2}$, ...: displayed parameters,
- $m_1$, $m_2$...: displayed measures,
- Aggregate($m_1$), Aggregate($m_2$)...: aggregation functions respectively associated to measure $m_1$, $m_2$...
- Pred: selection predicate on the fact and/or dimension(s) to limit the set of analyzed values.

The following example specifies the definition of a multidimensional table for analyzing the average temperature by month and department in the Rhône-Alpes region:

TM = (Temperature,
        <(Dates, Hmonth, <Year, MonthN>),
        (Geography, Hgeo_Scien, <Region, Department>)>,

---

[3] The customized aggregation function 'AVG_W' calculates a weighted average. It receives a parameter (TYPE DATA_WEIGHTED AS OBJECT (value NUMBER, weight NUMBER)) that consists in the data and its associated weight.

$<\{(2, \text{Avg(Tem\_Avg)}, \{\}, \{\}, \{\}, 0),$
$\quad (1, \text{Avg(Tem\_Avg)}, \{\text{Geography}\}, \{\text{Hgeo\_Scien}\}, \{\text{City}\}, 0)\}>,$
$\quad \text{GEOGRAPHY.Region} = \text{'Rhône-Alpes'})$

The graphical representation of this specification is shown in Fig. 3 (b):

| Temperature AVG(Tem_Avg) | | | Geography\|Hgeo_Scien | | |
|---|---|---|---|---|---|
| | | | Region | Rhône-Alpes | |
| | | | Department | Rhône | Isère |
| Dates\| Hmonth | Year | MonthN | | | |
| | 2012 | 2012-1 | | 1 | 2 |
| | | 2012-2 | | 5 | 4 |
| Geography.Region = 'Rhône-Alpes' | | | | | |

(a) classic

| Temperature <2> AVG(Tem_Avg) | | | Geography\|Hgeo_Scien | | |
|---|---|---|---|---|---|
| | | | Region | Rhône-Alpes | |
| | | | Department <1> AVG(Tem_Avg) | Rhône | Isère |
| Dates\| Hmonth | Year | MonthN | | | |
| | 2012 | 2012-1 | | 1 | 2 |
| | | 2012-2 | | 5 | 4 |
| Geography.Region = 'Rhône-Alpes' | | | | | |

(b) extended

**Fig. 3.** Graphical representation of a multidimensional table

In order to adapt the visualization of the MT to display several aggregation functions that can be used by a unique measure Fig. 3(b), the MT allows displaying the aggregation functions (along with their inputs, their execution orders and their constraints) used to obtain the displayed elements.

- The general function is displayed instead of the corresponding measure,
- The multiple dimensional function is displayed besides the name of the corresponding dimension,
- The multiple hierarchical function is displayed besides the name of the corresponding hierarchy,
- The differentiated function is displayed besides the name of the corresponding parameter,
- The function inputs are displayed between parentheses '()' after the function name,
- The function execution order is displayed between '<>' before the function name,
- The function constraints are displayed after the inputs, at the end of the function.

This visualization is based on the simplification of the functions as much as possible:

- Simplifying aggregation constraints: if a function is not constrained (the constraint value is 0), the MT does not display this constraint Fig. 3
- Simplifying execution orders: if all displayed aggregation functions have the same execution order, the MT hides all these execution orders. For example, if we analyze the average temperatures by department (using the 'Hgeo_simp' hierarchy) and by month, the functions used (the general 'AVG(Tem_Avg)' and the multiple hierarchical 'Select_center(Adm_lev, Tem_Avg)') have the same execution order (2). The resulting MT is shown in Fig. 4,

- Reduce the number of displayed functions:
  - If all the displayed parameters on a dimension have a differentiated function, the MT displays neither the multiple hierarchical function nor the multiple dimensional function on the considered dimension,
  - If a displayed hierarchy has an aggregation function, the MT does not display the multiple dimensional function of the considered dimension,
  - If the two displayed dimensions have a multiple dimensional function or a multiple hierarchy function for the displayed hierarchy or even a differentiated function for each displayed parameter, the MT does not display the general function.

| Temperature AVG(Tem_Avg) | | | Geography\|Hgeo_Simp Select_center(Adm_lev, Tem_Avg) | | |
|---|---|---|---|---|---|
| | | | Region | Rhône Alpes | |
| | | | Department | Rhône | Isère |
| Dates\| Hmonth | Year | MonthN | | | |
| | 2012 | 2012-1 | | 2 | 2 |
| | | 2012-2 | | 4 | 4 |
| Geography.Region = 'Rhône-Alpes' | | | | | |

**Fig. 4.** MT with simplified execution orders

# 4   Experiments

Our proposal is implemented in the prototype 'OLAP-Multi-Functions'. We use Java 7 on top of Oracle 11g DBMS. It allows the definition of a constellation with multiple and differentiated aggregations, as well as visualizing and querying multidimensional data. Aggregation functions are described in a meta-schema. This meta-schema also describes the structures of the multidimensional schema (facts, dimensions and hierarchies). To oversee the analysis, the prototype has a generator of SQL queries. The analyst selects the desired measure and aggregation levels. The prototype translates interactions by generating an executable SQL script in the context of a R-OLAP implementation.

In this section, we study the additional execution time required by the extended operators compared to classical operators.

**Collection:** We use the example shown in Fig. 1, where temperatures are recorded eight times a day (every three hours). Size grouping for the geography dimension is set to 5. This means that each instance of a higher level corresponds to five instances of lower level (for example, each department has five cities).

**Protocol:** We observe the execution time of the five queries above in accordance with the number of tuples of the fact (from two to eight millions). The fifth query includes a customized function 'Avg_w' that affects the execution time because the function is not optimized contrary to the standard functions (Sum, Avg, Count, Max, Min); therefore, we also study a sixth query (R6) identical to the fifth query but using a standard function: 'Avg' instead of 'Avg_w'.

**Results:** Figure 5 shows the curves corresponding to the six queries. The time required to execute query R3 (the basic analysis of the average temperatures of departments by month for the extended operator) is greater than the time required to execute query R1 (the basic analysis using the classical operator) because of the complexity of the query (using several functions) that uses the extended operator.

The time required to execute the queries R2 and R4 (that benefit from the intermediate values of the results Result1 and Result2 of the previous queries R1 and R3 respectively) is remarkably low (about 0.1 seconds) because the data has previously been highly aggregated. Here, we do not notice any difference between the classical operator and the extended operator.

The time required to execute the query R5 (that uses the extended operator in the absence of the possibility of using intermediate values) is greater than the time required to execute R3 (the basic analysis for the extended operator). However, a large part of that time is due to the fact that the customized function 'Avg_W' is not optimized, this is what we see clearly in the difference between the time required to execute R5 and R6 These results show that the additional time required to execute extended operators is relatively large, thus it is necessary to optimize queries and the used functions and benefit as much as possible of previously calculated values.

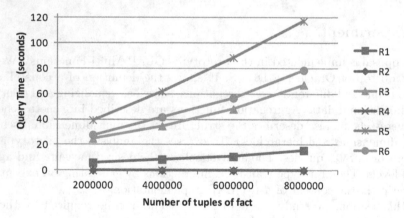

**Fig. 5.** Execution time of 6 queries according to the number of aggregated tuples

## 5   Conclusion

In this paper, we have studied the impact of our conceptual data representation model on the multidimensional table and the associated multidimensional algebra. This model allows associating to each measure several aggregation functions according to the dimensions, hierarchies and parameters. The non-commutative aggregation functions necessitate the use of execution orders as these later may influence the internal mechanism of OLAP operators (especially Rotate and RollUp). In order to adapt the visualization of the multidimensional table to our multifunction model, the multidimensional table allows displaying the aggregation functions with their inputs, execution orders and constraints; however, the multidimensional table simplifies the presentation of the functions in terms of aggregation constraints, execution order and number of displayed functions.

We are developing our prototype to benefit from intermediate values and temporary results of OLAP operators to improve the performance. We are also considering optimizing the extended operators algorithms and performing experiments based on our prototype.

# References

[1] Abelló, A., Samos, J., Saltor, F.: YAM2: A multidimensional conceptual model extending UML. Information Systems 31, 541–567 (2006)

[2] Boulil, K., Bimonte, S., Pinet, F.: Un modèle UML et des contraintes OCL pour les entrepôts de données spatiales. De la représentation conceptuelle à l'implémentation. Ingénierie des Systèmes d'Information (ISI) 16(6), 11–39 (2011) (in French)

[3] Codd, E.F.: Providing OLAP (on-line analytical processing) to user analysts: an IT mandate. Technical Report, E.F. Codd and Associates (1993)

[4] Golfarelli, M., Maio, D., Rizzi, S.: Conceptual Design of Data Warehouses from E/R Schemes. In: Intl. Conf. HICSS 1998, vol. 7, pp. 334–343 (1998)

[5] Gray, J., Bosworth, A., Layman, A., Pirahesh, H.: Data Cube: A Relational Aggregation Operator Generalizing Group-By, Cross-Tab, and Sub-Total. In: Intl. Conf. ICDE, vol. 96, pp. 152–159 (1996)

[6] Gyssens, M., Lakshmanan, L.V.S.: A Foundation for Multi-Dimensional Databases. In: Intl. Conf. VLDB., vol. 97, pp. 106–115 (1997)

[7] Harinath, S., Zare, R., Meenakshisundaram, S., Carroll, M., Guang-Yeu Lee, D.: Professional Microsoft SQL Server Analysis Services 2008 with MDX. Wiley Publishing, Indianapolis (2009)

[8] Hassan, A., Ravat, F., Teste, O., Tournier, R., Zurfluh, G.: Differentiated Multiple Aggregations in Multidimensional Databases. In: Cuzzocrea, A., Dayal, U. (eds.) DaWaK 2012. LNCS, vol. 7448, pp. 93–104. Springer, Heidelberg (2012)

[9] Pedersen, T., Jensen, C., Dyreson, C.: A foundation for capturing and querying complex multidimensional data. Information Systems 26(5), 383–423 (2001)

[10] Prat, N., Wattiau, I., Akoka, J.: Representation of aggregation knowledge in OLAP systems. In: The 18th European Conference on Information Systems, ECIS (2010)

[11] Ravat, F., Teste, O., Tournier, R., Zurfluh, G.: Algebraic and graphic languages for OLAP manipulations. International Journal of Data Warehousing and Mining 4(1), 17–46 (2008)

[12] Ravat, F., Teste, O., Tournier, R., Zurfluh, G.: Graphical Querying of Multidimensional Databases. In: Ioannidis, Y., Novikov, B., Rachev, B. (eds.) ADBIS 2007. LNCS, vol. 4690, pp. 298–313. Springer, Heidelberg (2007)

[13] Vassiliadis, P., Skiadopoulos, S.: Modelling and Optimisation Issues for Multidimensional Databases. In: Wangler, B., Bergman, L.D. (eds.) CAiSE 2000. LNCS, vol. 1789, pp. 482–497. Springer, Heidelberg (2000)

# Versatile XQuery Processing in MapReduce

Caetano Sauer, Sebastian Bächle, and Theo Härder

University of Kaiserslautern
P.O. Box 3049, 67653 Kaiserslautern, Germany
{csauer,baechle,haerder}@cs.uni-kl.de

**Abstract.** The MapReduce (MR) framework has become a standard tool for performing large batch computations—usually of aggregative nature—in parallel over a cluster of commodity machines. A significant share of typical MR jobs involves standard database-style queries, where it becomes cumbersome to specify *map* and *reduce* functions from scratch. To overcome this burden, higher-level languages such as HiveQL, PigLatin, and JAQL have been proposed to allow the automatic generation of MR jobs from declarative queries. We identify two major problems of these existing solutions: (i) they introduce new query languages and implement systems from scratch for the sole purpose of expressing MR jobs; and (ii) despite solving some of the major limitations of SQL, they still lack the flexibility required by *big data* applications. We propose *BrackitMR*, an approach based on the XQuery language with extended JSON support. XQuery not only is an established query language, but also has a more expressive data model and more powerful language constructs, enabling a much greater degree of flexibility. From a system design perspective, we extend an existing single-node query processor, *Brackit*, adding MR as a distributed coordination layer. Such heavy reuse of the standard query processor not only provides performance, but also allows for a more elegant design which transparently integrates MR processing into a generic query engine.

## 1 Introduction

MapReduce [4] arose as a key technology in the context of *big data*, a term whose definition depends not only on the subjective interpretation of "big", but also on the problem context. Most big-data technologies consist of systems which are designed from scratch to fulfill technical requirements imposed by the large size of the data, which in a particular point in time are not fulfilled by prevalent technologies. In the case of MR, the related prevalent technology to which it is usually compared consists of parallel database systems for analytics and data warehousing. In contrast to such systems, MR hits an "abstraction sweet spot", because it only abstracts the distribution and fault tolerance aspects of the execution, leaving data formats and operations performed on individual items *under complete responsibility* of the developer. Furthermore, MR is concerned solely with the execution of tasks and not the management of stored data— a key difference to database systems, which exclusively control all read and

B. Catania, G. Guerrini, and J. Pokorný (Eds.): ADBIS 2013, LNCS 8133, pp. 204–217, 2013.

write operations on the data. Because it is only focused on task execution, a large portion of the query processing logic present in a DBMS must in fact be implemented from scratch by the programmer.

A significant share of typical MR jobs consists of a composition of traditional database query operations such as filtering, grouping, sorting, aggregating, and joining. Therefore, it makes sense to push the abstraction further and introduce query languages that allow expressing such tasks in a declarative manner. A great advantage of this approach is that it allows the reuse of *existing data processing operators* as well as *query rewrite* and *optimization logic*.

It is clear that database systems and SQL can be used for that end, but the requirements of *big data* applications make them prohibitive for two main reasons: First, large-scale data warehouse systems are complex, expensive, and hard to customize towards specific end-user needs, especially for applications outside the traditional business domain. Second, and perhaps most critical, they make use of SQL, which enforces a flat, normalized representation of data. This restriction is significant for application domains like scientific computing and the Web 2.0. What is even worse for such scenarios, which often involve *ad-hoc* analysis procedures, is that the exact schema must be specified upfront, which contributes further to the inflexibility and high setup costs of such systems.

## 1.1   Related Work

One of the earliest proposals for a query language for MapReduce was from Hive [11], a data warehouse system with an SQL-like interface. Its evaluation engine represents the basic mechanism for converting a computational model based on query operators into that of MR. However, being designed to resemble traditional relational systems, it suffers from all the major limitations of SQL and the relational model. Its main focus is to provide a relational data warehouse as similar as possible to existing systems, but relying on MR for the execution layer. Pig [7] introduces a dataflow-oriented language, which aims to be easier to use than SQL for users accustomed with imperative programming. It allows computations to be described using variables which hold references to collections of data items. Operations are then applied to these variables and stored on a new variable. By collecting data dependencies of such variables, a directed acyclic graph (DAG) of operators is built, which serves essentially the same purpose as a query plan. JAQL [3] came up as a more advanced approach, relying heavily on functional-programming features such as higher-order functions. JAQL offers a very elegant solution from the language perspective, it provides a flexible JSON-based data model with support for partial schema specifications. The performance of the three related systems was measured in [10].

The work in [9] provides more detailed qualitative analysis of the four query languages, presenting basic language requirements for flexible query processing in the big data domain. As the authors observe, HiveQL and PigLatin do not fully overcome the deficiencies of SQL, and although no query language leads in all criteria, JAQL and XQuery clearly provide a higher degree of flexibility.

## 1.2   Contribution

We present BrackitMR, an XQuery processor extended with MR capabilities. XQuery is a wide-spread, generic data-processing language which can be used in a vast range of applications, from traditional DBMS processing to document management, data integration, and message processing. It supports the processing of untyped data, allowing schema information to be added at later stages, as well as partial schema definitions, which enable gradual schema refinement. Furthermore, it provides a hierarchical data model with support for path-based queries, allowing efficient storage and processing of denormalized data. Our prototype extends the XQuery data model with JSON, thereby providing concise representation and efficient processing not only for XML, but also for relational datasets. The main advantage of our approach is that it provides a more flexible data model while still reaching the same performance as the state-of-the art approaches for strictly relational data.

BrackitMR is also a flexible tool for large-scale query processing, because it builds upon an existing single-core query processor, Brackit, adding MR as a distributed coordination layer. This means that all available query rewrite and optimization techniques are simply reused, which is a big advantage, because virtually all "sequential" optimizations also hold on the distributed setting (e.g., push selective operations to the beginning, exploit existing sort orders, etc.). Furthermore, the execution of query fragments within nodes of a cluster is carried out entirely by the Brackit engine. This approach transparently integrates MR into an existing data-processing environment, giving the flexibility to choose which parts of a computation are shipped to MR and which are executed locally, in the standard query engine.

A further advantage of Brackit is that it provides an extensible framework for specifying storage modules. The query processing logic is decoupled from specific storage implementations, and a well-defined interface is offered for communication between these two components. This interface allows storage-specific optimizations such as index scans and predicate push-down to be implemented and integrated transparently in the optimizer.

The flexibility provided by BrackitMR, both at the query language and system architecture levels, also contributes for filling the gap between MapReduce and database systems, moving towards a unified solution which embraces the best from both worlds.

The remainder of this paper is organized as follows. Section 2 introduces our extended version of the XQuery data model and describes our generic *collection framework*, which enables the integration of various kinds of data sources into a single query processing environment. Section 3 describes the process of mapping an XQuery plan into the MR programming model. Section 4 provides an experimental performance evaluation of BrackitMR. We demonstrate that BrackitMR is as fast as the competing approaches when it comes to relational processing. Finally, Section 5 concludes this paper.

# 2    Data Model

## 2.1    Motivation

One of the major complaints of users against XQuery is that its XML-based data model, albeit optimal for document-centric scenarios, is too cumbersome for data-centric applications. The flexibility of the XML format for representing data, and of XQuery for processing it, comes with a heavy impact on query conciseness and evaluation performance in scenarios where data is purely or partially relational. Such limitations contributed to the adoption of JSON as a format for representing simple collections of objects, trimming away the complexities of document-centric XML. JSON was quickly adopted in Web 2.0 applications, which often interact with relational databases or external service APIs using asynchronous HTTP requests (i.e., Ajax applications).

The major drawback of the JSON data model is that there is currently no standard query language that supports it. Driven by the popularity of *NoSQL* databases, established techniques of database query processing are mostly ignored in such systems, and complex queries are often implemented from scratch in the application layer. To overcome this drawback, several NoSQL products like MongoDB currently provide rather primitive query constructs. For complex analytical queries, developers must implement ad-hoc glue code to integrate such databases with MapReduce-based query engines like Pig and Hive.

Another obvious drawback of such JSON-based approaches is the lack of support for XML, which is fundamental for Web applications. Brackit overcomes these problems by simply extending the XQuery data model with support for JSON objects and arrays, an approach which is also proposed in the JSONiq specification [8]. The integration of JSON values in an XQuery engine is very simple, but it results in a tremendous practical value, as it integrates document- and data-centric query processing in a single environment.

A further advantage that contributes to the versatility of BrackitMR is its *collection framework*. It defines a common interface for the communication between query engine and different storage modules, thus enabling the transparent processing of several data sources using a unified language. The interface not only allows data to be read from and written into various data sources, but also supports basic storage-related optimizations such as filter and projection push-down in a generic manner.

## 2.2    JSON Support

The XQuery data model (XDM) is centered around *sequences*, which are ordered collections of *items*. Items can be either XML nodes or atomic values, but they cannot be sequences again, which means that nested sequences are not supported. A fundamental property of XQuery, which is the basis of its composability, is that an item is indistinguishable from a singleton sequence containing it. This means that all values in XQuery are sequences, and that every expression returns a sequence when evaluated.

To integrate JSON in our data model, we simply add two new kinds of items: *objects* and *arrays*. Objects are simple record structures which map attribute names into values. In XDM, such values are themselves sequences, thus enabling nested data structures. Like SQL, JSON defines a special NULL value, which has no equivalent concept in XQuery[1], and therefore we also define a special null value. Arrays are similar to sequences in which they are ordered collections of items, but because arrays are a kind of item, they can therefore be nested.

Note that our scheme simply "reuses" the atomic and XML values from the original data model. This means that our objects and arrays are more powerful than those of the original JSON specification, which support only numbers (with no distinction between integer, decimal, double, etc.), strings and Boolean values. XQuery, on the other hand, provides the complete type hierarchy of XML Schema, which also includes user-defined types. This also implies that XML nodes can be embedded inside objects and arrays, which gives great flexibility for mixed document- and data-centric workloads.

Figure 1 illustrates a query over two relational tables from the TPC-H benchmark, which are accessed as collections of JSON objects. For comparison, we show the equivalent SQL query on the right-hand side.

```
for $l in collection('lineitem')          SELECT l.returnflag AS retflag,
for $o in collection('orders')                 avg(o.totalprice) AS avg_price
where $l=>orderkey = $o=>orderkey         FROM lineitem l, orders o
  and $l=>shipdate >= '1995-01-01'        WHERE l.orderkey = o.orderkey
let $rf := $l=>returnflag                    AND l.shipdate >= '1995-01-01'
group by $rf                              GROUP BY l.returnflag
return { retflag: $rf,
        avg_price: avg($o=>totalprice) }
```

**Fig. 1.** Example of XQuery expression with JSON support and equivalent SQL query

The query performs a join between the tables *lineitem* and *orders*, whereas the former is filtered by a predicate on its *shipdate* attribute. The tuples are then grouped by the *returnflag* attribute, which according to XQuery semantics must first be extracted into its own variable $rf before grouping. Finally, each return flag is returned together with the average of its associated order prices. In this query, the use of the JSON data model is demonstrated in expressions accessing object attributes with the => operator. Furthermore, the query returns JSON objects using a constructor syntax that is equivalent to the string representation of JSON objects.

For brevity, we rely on this simple example to demonstrate the basic capabilities of JSON support in XQuery. For a more precise specification, we refer to the JSONiq language [8], which is implemented in the Zorba XQuery processor.

---

[1] In XQuery, the empty sequence is usually employed to denote the absence of a value, but this is not semantically equivalent to NULL. An empty sequence is an empty, but existing value, while NULL indicates the absence of a value.

## 2.3   Collection Framework

The main design goal of the Brackit[2] query engine is to provide a common framework for query compilation, optimization, and execution, abstracting away from specific storage modules. This abstraction is realized in the *collection framework*, which provides a Java interface that is implemented by specific storage modules. The query evaluation engine interacts with collection instances when the function `collection` is invoked, as in the example of Figure 1. In the XQuery specification, this function must return XML elements, but our implementation relaxes this constraint by allowing general XDM items to be returned. An item is represented by a Java interface as well, and hence storage modules are free to represent items in an efficient manner. A relational tuple, for instance, would be best represented internally as an array of atomic values, but it would still behave transparently, from the query engine perspective, as a JSON object.

In order to be found by the query engine, a collection must be registered in the metadata catalog, which in XQuery is modelled by the static context [12]. Because XQuery provides no standard mechanism for registering collections, our implementation introduces the `declare collection` primitive. It takes four parameters: the collection name as a string, the name of the class which implements the collection interface, and an optional URI for locating the collection within the implemented module, and the type of its items. Consider, for example, the following declaration of the collection *lineitem* of Figure 1:

```
declare collection lineitem of CSVCollection
  at hdfs://lineitem.csv as object(type:lineitem)*;
```

It declares a collection called *lineitem*, which is instantiated using the class *CSVCollection*, an implementation for CSV files where each line is treated as a JSON object. The `at` keyword specifies a path, in this case in HDFS (i.e., Hadoop distributed file system), where the file can be located. The `as` keyword specifies the sequence type which is returned by the collection. The `object` keyword specifies that the items returned are JSON objects, each having the type `type:lineitem`, which is declared separately. The `*` symbol indicates that the collection contains zero or more items.[3] Because the original XQuery does not support JSON types, we must extend it with a type declaration primitive or schema import mechanisms. For brevity, we abstract the declaration of JSON types from our discussion. We refer to JAQL [3], which served as inspiration for JSON type declaration in our system.

Note that the type argument can be any valid sequence type, and thus we can use abstract types like `item()*` to leave the schema unspecified. Partial schema specifications can be implemented in XML Schema using complex types, and in JSON using the JAQL notation with wild-card markers like `*` and `?`.

---

[2] http://www.brackit.org

[3] The use of such cardinality symbols may seem unnecessary, but it is part of the universal sequence-type syntax of XQuery, which is also used in function arguments and results.

# 3   Compilation and Execution

## 3.1   Query Plans

A query is compiled in Brackit into a tree of expressions which is evaluated in a bottom-up manner. A FLWOR expression is the standard construct used for processing bulk data, and hence we focus on the compilation of FLWOR expressions only. When compiled, a FLWOR expression is represented by a tree of operators, just like a relational query plan. For an overview of how such plans are evaluated, we refer to the plan on the left-hand side in Figure 2, which is generated for the query in Figure 1.

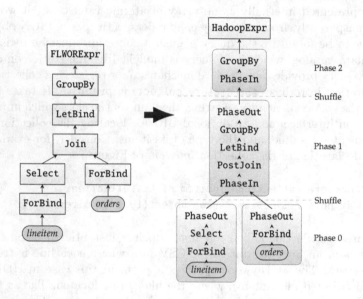

**Fig. 2.** Example of a query plan and its translation into task functions

At the top, we have a FLWOR expression node, which is responsible for evaluating the operators under it and delivering the final expression in the return clause as a result. Each operator in the plan consumes and produces a tuple of bound variables, in accordance to the semantics of FLWOR clauses [12]. These tuples, however, are not part of the XQuery data model, but internal structures which serve the exclusive purpose of evaluating operators. The FLWORExpr node, therefore, serves as a bridge between the iterator-based, set-oriented model of operators (i.e., the traditional model of relational query processing) and the sequence-based model of XQuery expressions.

The query starts with ForBind operators, which evaluate the collection function and bind each delivered item to a given variable, generating the initial tuples. Its functionality resembles that of a relational table scan. The input tuple of an operator contains the currently available variable bindings, whose values are used to evaluate expressions attached to each operator. Depending on the result of the evaluation, tuples are modified, discarded, grouped, or reordered. This corresponds to the classical mechanism used for query evaluation in relational systems, as established by the iterator model [5].

Each operator in the query plan is basically derived from a corresponding clause in the FLWOR expression. The `Join` operator is derived by our query rewrite engine from a sequence of two `ForBind` operators followed by a predicate on both their inputs. For a detailed description of each operator's semantics, we refer to the FLWOR expression specification in the XQuery standard [12].

## 3.2    Generalized MapReduce Model

The generation of MR jobs from query plans follows the same basic mechanism used in Hive, Pig, and JAQL. In previous work, we have discussed general mechanisms for compiling query languages into the MR programming model [9]. Our discussion of the mapping from XQuery to MR thus relies on the Generalized MapReduce model (GMR) introduced there. Nevertheless, we provide a brief explanation of the GMR model for completeness. It is derived from the original MR model by generalizing a few aspects of a job specification. Despite being a generalization, we emphasize that the model can be fully implemented in Hadoop, and therefore it is the same basic model used in Hive, Pig, and JAQL. The goal of our generalizations is simply to provide a more concise model for discussing the mapping of query plans into MR jobs.

The first, and most important, generalization of GMR is the use of *task functions* instead of the original map and reduce functions. As noted in [6], the Mapper and Reducer tasks—which iterate over the input key-value pairs and invoke the map or reduce function on each of them—are both implementations of the traditional `map` higher-order function of functional programming languages. The only difference between the map and reduce functions is that a reduce is applied on a list of values rather than on a single value, but if we consider a list as a proper value, the functions are essentially the same. Thus, we can specify a job as a sequence of Mapper-Shuffle-Mapper stages. Furthermore, we allow the user to specify the function which is executed by a task for each partition of the input, which we refer to as *task function*. This is in contrast to the original map and reduce functions, which are called for each key-value pair in an input partition. With this generalization, MR jobs can be specified as a pair of generalized task functions—one executed before and the other after the Shuffle stage.

The further generalizations enable the specification of jobs as arbitrary trees of task functions. First, we allow jobs to contain multiple map functions and a single reduce, or, equivalently in our generalization, multiple task functions before a Shuffle stage. The intermediate Shuffle keys are then grouped regardless of which task function produced them, and a tag is appended to allow separating the key-value pairs again on the reduce side. This is a standard technique for performing joins in Hadoop, discussed in mored detail in [13]. Furthermore, we can allow arbitrarily long sequences of task functions interleaved with Shuffle stages. This is necessary to implement complex operations that require multiple MR jobs. In the original MR model, this can be simulated by executing jobs with empty map functions, which simply propagate the key-value pairs generated by a preceding reduce to another Shuffle stage. In our GMR model, this finally allows the specification of trees of task functions, where between each level of the tree there is a Shuffle stage.

The GMR model provides an ideal level of abstraction for discussing query processing in MR, because it has greater resemblance to the model of query plans and the iterator model. On the other hand, the model is not over-generalized, such that the MR nature would be lost. It can be fully implemented in Hadoop, and it also applies to query compilation in Pig, Hive, and JAQL. For further details on GMR, we refer to [9].

### 3.3 Mapping Query Plans to GMR

The use of the GMR model makes it very natural to map query plans into MR jobs. The process is based on the classification of operators into *blocking* and *non-blocking*, and it is illustrated in Figure 2 for our example query. The idea is to convert all blocking operators (in our case Join, GroupBy, and OrderBy) into an MR Shuffle, which is the only blocking operation provided by the MR framework. The Shuffle performs a grouping operation on multiple inputs, followed by a repartition of the key-value pairs across the worker nodes in the cluster. It is generic enough to implement all standard blocking operations, but it requires, in some cases, special non-grouping post-processing operators which restore the semantics of the original operator after a Shuffle.

A sequence of non-blocking operators, on the other hand, is "packed" into a task function, as illustrated in Figure 2. Therefore, the execution of a task function across worker nodes, which we refer to as a *phase*, consists of compiling and executing the operators using the standard Brackit query processor, based on the iterator model. In order to generate proper key-value pairs from tuples and to build tuples back after shuffling, the special operators PhaseOut and PhaseIn are used. A key in this case is simply a subset of the tuple's fields (i.e., a *sub-tuple*), whereas a value is composed of the remaining fields.

To clarify the compilation and execution process, we make use of the example in Figure 2, starting at the root node FLWORExpr. The FLWOR expression itself is compiled into a HadoopExpr, which executes the contained query plan in the MR framework and then brings the generated results back to the client machine, serving as a "bridge" between local and distributed evaluation. Note that this approach allows a hybrid query evaluation mechanism in which only the heavy parts of an XQuery program—namely those that access collections on distributed storage—are shipped for execution in MR. This means that a HadoopExpr may, for example, compute aggregations on a large dataset and return the results as a sequence to the parent expression, which may generate an XML report or store the results in a local database.

The evaluation of the plan starts at the leaf task functions, which must start with a ForBind operator. MR jobs usually exploit the data parallelism of the input datasets, by processing each key-value pair independently. This is exactly the functionality of the `for` clause in a FLWOR expression, namely to iterate over a sequence of items, bind it to a variable, and execute the following clauses once for each operator. The non-blocking operators are then evaluated until a `PhaseOut` is reached. It extracts specific fields of the input tuple into a sub-tuple which will be used as key in the Shuffle stage. The fields to be extracted depend of course on the blocking operator being executed. In the example, the two leaf task functions serve as input to a `Join`, and hence the keys extracted are the join keys, namely the *orderkey* attribute. Note that the variables $l and $o contain the whole JSON objects, so their *orderkey* attributes have to be extracted into their own variables using a `LetBind`, which we omitted here for space reasons.

The task function in the second phase of our example starts with a `PhaseIn` operator, which will rebuild the original tuples from the key and value sub-tuples. The `PostJoin` separates the input tuples based on the tag value and joins them locally. Our implementation is based on a hash-join algorithm, and so the tuples are repartitioned by the Shuffle based on a hash value of the keys. Furthermore, tuples within each partition are sorted by the tag value, so that all tuples from the right input come first and are used to build the hash table.

After the `Join`, the `LetBind` operator is executed and its outputs are fed into a `GroupBy` operator. Note, despite being a blocking operator, the `GroupBy` is executed inside the task function. It pre-aggregates values locally before the global grouping that occurs in the Shuffle. This functionality corresponds to the MR *combine* function, but instead it reuses the `GroupBy` operator provided by the Brackit engine. The local `GroupBy` is a *partition-wide* blocking operator [9], meaning that it blocks the evaluation of a single partition only. In general, non-blocking as well as partition-wide blocking operators can be executed inside a task function. Note that the `PostJoin` operator also belongs to this class.

The last task function then groups the pre-aggregated tuples to compute the global aggregations, and another `GroupBy` is used for that end. Because it is the last operator in the evaluation, its results are written to distributed storage and retrieved by the `HadoopExpr` instance. It is also possible to directly transmit the output tuples back to the client, but this would require some synchronization mechanism in order to keep the sort order of the output.

Note, because Hadoop does not directly implement the GMR model, the task function tree of Figure 2 requires one MR job for the execution of phases 0 and 1, inside Mapper and Reducer tasks, respectively, and a second job with an empty Mapper is then required to process phase 2.

### 3.4 Processing XML Data

So far, we have discussed examples based only on relational datasets modelled as JSON collections. However, one major advantage of BrackitMR towards the related approaches is its native support for XML data. This feature is primordial in scenarios like Web page crawling, generation of HTML reports or SVG graphs, and many other typical scenarios in the Web 2.0. Our generic collection framework combined with the transparent integration of JSON in XDM enables queries which mix JSON and XML data sources, and because arbitrary items can be nested inside JSON objects, queries can also generate JSON documents or relational tables which contain embedded XML data.

BrackitMR, however, only supports XML data stored in collections, because they exhibit the degree of data parallelism required in MR processing. If large distributed XML documents are to be processed in MR, we need to provide the query engine with knowledge about how XML fragments are partitioned. Such partitioning strategies heavily depend on the particular structure of each document, and the absence of schema information complicates the problem even further. Even if the partitioning scheme is provided, the query still needs to be checked at compile time to detect paths that cross partition boundaries. Given the management complexity, large XML data sources rarely occur as a single document, but this does not represent a major limitation for BrackitMR. If we consider the domain of MR processing, this becomes an even smaller concern, because typical MR tasks like log processing or document crawling usually deal with large collections of small and independent items.

## 3.5   Limitations

XQuery is in fact a Turing-complete functional programming language, and thus it provides general recursive functions, arbitrarily nested FLWOR expressions, as well as several constructs that rely on strict sequential processing. Because MR provides a simple programming model, suited only for data-parallel computations, it is not possible to cover the complete XQuery standard in BrackitMR. Obvious limitations are, for instance, expressions that depend on sequential evaluation, like the count clause in FLWOR expressions or the position variable binding using the at keyword inside for clauses.

A further limitation of our current prototype is related to dependent subqueries. When evaluated naively, a nested-loops computation is required, which is catastrophic in the MR scenario, given the extremely high latencies intrinsic to the batch processing model. The Brackit query engine extracts arbitrarily nested sub-queries into a single unnested pipeline, using a technique referred to as *pipeline unnesting*. Using this feature, described in detail in [2], nested query semantics is simulated by left outer joins and grouping operations on additional *count variables*, which keep track of the position of a tuple within an iteration. Furthermore, it requires operators to keep track of "empty" iterations that emerge when sub-queries do not deliver any tuples. The unnesting feature is currently not supported in BrackitMR, but because the rewrite rules of the standard Brackit engine can be simply reused, it is not a conceptual limitation.

A further use case in XQuery is the use of recursive functions. Such functions which access collections require an evaluation model that supports fixed point computations. The technique required to implement recursive queries is essentially the same used in recursive SQL, and one approach for XQuery was proposed in [1]. Our current prototype does not implement the technique. However, none of the related approaches support recursive queries, and so the use case remains a corner stone for MR processing.

Aggregate task:

```
for $l in collection('lineitem')
let $month :=
      substring($l=>shipdate, 1, 7)
group by $month
return { month: $month,
      price: sum($l=>extendedprice) }
```

Join task:

```
for $l in collection('lineitem'),
    $o in collection('orders')
where $l=>orderkey eq $o=>orderkey
  and $l=>shipdate gt '1994-12-31'
  and $o=>totalprice gt 70000.00
return { o: $o=>orderkey,
         l: $l=>linenumber }
```

TPC-H Q3 Task:

```
for $c in collection("customer"),
    $o in collection("orders"),
    $l in collection("lineitem")
where $c=>custkey = $o=>custkey
  and $l=>orderkey = $o=>orderkey
  and $o=>orderdate < "1995-09-15"
  and $l=>shipdate > "1995-09-15"
let $orderkey := $l=>orderkey,
    $orderdate := $o=>orderdate,
    $shippriority := $o=>shippriority,
    $discounted := $l=>extendedprice
      * (1 - $l=>discount)
group by $orderkey,
    $orderdate,
    $shippriority
let $revenue := sum($discounted)
order by $revenue
return { order_key: $orderkey,
         revenue: $revenue,
         order_date: $orderdate,
         ship_priority: $shippriority }
```

**Fig. 3.** Queries used in the experiments

# 4    Experiments

To measure the performance of BrackitMR, we ran experiments based on the TPC-H dataset with a size of 10 GB, stored in plain CSV files. The experiments were run on a small cluster with 5 worker nodes and a separate master. In order to ensure similar conditions, we have tuned the Hadoop jobs to use the same number of Mapper and Reducer tasks. We also disabled compression of data shipped between Mapper and Reducer tasks.

We compared the execution times with Pig and Hive. JAQL was not included because its development was moved to a proprietary data warehouse system, and thus its open-source release is not being maintained anymore. It also depends on a discontinued version of Hadoop. However, published measurements comparing JAQL with Pig and Hive have shown that it is outperformed in all tests [10].

We used three queries in our experiments, shown in Figure 3. The first one is an aggregation task on the *lineitem* table, which computes the sum of item prices for each month. The second is a Join task, which filters and joins the *lineitem* and *orders* tables. Last, we ran a slightly modified version of the official TPC-H query number 3. Figure 4 shows the queries expressed in XQuery and the measured execution times in seconds.

The optimization techniques implemented in BrackitMR make use of the collection framework to push down filters and projections. After close inspection of the data produced at each phase of the computation, we conclude that the same

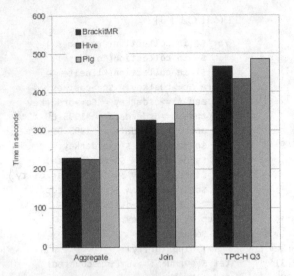

| | Brackit | Hive | Pig |
|---|---|---|---|
| Aggregate | 228 | 225 | 339 |
| Join | 326 | 318 | 366 |
| TPC-H Q3 | 466 | 433 | 486 |

Measured times in seconds

**Fig. 4.** Query execution times in seconds

techniques are employed by Hive. Pig, however, does not perform automatic projection, and so the queries were manually modified so that unused columns were discarded.

The experiments show that Hive is the fastest system for the three tasks, followed closely by BrackitMR and then by Pig. However, the difference between BrackitMR and Hive for each of the queries was 1.3%, 2,5%, and 7.6%, respectively. Thus, it empirically confirms our claim that despite having a more generic data model and more expressiveness in the query language, BrackitMR does exhibit the same performance as the state-of-the art approaches.

## 5    Conclusion

We have developed BrackitMR, an extension of the Brackit query engine which executes XQuery FLWOR expressions in the MR framework. In comparison to existing approaches for query processing in MR, our system relies on an already established, flexible query language. Despite having similar characteristics, such as a semi-structured data model and a more flexible means to compose operations, languages like Hive or PigLatin are currently only used in the context of MR, whereas XQuery is widespread on varying application scenarios from database systems to the Web. Our approach of query engine reuse represents an elegant solution, which simplifies the MR computational model, making greater use of long-established query processing logic.

As a model for distributed query processing, MR actually has significant drawbacks, especially if we consider complex queries with multiple non-blocking operators. The main reason is that it simulates the GMR model using identity Mapper tasks. These tasks represent a major performance bottleneck, because the output of a Reducer phase is always written to the distributed file system, which in most scenarios has a replication factor of three. This output is then fed into the identity Mapper tasks, which simply write the whole data unmodified to the local file systems of the worker nodes. Only then a Shuffle phase can start to group data and perform the operation required by the non-blocking operator. Note that a much more efficient variant would allow the Reducer output to stay within local file system boundaries and be fetched directly by the shuffle tasks.

BrackitMR is at an early stage, and the goal of this paper was simply to show the flexibility potential of XQuery combined with our modular architecture. A crucial requirement, however, is that the higher flexibility does not incur a higher cost in performance. Our goal is that whenever the conditions for efficient data processing are met—in this case, relational structures with full schema information—the query engine must perform as fast as an approach designed specifically for those conditions. We believe this has been achieved so far, as reported in our experiments.

# References

1. Afanasiev, L., Grust, T., Marx, M., Rittinger, J., Teubner, J.: An Inflationary Fixed Point Operator in XQuery. In: ICDE Conference, pp. 1504–1506. IEEE (2008)
2. Bächle, S.: Separating Key Concerns in Query Processing – Set Orientation, Physical Data Independence, and Parallelism. Ph.D. thesis, University of Kaiserslautern, Germany (2012)
3. Beyer, K.S., Ercegovac, V., Gemulla, R., Balmin, A., Eltabakh, M.Y., Kanne, C.C., Özcan, F., Shekita, E.J.: Jaql: A Scripting Language for Large-Scale Semistructured Data Analysis. PVLDB 4(12), 1272–1283 (2011)
4. Dean, J., Ghemawat, S.: MapReduce: A Flexible Data Processing Tool. Commun. ACM 53(1), 72–77 (2010)
5. Graefe, G.: Query Evaluation Techniques for Large Databases. ACM Comput. Surv. 25(2), 73–170 (1993)
6. Lämmel, R.: Google's MapReduce Programming Model – Revisited. Sci. Comput. Program. 70(1), 1–30 (2008)
7. Olston, C., Reed, B., Srivastava, U., Kumar, R., Tomkins, A.: Pig Latin: A Not-So-Foreign Language for Data Processing. In: SIGMOD Conference, pp. 1099–1110 (2008)
8. Robie, J., Brantner, M., Florescu, D., Fourny, G., Westmann, T.: JSONiq: XQuery for JSON, JSON for XQuery, pp. 63–72 (2012)
9. Sauer, C., Härder, T.: Compilation of Query Languages into MapReduce. Datenbank-Spektrum 13(1), 5–15 (2013)
10. Stewart, R.J., Trinder, P.W., Loidl, H.-W.: Comparing High Level MapReduce Query Languages. In: Temam, O., Yew, P.-C., Zang, B. (eds.) APPT 2011. LNCS, vol. 6965, pp. 58–72. Springer, Heidelberg (2011)
11. Thusoo, A., Sarma, J.S., Jain, N., Shao, Z., Chakka, P., Zhang, N., Anthony, S., Liu, H., Murthy, R.: Hive – A Petabyte Scale Data Warehouse using Hadoop. In: ICDE Conference, pp. 996–1005 (2010)
12. W3C: XQuery 3.0: An XML Query Language (2011), http://www.w3.org/TR/xquery-30/
13. White, T.: Hadoop - The Definitive Guide: Storage and Analysis at Internet Scale, 2nd edn. O'Reilly (2011)

# FibLSS: A Scalable Label Storage Scheme for Dynamic XML Updates*

Martin F. O'Connor and Mark Roantree

Interoperable Systems Group, School of Computing,
Dublin City University, Dublin 9, Ireland
{moconnor,mark}@computing.dcu.ie

**Abstract.** Dynamic labeling schemes for XML updates have been the focus of significant research activity in recent years. However the label storage schemes underpinning the dynamic labeling schemes have not received as much attention. Label storage schemes specify how labels are physically encoded and stored on disk. The size of the labels and their logical representation directly influence the computational costs of processing the labels and can limit the functionality provided by the dynamic labeling scheme to an XML update service. This has significant practical implications when merging XML repositories such as clinical studies. In this paper, we provide an overview of the existing label storage schemes. We present a novel label storage scheme based on the Fibonacci sequence that can completely avoid relabeling existing nodes under dynamic insertions. Theoretical analysis and experimental results confirm the scalability and performance of the Fibonacci label storage scheme in comparison to existing approaches.

## 1   Introduction

There has been a noticeable increase in research activity concerning dynamic labeling schemes for XML in recent years. As the volume of XML data increases and the adoption of XML repositories in mainstream industry becomes more widespread, there is a requirement for labeling schemes that can support updates. While read-only XML repositories such as data warehouses have seen significant optimization using views and query adaptation [13], and novel approaches to multi-dimensional modeling [6] faciliate complex rollup and drill-down operations, these efforts do not tackle issues of major changes to the underlying XML documents.

A major obstacle in the provision of an XML update service is the limited functionality provided by existing dynamic labeling schemes. There are a number of desirable properties that characterize a *good* dynamic labeling scheme for XML [16], such as the ability to determine ancestor-descendant, parent-child, and sibling-order relationships between nodes from the labels alone; the generation

---

* The research leading to these results has received funding from the European Union Seventh Framework Programme (FP7/2012) under grant agreement no. 304979.

B. Catania, G. Guerrini, and J. Pokorný (Eds.): ADBIS 2013, LNCS 8133, pp. 218–231, 2013.

of compact labels under arbitrary dynamic node insertions; and the ability to support the reuse of deleted node labels. In this paper, we address the problem of storing scalable binary encoded bit-string dynamic labeling schemes for XML. By scalable, we mean the labeling scheme can support an arbitrary number of node insertions and deletions while completely avoiding the need to relabel nodes. As the size of the databases grow from Gigabytes to Terabytes and beyond, the computational costs of relabeling nodes and rebuilding the corresponding indices becomes prohibitive, not the mention the negative impact on query and updates services while the indices are under reconstruction.

## 1.1  Motivation

The In-MINDD FP7 project is funded by the European Commission to investigate means to decrease dementia risk and delay the onset of dementia by combining areas of social innovation, multi-factorial modelling and clinical expertise [8]. The project aims to quantify dementia risk and deliver personalised strategies and support to enable individuals to reduce their risk of dementia in later life. One of the main tasks is to integrate longitudinal studies such as the Maastricht Ageing Study (MAAS) [9], construct XML views, and integrate the views for various clinical studies. However, this integration process requires the threading of XML elements from one study into another, requiring many and frequent relabeling of nodes. The benefit of XML views is their highly interoperable qualities but their usage presents the problem of XML updates.

There are only two reasons that cause a dynamic labeling scheme to relabel nodes when updating XML. The first reason is that the node insertion algorithms of the dynamic labeling scheme do not permit arbitrary dynamic node insertions without relabeling. For example, when a new node is inserted into an XML tree, the DeweyID labeling scheme [21] requires the relabeling of all *following-sibling* nodes (and their descendants). The second reason that causes a dynamic labeling scheme to relabel nodes is due to the overflow problem.

*The Overflow Problem.* The Overflow Problem concerns the label storage scheme used to encode and store the labels on disk or any physical digital medium and affects both fixed-length and variable-length encodings. It should be clear that all fixed length label storage schemes are subject to overflow once all the assigned bits have been consumed by the update process and consequently require the relabeling of all existing nodes. It is not so obvious that variable-length encodings are also subject to the overflow problem. Variable length labels require the size of the label to be stored in addition to the label itself. Thus, if many nodes are inserted into the XML tree, then at some point, the original fixed length of bits assigned to store the size of the label will be too small and overflow, requiring all existing nodes to be relabeled. This problem has been named the overflow problem in [11].

We hold the position that a modern dynamic labeling scheme for XML should **not** be subject to the overflow problem. All dynamic labeling schemes subject to the overflow problem must relabel existing nodes after a certain number of updates have been performed. In our previous work [17], we highlighted that there are only two existing dynamic labeling schemes that can completely avoid the need to relabel nodes, namely QED [11] and SCOOTER [17]. All other dynamic labeling schemes for XML must relabel existing nodes after an arbitrary number of node insertions due to either limitations in the node insertions algorithms or limitations in the label storage scheme employed by the dynamic labeling scheme.

## 1.2 Contribution

In this paper, we provide a comprehensive review of the existing state-of-the-art in label storage schemes employed by XML dynamic labeling schemes. We present a novel label storage scheme that exploits the properties of the Fibonacci sequence to encode and decode node labels of any arbitrary size. The Fibonacci label storage scheme is scalable - it will never require a dynamic labeling scheme for XML to relabel nodes regardless of arbitrary or repeated dynamic node insertions and deletions. The Fibonacci label storage scheme offers comparable storage costs with the best existing approaches and in particular, is well suited for large data volumes. It also offers the best performance in computational processing costs compared to existing approaches. We provide both theoretical analyses and experimental evaluations to validate our approach.

This paper is structured as follows: in §2, we review and analyze the state-of-the-art in label storage schemes for XML, with a particular focus on scalability. In §3, we present the Fibonacci label storage scheme and the properties that underpin it. We present the algorithms for encoding and decoding a node label and provide a detailed explanation of the encoding transformation. In this section, we also provide a theoretical analysis of the growth rate of the Fibonacci encoded labels. In §4, we provide experimental evaluations of our approach in terms of execution time and total label storage costs and analyze the results. Finally in §5, our conclusions are presented.

## 2    Related Research

A key consideration for all dynamic labeling schemes for XML is how they choose to physically encode and store their labels on disk. All digital data is ultimately stored as binary, but the logical representation of the label on disk directly influences the size of the label on disk and the computational cost to encode/decode from the logical to the physical representation. In this section, we provide an overview of label storage schemes. All existing approaches to the storage of dynamic (variable-length) labels fall under four classifications: length fields, control tokens, separators and prefix-free codes. We employ the same four classifications as those presented in [7].

### 2.1    Length Fields

The concept underlying length fields is to store the length of the label immediately before the label itself. The naive approach is to assign a fixed-length bit code to indicate the length of the label. In a dynamic environment, after a certain number of node insertions, the label size will grow beyond the capacity indicated by the fixed-length bit code and consequently a larger fixed-length bit code will have to be assigned and all existing labels will have to be relabeled according to the new larger fixed-length bit code. One could initially assign a very large fixed-length bit code to minimize the occurrence of the relabeling process, but that would lead to significant wastage in storage for all relatively small labels. In [7], they present several different variations of variable-length bit codes to indicate the size of the label but the authors acknowledge that all of the variable-length approaches lead to either relabeling of existing nodes or involve significant wastage of storage space.

## 2.2  Control Tokens

The concept underlying control tokens is similar to length fields, except rather than storing the length of the label, tokens are used instead to indicate or *control* how the subsequent bit sequence is to be interpreted. We now provide a brief overview of UTF-8 [26] which is a multi-byte variable encoding that uses control tokens to indicate the size of a label.

UTF-8 is employed by the DeweyID [21] and Vector labeling schemes [24]. Originally, UTF-8 was designed to represent every character in the UNICODE character set, and to be backwardly compatible with the ASCII character set.

Referring to Table 1, any number between 0 and 127 ($2^7 - 1$) inclusive, may be represented using 1 byte. The first bit sequence in the label is the control token(s). If the first bit is the control token "0", it indicates the label length is 1 byte. If the first bit is the control token "1", then the number of bytes used to represent the label is computed by counting the number of consecutive control token "1" bits until the control token "0" bit is encountered. The first two bits of the second and subsequent bytes always consist of the bit sequence "10" as illustrated in Table 1.

**Table 1.** UTF-8 Multi-byte Encoding using Control Tokens

| Value | Byte1 | Byte2 | Byte3 | Byte4 | Byte5 | Byte6 |
|-------|-------|-------|-------|-------|-------|-------|
| $0 - (2^7 - 1)$ | 0xxxxxxx | | | | | |
| $2^7 - (2^{11} - 1)$ | 110xxxxx | 10xxxxxx | | | | |
| $2^{11} - (2^{16} - 1)$ | 1110xxxx | 10xxxxxx | 10xxxxxx | | | |
| $2^{16} - (2^{21} - 1)$ | 11110xxx | 10xxxxxx | 10xxxxxx | 10xxxxxx | | |
| $2^{21} - (2^{26} - 1)$ | 111110xx | 10xxxxxx | 10xxxxxx | 10xxxxxx | 10xxxxxx | |
| $2^{26} - (2^{31} - 1)$ | 1111110x | 10xxxxxx | 10xxxxxx | 10xxxxxx | 10xxxxxx | 10xxxxxx |

*Example 1.* To encode the DeweyID label 1.152 in UTF-8, we first determine how many bytes each component requires, convert each component to binary and finally encode using the appropriate number of bytes. 1 is less than 127 ($2^7 - 1$), hence the UTF-8 encoding of 1 is 0 0000001. 152 is between 128 ($2^7$) and

2047 ($2^{11} - 1$), and 152 in binary is 10011000, hence the UTF-8 encoding of 152 requires two bytes and is 110 00010 10 011000 (the spaces are present for readability only). Finally, the full UTF8 encoding of the label 1.152 is 0 0000001 110 00010 10 011000.

The primary limitation of control tokens relate to the requirement to predefine a fixed-length step governing the growth of the labels under dynamic insertions. The fixed-length step cannot dynamically adjust to the characteristics of the XML document or the type of updates to be performed. Furthermore, from the point of view of scalability, UTF-8 cannot encode a number larger than $2^{31} - 1$.

### 2.3  Separators

Whereas control tokens are used to interpret and give meaning to the sequence of bits that immediately follow the token, a separator reserves a predefined bit sequence to have a particular meaning. Consequently, regardless of where the predefined bit sequence occurs, it must be interpreted as a separator. The QED [11] and SCOOTER [17] schemes are the only dynamic labeling schemes to date that employ the separator storage scheme to encode their labels. We now describe the label storage scheme employed by the QED labeling scheme and omit the SCOOTER labeling scheme as the label storage scheme adopted is conceptually very similar.

In QED, a quaternary code is defined as consisting of four numbers *0*, *1*, *2*, *3* and each number is stored with two bits, i.e.: *00*, *01*, *10*, *11*. The number *0* (and bit sequence *00*) is reserved as a separator and only *1*, *2*, and *3* are used in the QED code itself. Therefore, any positive integer can be encoded in the base 3 and represented as a quaternary code. For example, the DeweyID label 2.10.8 can be represented in the base 3 as 2.101.22 and can be encoded using quaternary codes and stored on disk as 11 00 100110 00 1111 (the spaces are present for readability only).

The primary advantage of separator storage schemes over control token schemes is that no matter how big the individual components of a label grow, the separator size remains constant. In the case of quaternary code, the separator size will always be 2-bits no matter how large the label grows. A disadvantage suffered by separator storage schemes compared to control token schemes is that control tokens permit a fast byte-by-byte or bit-by-bit comparison operation [7] and consequently facilitate fast query performance when labels have comparable lengths.

### 2.4  Prefix-Free Codes

Prefix-free codes [4] are fixed-length or variable-length numeric codes that are members of a set which have the distinct property that no member in that set is a prefix to any other member in that set. For example, the set m={1,2,3,4} is a prefix set, however the set n={1,2,3,22} is not a prefix set because the member "2" is a prefix of the member "22".

The ORDPATH [18] dynamic labeling scheme uses prefix-free codes as its label storage scheme. The authors present two prefix-free encoding tables; we present their first encoding table in Table 1 and omit the second table as it is conceptually very similar.

**Table 2.** ORDPATH Variable-length Prefix-free codes and Value range

| Prefix-free Code | Number of bits | Value range |
|---|---|---|
| 0000001 | 48 | $[-2.8 \times 10^{14}, -4.3 \times 10^{9}]$ |
| 0000010 | 32 | $[-4.3 \times 10^{9}, -69977]$ |
| 0000011 | 16 | $[-69976, -4441]$ |
| 000010 | 12 | $[-4440, -345]$ |
| 000011 | 8 | $[-344, -89]$ |
| 00010 | 6 | $[-88, -25]$ |
| 00011 | 4 | $[-24, -9]$ |
| 001 | 3 | $[-8, -1]$ |
| 01 | 3 | $[0, 7]$ |
| 100 | 4 | $[8, 23]$ |
| 101 | 6 | $[24, 87]$ |
| 1100 | 8 | $[88, 343]$ |
| 1101 | 12 | $[344, 4439]$ |
| 11100 | 16 | $[4440, 69975]$ |
| 11101 | 32 | $[69976, 4.3 \times 10^{9}]$ |
| 11110 | 48 | $[4.3 \times 10^{9}, 2.8 \times 10^{14}]$ |

*Example 2.* To encode the ORDPATH label 1.152, we must encode each of the components in the label individually. 1 lies in the value range $[0, 7]$ and hence will be represented using three bits (001) and have the prefix code 01. Thus, the full representation of 1 is 01 001. 152 lies in the value range $[88, 343]$ and hence will be represented using 8 bits and have the prefix code 1100. Note, 152 in binary is the 8 digit number 10011000 but ORDPATH uses the binary representation of 64 to represent this number. 64 is obtained by subtracting the start of the value range from the number to be encoded, that is $152 - 88 = 64$. The binary representation of 64 is 01000000 (using 8 bits). Hence the full representation of 152 is 1100 01000000. Finally the full representation of the ORDPATH label 1.152 is 01 001 1100 01000000 (the spaces are present for readability only).

The ORDPATH prefix-free label storage schemes often require less bits to represent a label that the UTF-8 control token scheme - recall the label 1.152 requires 24-bits to be represented in UTF-8 but only 17 bits using ORDPATH prefix-free codes. However, the ORDPATH prefix free codes have higher computational costs in order to decode a label.

## 2.5  Critique of Label Storage Schemes

In this section, we outlined the four approaches underlying the implementation of all existing label storage schemes for XML to date: length fields, control token, separators, and prefix-free codes. No single approach stands out, each has their own advantages and limitations. Fixed length fields are ideal for static data and variable length fields are ideal for data that is rarely updated. Control token storage schemes *may* facilitate fast byte-by-byte label comparison operations if the properties of the labeling scheme is designed to take advantage of such operations. However the control token storage schemes proposed to date are not compact. The separator storage schemes offer compact label encoding however, the entire label must be decoded bit-by-bit in order to identify each individual component in the label. Lastly, prefix-free codes *may* also permit fast byte-by-byte comparisons but require a pre-computed prefix-free code table to encode and decode labels and a more complex encode/decode function that leads to higher label comparison computational costs.

The key problem we seek to address in this paper is the provision of scalability - that is a label storage scheme that will *never* require the labeling scheme to relabel existing nodes under any arbitrary combination of node insertions and deletions. The SCOOTER and QED labeling schemes are the only labeling schemes to successfully provide this feature by employing the separator label storage scheme in conjunction with node insertion algorithms that do not require the relabeling of existing nodes. Control tokens and prefix-free label storage schemes have been deployed by dynamic labeling schemes that numerically or alphanumerically encoded their labels (DeweyID [21], ORDPATH [18], DLN [2], LSDX [3], Vector [24], DDE [25], however, none of these labeling schemes are scalable as illustrated in [17]). In contrast, length field and separator label storage schemes have been deployed by bit-string dynamic labeling schemes. However, all binary encoded bit-string dynamic labeling schemes (ImprovedBinary [10], CDBS [12], EXEL [14], Enhanced EXEL [15]) are unable to avail of the separator storage scheme (because a bit sequence is reserved as a separator) and are not scalable. Lastly, all length field label storage schemes are subject to relabeling after an arbitrary large number of node insertions. Consequently, there does not exist a label storage scheme that enables binary encoded bit-string dynamic labeling schemes to completely avoid the relabeling of nodes. We address this problem now.

## 3     Fibonacci Label Storage Scheme

The Fibonacci label storage scheme may be employed by any binary encoded bit-string dynamic labeling scheme and enables the labeling scheme to completely avoid the relabeling of nodes. The Fibonacci label storage scheme is a hybrid of the control token and length field classifications. Before we describe the label storage scheme, we provide a brief overview of the Fibonacci sequence [22] and the Zeckendorf representation [23].

**Definition 1.** *Fibonacci Sequence.*
*The Fibonacci sequence is given by the recurrence relation $F_n = F_{n-1} + F_{n-2}$ with $F_0 = 0$ and $F_1 = 1$ such that $n \geq 2$.*

The first 10 terms of the Fibonacci sequence are: 0, 1, 1, 2, 3, 5, 8, 13, 21, 34. Each term in the sequence is the sum of the previous two terms.

**Definition 2.** *Zeckendorf Representation.*
*For all positive integers n, there exists a positive integer N such that*

$$n = \sum_{k=0}^{N} \epsilon_k F_k \quad \text{where } \epsilon_k \text{ is 0 or 1, and } \epsilon_k * \epsilon_{k+1} = 0.$$

This may be more informally stated as: every positive integer n has a unique representation as the sum of one or more distinct non-consecutive Fibonacci numbers. It should be noted that although there are several ways to represent a positive integer $n$ as the sum of Fibonacci numbers, only one representation is the Zeckendorf representation of $n$. For example, the positive integer 111 may be represented as the sum of Fibonacci numbers in the following way:

1. $111 = 89 + 21 + 1$
2. $111 = 55 + 34 + 13 + 8 + 1$
3. $111 = 89 + 13 + 5 + 3 + 1$

However, only the first expression is the Zeckendorf representation of 111, because the second expression contains two consecutive Fibonacci terms (55 + 34) as does the third expression (5 + 3). We will exploit the property that no two consecutive Fibonacci terms occur in the Zeckendorf representation of a positive integer to construct the Fibonacci label storage scheme.

### 3.1 Encoding and Decoding the Length of the Label

We begin with a simple example providing an overview of the encoding process for the Fibonacci label storage scheme before we present our algorithms. Given a binary encoded bit-string label $N_{new} = 110101$,

- We first determine the length of $N_{new}$. It has 6 bits.
- We then obtain the Zeckendorf representation of the length of the label. The Zeckendorf representation of 6 is 5 + 1.
- We then encode the Zeckendorf representation of the label length as a Fibonacci coded binary string. Specifically, starting from the Fibonacci term $F_2$ (recall $F_0 = 0$ and $F_1 = 1$), if the Fibonacci term $F_{k+1}$ occurs in the Zeckendorf representation of the label length, then the $k^{th}$ bit in the Fibonacci coded binary string is set to "1". If the Fibonacci term $F_{k+1}$ does *not* occur in the Zeckendorf representation, then the $k^{th}$ bit in the Fibonacci coded binary string is set to "0".
- For example, the first term $F_2$ (1) occurs in the Zeckendorf representation of 6 and thus, the first bit in the binary string is "1". The second term $F_3$ (2) does not occur in the Zeckendorf representation of 6 and thus, the binary string is now "10". The third term $F_4$ (3) does not occur in the Zeckendorf representation of 6 and thus, the binary string is now "100". The

**Algorithm 1.** EncodeLabelLength.

```
/* Encode n to Fibonacci coded binary string of Zeckendorf representation of n.    */
input  : n - a positive integer representing the length of a label.
output: fibStr - a Fibonacci coded binary string of the Zeckendorf representation of n.
1  begin
2      F_0 ← 0;
3      F_1 ← 1;
4      F_start ← F_0 + F_1;
5      F_end ← the largest Fibonacci number ≤ n;
6      fibArray ← the Fibonacci sequence from F_start to F_end inclusive;
7      fibStr ← "1";
8      for (i=length(fibArray); i=1; i−−) do
9          if (n ≥ fibArray[i]) then
10             fibStr ← "1" ⊕ fibStr;
11             n ← n − fibArray[i];
12         else
13             fibStr ← "0" ⊕ fibStr;
14         end
15     end
16     return fibStr;
17 end
```

fourth term $F_5$ (5) occurs in the Zeckendorf representation of 6 and thus, the binary string is now "1001". There are no more terms in the Zeckendorf representation of the length of $N_{new}$ (6 bits), therefore stop.

- The Fibonacci coded binary string of the Zeckendorf representation will never contain two consecutive "1" bits, precisely because it is a Fibonacci encoding of an Zeckendorf representation. Also, given that the construction of the Fibonacci coded binary string stops after processing the last term in the Zeckendorf representation, we are certain the last bit in the Fibonacci coded binary string must be "1". We subsequently append an extra "1" bit to the end of the Fibonacci coded binary string to act as a control token or delimiter. Thereafter, we know the only place two consecutive "1" bits can occur in the Fibonacci coded binary string is at the end of the string. Thus the binary string is now "10011".

- The Fibonacci label storage scheme adopts a length field storage approach, which means we encode and store the size of the label immediately before the label itself. The last two bits of the Fibonacci coded binary string will always consist of two consecutive "1" bits and they act as a control token separating the length field of the label from the label itself. To complete our example, the label $N_{new}$ (110101) is encoded and stored using the Fibonacci label storage scheme as 10011 110101 (the space is provided as a visual aid).

It can be seen from above that the Fibonacci label storage scheme is a hybrid of the control token and length field label storage schemes. In [1] and [5], the authors exploit a Fibonacci coding of the Zeckendorf representation of variable-length binary strings for synchronization and error correction during the transmission of codes. However, to the best of our knowledge, Fibonacci coded binary strings have never been proposed as a foundation for a label storage scheme nor have they been proposed to provide scalabilty to dynamic labeling schemes.

---

**Algorithm 2.** DecodeLabelLength.

```
/* Decode a Fibonacci coded binary string of a Zeckendorf representation to n.    */
input  : fibStr - a Fibonacci coded binary string of the Zeckendorf representation of n.
output: n - a positive integer representing the length of a label.
1 begin
2  |   F₀ ⟵ 0;
3  |   F₁ ⟵ 1;
4  |   F_start ⟵ F₀ + F₁;
5  |   fibCount ⟵ length(fibStr);
6  |   fibArray ⟵ the first fibCount terms of the Fibonacci sequence from F_start inclusive;
7  |   n ⟵ 0 ;
8  |   for (i=1; i < length(fibArray); i++) do
9  |   |   if (fibStr[i] == "1") then
10 |   |   |   n ⟵ n + fibArray[i];
11 |   |   end
12 |   end
13 |   return n;
14 end
```

---

Algorithm 1 outlines the label length encoding process. It receives as input a positive integer $n$ representing the label length and outputs a Fibonacci coded binary string of the Zeckendorf representation of $n$. Algorithm 2 outlines the label length decoding process which is the reverse transformation of algorithm 1.

### 3.2 Fibonacci Label Storage Scheme Size Analysis

In Table 3, we illustrate the relationship between the growth in label size and the corresponding growth in the quantity of labels that may be encoded. Given a label encoding length $n$, the quantity of labels that may be encoded with length $n$ is equal to the Fibonacci term $F_{n-1}$. In [19], the authors prove that the average value of the $n^{th}$ term of a sequence defined by the general recurrence relation $G_n = G_{n-1} +- G_{n-2}$ increases exponentially. *Therefore, as the number of labels to be encoded using the Fibonacci label storage scheme increases exponentially, the corresponding growth in the size of the Fibonacci coded binary string is linear.* This demonstrates that when processing a large quantity of labels the Fibonacci label storage scheme scales gracefully.

## 4 Evaluation

In this section, we evaluate the Fibonacci label storage scheme by comparing it with three other label storage schemes, namely ORDPATH Compressed binary format, UTF-8 and the Separator label storage schemes. All label storage schemes were implemented in Java version 6.38 and all experiments were carried out on a 2.66Ghz Intel(R) Core(TM)2 DUO CPU with 4GB of RAM. The experiments were performed 11 times, the time from the first run was discarded and the results of the subsequent 10 experiments averaged. For all experiments, the unit of storage is in *bits* and the unit of time is in milliseconds (ms). The ORDPATH prefix-free code tables and the array of Fibonacci numbers from 1 to N

**Table 3.** Fibonacci Encoded Label Length Growth Rate

| Growth Counter | Label Encoding Length | Num of labels |
|:---:|:---:|:---:|
| 1 | 2 | 1 |
| 2 | 3 | 1 |
| 3 | 4 | 2 |
| 4 | 5 | 3 |
| 5 | 6 | 5 |
| 6 | 7 | 8 |
| 7 | 8 | 13 |
| 8 | 9 | 21 |
| 9 | 10 | 34 |
| 10 | 11 | 55 |
| 11 | 12 | 89 |
| 12 | 13 | 144 |
| 13 | 14 | 233 |
| 14 | 15 | 377 |
| $\vdots$ | $\vdots$ | $\vdots$ |
| n | n + 1 | $F_n$ |

are computed once (in advance), and not each time a label is encoded/decoded, so as to reflect a real-world implementation scenario.

The Fibonacci and Separator label storage schemes were designed to encode bit-string labels, whereas the ORDPATH and UTF-8 label storage schemes were designed to encode integer-based labels. Consequently, to ensure an equitable and fair experimental evaluation, all four label storage schemes encode the positive integers from 1 to $10^n$ where n has the values from 1 to 6 inclusive. Given that the Fibonacci and Separator label storage schemes expect a bit-string label to encode, the integer is converted from base 10 to base 2 (binary) and the binary string representation of the integer is encoded. In Figure 1, we illustrate the storage costs for all four label storage schemes using labels derived from the integer encodings from 1 to $10^6$ (Note: a logarithmic scale is used in the illustration). The ORDPATH compressed binary format provides a choice of two encoding tables to use; we present both encodings to enable a comprehensive evaluation and analysis. In this remainder of this section, "FIB" is used to denote the Fibonacci label storage scheme.

ORDPATH2 provides the most compact storage representation when encoding less than 10 integers. UTF-8 provides the most compact storage representation when encoding $10^2$ integers and SEPARATOR when encoding $10^3$ through $10^5$ inclusive. When encoding $10^6$ integers, FIB provides the most compact storage representation. This result is in line with our theoretical analysis in §3.2 which observed that as the number of labels to be encoded using FIB increases exponentially, the corresponding growth rate in the size of the Fibonacci coded binary string is linear. Hence, although the performance of FIB is average for small to medium sized labels, FIB provides a highly compact storage representation for large labels. However, unlike ORDPATH and UTF-8, FIB is not subject to the overflow problem and will never require existing labels to be relabeled.

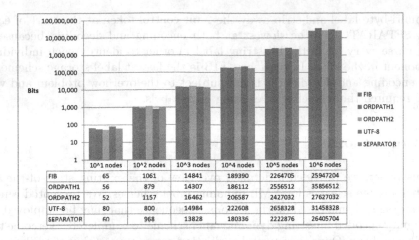

**Fig. 1.** Storage Costs of Encoding Integers for the Label Storage Schemes

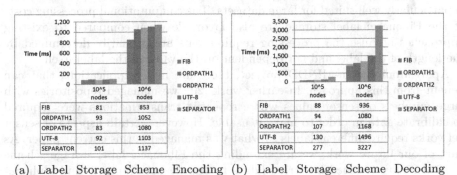

(a) Label Storage Scheme Encoding Times

(b) Label Storage Scheme Decoding Times

**Fig. 2.** Integer Encoding and Decoding times for Label Storage Schemes

In Figure 2, we illustrate the computational processing times to encode and decode $10^5$ and $10^6$ integer labels. The times for $10^4$ (or less) integer encodings are not shown because they are single digit results with negligible differences between them. FIB is the fastest label storage scheme at both encoding and decoding. FIB is the fastest because as a length field label storage scheme it only has to encode and decode the length of the label. The actual bit-string label is stored immediately after the Fibonacci coded binary string and can be read and written without having to process each individual bit. All of the other label storage schemes must process the entire label to generate their encoding. ORDPATH1 has similar encode and decode computational processing costs. OR-DPATH2 encodes more quickly than it decodes because the key size that maps to the range of the integer encoding grows more quickly than the encoding table employed by ORDPATH1. UTF-8 decodes approximately 30 percent slower than it encodes because when decoding, it must parse each individual byte in the multi-byte label and strip away the 2-bit control token at the start of each byte. SEPARATOR is the slowest at both encoding and decoding because it must parse every bit in the bit-string label in order to identify each individual component in the label. In summary, FIB is the fastest label storage scheme at both encoding and decoding, it is not subject to the overflow problem and will never required the relabeling of existing node labels.

## 5    Conclusions

In this paper, we provided a detailed overview of the existing state-of-the-art in label storage schemes for XML dynamic label schemes. We presented a new label storage scheme based on the Fibonacci sequence that may be employed by any binary encoded bit-string dynamic label scheme to completely overcome the overflow problem. Our experimental evaluation demonstrated that the Fibonacci label storage scheme offers storage costs comparable to existing approaches and is particularly well suited for large datasets. The computational processing costs of the Fibonacci label storage are also favorable when compared to existing approaches because the processing requirements are primarily determined by the length of the label and not dependent on the value of the label itself.

Apart from the In-MINDD project, there are many applications that can benefit from this approach. In earlier work [20], we managed repositories with large numbers of sensor values where repeated transformations were required to calibrate data in order to make it usable. However, certain aspects of sensor networks require different labels to what was managed in this current paper. As part of our future work, we are extending the Fibonacci label storage scheme beyond binary encoded bit-string labels to encode both numeric and alphanumeric labels. The goal is to provide an alternative label storage scheme that may be employed by all dynamic labeling schemes, offering a compact storage representation while minimizing processing costs.

## References

1. Apostolico, A., Fraenkel, A.S.: Robust Transmission of Unbounded Strings Using Fibonacci Representations. IEEE Transactions on Information Theory 33(2), 238–245 (1987)
2. Böhme, T., Rahm, E.: Supporting Efficient Streaming and Insertion of XML Data in RDBMS. In: DIWeb, pp. 70–81 (2004)
3. Duong, M., Zhang, Y.: LSDX: A New Labelling Scheme for Dynamically Updating XML Data. In: ADC, pp. 185–193 (2005)
4. Elias, P.: Universal Codeword Sets and Representations of the Integers. IEEE Transactions on Information Theory 21(2), 194–203 (1975)
5. Fraenkel, A.S., Kleinb, S.T.: Robust Universal Complete Codes for Transmission and Compression. Discrete Applied Mathematics 64(1), 31–55 (1996)
6. Gui, H., Roantree, M.: A Data Cube Model for Analysis of High Volumes of Ambient Data. Procedia CS 10, 94–101 (2012)

7. Härder, T., Haustein, M.P., Mathis, C., Wagner, M.: Node Labeling Schemes for Dynamic XML Documents Reconsidered. Data Knowl. Eng. 60(1), 126–149 (2007)
8. In-MINDD - INnovative, Midlife INtervention for Dementia Deterrence (2013), online Resource http://www.inmindd.eu/
9. Jolles, J., Houx, P., van Boxtel, M., Ponds, R.: Maastricht Aging Study: Determinants of Cognitive Aging. Neuropsych Publishers (1995)
10. Li, C., Ling, T.-W.: An Improved Prefix Labeling Scheme: A Binary String Approach for Dynamic Ordered XML. In: Zhou, L.-Z., Ooi, B.-C., Meng, X. (eds.) DASFAA 2005. LNCS, vol. 3453, pp. 125–137. Springer, Heidelberg (2005)
11. Li, C., Ling, T.W.: QED: A Novel Quaternary Encoding to Completely Avoid Re-labeling in XML Updates. In: CIKM, pp. 501–508 (2005)
12. Li, C., Ling, T.W., Hu, M.: Efficient Processing of Updates in Dynamic XML Data. In: ICDE, p. 13 (2006)
13. Liu, J., Roantree, M., Bellahsene, Z.: A SchemaGuide for Accelerating the View Adaptation Process. In: Parsons, J., Saeki, M., Shoval, P., Woo, C., Wand, Y. (eds.) ER 2010. LNCS, vol. 6412, pp. 160–173. Springer, Heidelberg (2010)
14. Min, J.-K., Lee, J., Chung, C.-W.: An Efficient Encoding and Labeling for Dynamic XML Data. In: Kotagiri, R., Radha Krishna, P., Mohania, M., Nantajeewarawat, E. (eds.) DASFAA 2007. LNCS, vol. 4443, pp. 715–726. Springer, Heidelberg (2007)
15. Min, J.-K., Lee, J., Chung, C.-W.: An Efficient XML Encoding and Labeling Method for Query Processing and Updating on Dynamic XML Data. Journal of Systems and Software 82(3), 503–515 (2009)
16. O'Connor, M.F., Roantree, M.: Desirable Properties for XML Update Mechanisms. In: EDBT/ICDT Workshops (2010)
17. O'Connor, M.F., Roantree, M.: SCOOTER: A Compact and Scalable Dynamic Labeling Scheme for XML Updates. In: Liddle, S.W., Schewe, K.-D., Tjoa, A.M., Zhou, X. (eds.) DEXA 2012, Part I. LNCS, vol. 7446, pp. 26–40. Springer, Heidelberg (2012)
18. O'Neil, P.E., O'Neil, E.J., Pal, S., Cseri, I., Schaller, G., Westbury, N.: ORDPATHs: Insert-Friendly XML Node Labels. In: SIGMOD Conference, pp. 903–908 (2004)
19. Rittaud, B.: On the Average Growth of Random Fibonacci Sequences. Journal of Integer Sequences 10(2), 3 (2007)
20. Roantree, M., Shi, J., Cappellari, P., O'Connor, M.F., Whelan, M., Moyna, N.: Data Transformation and Query Management in Personal Health Sensor Networks. J. Network and Computer Applications 35(4), 1191–1202 (2012)
21. Tatarinov, I., Viglas, S., Beyer, K.S., Shanmugasundaram, J., Shekita, E.J., Zhang, C.: Storing and Querying Ordered XML using a Relational Database System. In: SIGMOD Conference, pp. 204–215 (2002)
22. Wolfram‖Alpha: Fibonacci Numbers, Wolfram Alpha LLC edn. (December 2012), online Resource http://mathworld.wolfram.com/FibonacciNumber.html
23. WolframAlpha: Zeckendorf Representation, Wolfram Alpha LLC edn. (December 2012), online Resource http://mathworld.wolfram.com/ZeckendorfRepresentation.html
24. Xu, L., Bao, Z., Ling, T.-W.: A Dynamic Labeling Scheme Using Vectors. In: Wagner, R., Revell, N., Pernul, G. (eds.) DEXA 2007. LNCS, vol. 4653, pp. 130–140. Springer, Heidelberg (2007)
25. Xu, L., Ling, T.-W., Wu, H., Bao, Z.: DDE: From Dewey to a Fully Dynamic XML Labeling Scheme. In: SIGMOD Conference, pp. 719–730 (2009)
26. Yergeau, F.: UTF-8, A Transformation Format of ISO 10646, Request for Comments (RFC) 3629 edn. (November 2003)

# Exploiting the Relationship between Keywords for Effective XML Keyword Search

Jiang Li[1], Junhu Wang[1], and Maolin Huang[2]

[1] School of Information and Communication Technology,
Griffith University, Gold Coast, Australia
[2] Faculty of Engineering and Information Technology
The University of Technology, Sydney, Australia

**Abstract.** XML keyword search provides a simple and user-friendly way of retrieving data from XML databases, but the ambiguities of keywords make it difficult to *effectively* answer keyword queries. In this paper, we tackle the keyword ambiguity problem by exploiting the relationship between keywords in a query. We propose an approach which infers and ranks a set of likely search intentions. Extensive experiments verified the better effectiveness of our approach than existing systems.

## 1 Introduction

Keyword search in XML databases has been extensively studied recently. However, the search effectiveness problem is far from solved. One of the main causes of the problem is the ambiguity of keywords. In particular, a word can have multiple meanings[1]. Consider the XML tree in Fig. 1. The word *16* appears as a text value of *volume* and *initPage* nodes, and the word *issue* exists as an XML tag name and a text value of *title* node. When a user types in such keywords, it is hard to know which meaning of the keyword the user wants.

In this work, we propose to tackle keyword ambiguities and infer users' search intention by exploiting the relationship between different keywords in a query. The basic observation is that users seldom issue a query arbitrarily. Instead, most of the time they construct queries logically. They usually place closely related keywords at *adjacent* positions. For example, if the user intends to retrieve the articles about database from issue 16, he is more likely to submit the query {**issue 16** database} than the query {**16** database **issue**} because the keywords "16" and "issue" are closely related. This intuition motivates us to infer a keyword's meaning by *preferentially* evaluating the relationship between this keyword and its adjacent keywords. In the example above, if the query is {**issue 16** database}, it is more likely to infer "issue" as the tag name *issue* by considering this keyword together with its adjacent keyword "16" than considering this keyword with "database".

---

[1] We use word type to represent a word's meaning. See Definition 5 for the definition of word type.

B. Catania, G. Guerrini, and J. Pokorný (Eds.): ADBIS 2013, LNCS 8133, pp. 232–245, 2013.
© Springer-Verlag Berlin Heidelberg 2013

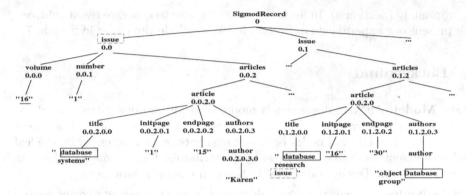

**Fig. 1.** A Sample XML data tree of SigmodRecord

We developed a system `XInfer` that infers a set of likely search intentions using the above intuition as well as keyword distribution in the data tree, and ranks them appropriately. For example, for the data tree in Fig. 1, if the user types the query {issue 16 database}, `XInfer` will infer that the most likely search intention as *articles in issue 16 whose title contains "database"*; if the query is {16 database issue}, it will infer the most likely search intention as *articles with initial page "16" whose title contains the words "database" and "issue"*; If the user query is {database issue 16}, then both of the afore-mentioned search intentions will be ranked highly, with the second one being ranked the highest.

**Related Work.** Works that are most closely related to ours include XReal [2], XBridge [3] and `XSeek` [4]. XReal uses statistics of the data (mainly term frequency) to find a *search-for node type (SNT)*, and XBridge [3] estimates the promising result types based on off-line synopsis of the XML tree (structural distribution and value distribution). `XSeek` tried to recognize the possible entities and attributes in the data tree, distinguish between search predicates and return specifications in the keywords, and return nodes based on the analysis of both XML data structures and keyword match patterns. Many other works on XML keyword search are based on variants of LCA (lowest common ancestor). One of the variants is maxMatch [5], which prunes irrelevant nodes from result subtrees obtained using the SLCA semantics. Due to page limit, we refer the readers to the full version of this paper[2] and the recent survey [6] for more details. To the best of our knowledge, no previous work has utilized the relationship between adjacent keywords when inferring the search intention.

*Organization.* After introducing the data model (Section 2), we design a formula to evaluate the relationship between two adjacent keywords without considering other keywords in the query, which takes into account the statistics and structural properties of a keyword's different meanings (Section 3). Then we propose the Pair-wise Comparison Strategy, which utilizes the inference results of pairs of adjacent keywords, to infer a set of likely search intentions and rank them

---

[2] Available from `http://www.ict.griffith.edu.au/~jw/report/xinfer.pdf`

appropriately (Section 4). In Section 5, we show how to generate result subtrees. We present our experiments in Section 6 and conclude the paper in Section 7.

## 2   Background

**Data Model.** An XML document is modeled as an unordered tree, called the *data tree*. Each *internal node* (i.e., non-leaf node) has a label, and each *leaf node* has a value. The internal nodes represent elements or attributes, while the leaf nodes represent the values of elements or attributes. Each node $v$ in the data tree has a unique Dewey code. Fig. 1 shows an example data tree.

**Entity Nodes.** In reality, an XML document is usually a container of related entities. For instance, Fig. 1 is a collection of *issue* and *article* entities. We use an approach similar to that of [4] to identify entity nodes.

**Definition 1.** *(Entity Node) Let t be a data tree. A node v in t is said to be a* simple node *if it is a leaf node, or has a single child which is a leaf node. A node v is said to be an* entity node *if: (1) it corresponds to a ∗-node in the DTD (if DTD exists), or has siblings with the same tag name as itself (if DTD does not exist), and (2) it is not a simple node. The* entity type *of an entity node e refers to the node type of e. A node v in t is called a* grouping node *if it has children of entity nodes only.*

Consider the data tree in Fig. 1, nodes *issue* (0.0) and *article* (0.0.2.0) are entity nodes. The nodes *SigmodRecord* (0) and *articles* (0.0.2) are grouping nodes.

**Keyword Query.** A keyword query is a finite set of keywords $K = \{w_1, ..., w_n\}$. Given a keyword $w$ and a data tree $t$, the search of $w$ in $t$ will check both the labels of internal nodes and values of leaf nodes for possible occurrences of $w$.

## 3   Inferring the Meaning of Two Adjacent Keywords

### 3.1   Preliminary Definitions

**Definition 2. (Node Type)** *Let* v *be a node in data tree* t. *The node type of* v *is the sequence of node labels on the path from the* root *to* v *if* v *is an internal node, and is denoted* $l_1.l_2.\cdots.l_n$, *where* $l_i$ *($1 \le i \le n$) is the label of the $i^{th}$ node on the path. If* v *is a leaf node, its node type is the node type of its parent. The length of a node type is the number of nodes on the path.*

In Fig. 1, the node type of *author* (0.0.2.0.3.0) is `SigmodRecord.issue.articles.article.authors.author`. For simplicity, we will use the tag name of a node to denote the node type when there is no confusion.

**Definition 3. (Ancestor Node Type)** *Given a node type* $T \equiv l_1.l_2.\cdots.l_n$, *we say* $l_1.l_2.\cdots.l_i$ *is an* ancestor node type *of* $T$, *for any* $i \in [1, n-1]$.

We use $T_1 \prec T_2$ to denote that $T_1$ is an ancestor node type of $T_2$.

**Definition 4.** *Let $T_1, \cdots, T_n$ be node types. The* longest common ancestor *of $T_1, \cdots, T_n$, denoted* NtLCA$(T_1, \cdots, T_n)$*, is a node type $V$ such that (1) $V \prec T_i$ for all $i \in [1, n]$; (2) there is no node type $U$ such that $V \prec U \prec T_i$ for all $i \in [1, n]$.*

We now define word type, which is used to represent the meaning of a word.

**Definition 5. (Word Type)** *Let $t$ be the data tree and $w$ be a word that occurs in $t$. A node type is said to be a word type of $w$ if some nodes of this node type directly* contain *$w$.*

Table 1 lists the word types of several words for the data tree in Fig. 1.

**Table 1.** Sample Word Types

| Word | No. | Word Type |
|------|-----|-----------|
| issue | 1 | SigmodRecord.issue |
| | 2 | SigmodRecord.issue.articles.article.title |
| 16 | 3 | SigmodRecord.issue.volume |
| | 4 | SigmodRecord.issue.articles.article.initPage |
| database | 5 | SigmodRecord.issue.articles.article.title |
| | 6 | SigmodRecord.issue.articles.article.authors.author |

**Definition 6. (Search Intention)** *Given a keyword query $K = \{w_1, \cdots, w_n\}$ and their corresponding word type sets $\{WT_1, \cdots, WT_n\}$, a search intention of this query is a tuple of word types $(wt_1, \cdots, wt_n)$, where $wt_i \in WT_i$ ($1 \le i \le n$).*

Once we have a search intention, we can find the result subtrees accordingly. The details of the result subtree will be given in Section 5.

### 3.2   Inferring the Word Types of Two Adjacent Keywords

If two adjacent keywords $w_i$ and $w_{i+1}$ are taken as a keyword query with two keywords, the number of possible search intentions is equal to $|WT_i| \times |WT_{i+1}|$. We design a formula to compute the *proximity score* between $wt_i$ and $wt_{i+1}$ which is used to evaluate how closely $wt_i$ and $wt_{i+1}$ are related, where $(wt_i, wt_{i+1}) \in WT_i \times WT_{i+1}$. The tuple $(wt_s, wt_t)$ that achieves the *largest proximity score* is considered as the desired search intention of $w_i$ and $w_{i+1}$.

Basically, the proximity score between two word types is affected by the distance between them and the statistics of them.

**Distance between Two Word Types.** The distance between two word types $wt_1$ and $wt_2$ is defined as the total number of edges from NtLCA$(wt_1, wt_2)$ to the ends of $wt_1$ and $wt_2$, which can be calculated using the following formula:

$$Dis(wt_1, wt_2) = len(wt_1) + len(wt_2) - 2 * clen(wt_1, wt_2) \qquad (1)$$

In the formula above, *len* is the length of a word type. $clen(wt_1, wt_2)$ is the length of the longest common ancestor of $wt_1$ and $wt_2$. Consider word types

2 and 6 in Table 1. According to Formula (1), the distance between them is $5 + 6 - 2 * 4 = 3$.

Intuitively, *the shorter the distance between two word types, the more closely these two word types are related*. In our work, we use $p^{Dis(wt_1,wt_2)}$ to formulate the intuition above, where $p$ is a tuning parameter which is used to determine how much penalty should be given to the distance between two word types. In our experiments we found setting $p$ to 0.87 achieves good results.

**Statistics of Word Types.** Another important factor that influences the proximity score is the statistics of word types. Given a pair of word types of two adjacent keywords, we formulate the influencing factor of the statistics of the two word types as follows:

$$Sta(wt_1, wt_2) = log_e(f_{w_1,wt_1}^{NtLCA(wt_1,wt_2)} + f_{w_2,wt_2}^{NtLCA(wt_1,wt_2)}) * R \tag{2}$$

where $R$ is a reduction factor which will be explained shortly. $f_{w,wt}^T$ is the number of T-typed nodes that contain the keyword $w$ with the word type $wt$ in their subtrees. The intuition (similar to XReal) behind the statistics is as follows:

*The more XML nodes of node type NtLCA($wt_1, wt_2$) contain the keywords with word types $wt_1$ and $wt_2$, the more likely these two word types are related through the nodes of node type NtLCA($wt_1, wt_2$) and this kind of relationship between $wt_1$ and $wt_2$ is desired by the user.*

We now explain the reduction factor $R$. Due to the tree structure, the number of nodes at higher levels is usually significantly less than the number of nodes at lower levels. This may bring unfairness when collecting statistics. Fewer nodes at higher levels are usually caused by design, which may not reflect the real occurrences of data. Therefore, we put a reduction factor of depth to the formula to reduce the unfairness. A straightforward reduction factor of depth is $\frac{1}{Dep(NtLCA(wt_1,wt_2))}$, but it reduces too much and too quickly when the depth increases. Instead, we use the following formula as the reduction factor:

$$R = \sqrt{\frac{1}{Dep(\text{NtLCA}(wt_1, wt_2))}} \tag{3}$$

With the two influencing factors above, the proximity score between two word types is defined as follows:

$$P(wt_1, wt_2) = p^{Dis(wt_1,wt_2)} * Sta(wt_1, wt_2) \tag{4}$$

**Desired Word Types.** Given two adjacent keywords $w_i$ and $w_{i+1}$, the preferred relationship between them achieves the largest proximity score among all $P(wt_i, wt_{i+1})$, where $wt_i \in WT_i$ and $wt_{i+1} \in WT_{i+1}$. The word types that form this relationship are the desired word types of $w_i$ and $w_{i+1}$. In the sequel, the largest proximity score between $w_i$ and $w_{i+1}$ will be denoted $HP(w_i, w_{i+1})$, i.e.,

$$HP(w_i, w_{i+1}) = max\{P(wt_i, wt_{i+1}) \mid wt_i \in WT_i, wt_{i+1} \in WT_{i+1}\} \tag{5}$$

# 4   Inferring Likely Search Intentions

## 4.1   One Keyword

If a query contains *only one* keyword, each word type of this keyword is considered as a search intention. We use the following formula to compute the likelihood that the word type $wt$ is desired by the user.

$$C(wt) = log_e(f_{k,wt}^{wt}) * \sqrt{\frac{1}{Dep(wt)}} \tag{6}$$

## 4.2   Two or More Keywords

When there are two or more keywords, we use a *pair-wise comparison strategy* (PCS) to infer a set of likely search intentions. We first explain the ideas  and then present the detailed algorithms.

**Inferring One Likely Search Intention.** Given the keywords $\{w_1, \cdots, w_n\}$ $(n > 1)$ and their corresponding word type sets $\{WT_1, \cdots, WT_n\}$, we infer a likely search intention as follows.

In the case $n = 2$, i.e., there are only two keywords $w_1$ and $w_2$ in the query, we will compute the largest proximity score between $w_1$ and $w_2$ (i.e., $HP(w_1, w_2)$) using formula (5), and choose the word types of $w_1$ and $w_2$ that achieve $HP(w_1, w_2)$ as the word types of $w_1$ and $w_2$.

In the case $n > 2$, we need to go through several iterations. In the first iteration, we scan the keywords from left to right pair by pair and compute $HP(w_i, w_{i+1})$ $(1 \leq i < n)$. We group $w_j$ and $w_{j+1}$ together if they achieve the largest $HP$ value, that is, if $HP(w_j, w_{j+1}) = max\{HP(w_i, w_{i+1}) \mid i \in [1, n-1]\}$. Whenever a pair of adjacent keywords $w_i, w_{i+1}$ are put into a group, the word types of $w_i$ and $w_{i+1}$ that achieve $HP(w_i, w_{i+1})$ will be chosen as the word types of $w_i$ and $w_{i+1}$ respectively, and we will go to the next iteration. In each subsequent iteration, we scan the keywords or groups of keywords from left to right, grouping a pair of two keywords, or a keyword and a group, into the same group by using the largest proximity score in a way similar to the first iteration, except that for each group (that contains more than one keyword), a unique word type of each keyword in the group has been chosen as the word type for that keyword (and this unique word type will not be changed later), thus the computation of the largest proximity score between a keyword and a group, or two groups will use the word types of each keyword in the group as shown in formulae 7 to 8 below. This process continues until all keywords joins a group (so that its word type can be determined).

For a keyword $w_i$ and a group $g = \{w_k, \ldots, w_s\}$ of keywords, suppose the word types of $w_k, \cdots, w_s$ are $wt_k, \cdots, wt_s$ respectively, then

$$HP(w_i, g) = HP(g, w_i) = max\{P(wt_i, wt) \mid wt_i \in WT_i, wt \in \{wt_k, \ldots, wt_s\}\} \tag{7}$$

For two neighboring groups $g_1 = \{w_j, \ldots, w_k\}$ and $g_2 = \{w_{k+1}, \ldots, w_t\}$ with word types $wt_j, \ldots, wt_k$ and $wt_{k+1}, \ldots, wt_t$ respectively, we define

$$\text{HP}(g_1, g_2) = P(wt_k, wt_{k+1}) \tag{8}$$

The HP score between two groups is not used to merge two groups. It is only used in the ranking of the returned search intentions, as we will discuss later.

**Inferring a Set of Likely Search Intentions.** To reduce the possibility of missing the real search intention, we generate a set of likely search intentions as described below.

Given the keywords $\{w_1, \ldots, w_n\}$ and their word type sets $\{WT_1, \ldots, WT_n\}$, we treat each word type in $WT_i$ as the only word type of $w_i$, and use PCS to infer a search intention which is considered as a *likely search intention*. In total, we will produce $\sum_{1 \leq i \leq n} |WT_i|$ likely search intentions if there are no duplicates[3], which is much smaller than $\prod_{1 \leq i \leq n} |WT_i|$ (the number of all possible search intentions). Applying PCS against each word type guarantees a good coverage of different word types in the query. In other words, every word type of each keyword exists in at least one likely search intention.

**Ranking Likely Search Intentions.** In order to rank the inferred likely search intentions, we define their ranking scores as follows.

**Definition 7.** *(Ranking Score) Given a query* $K = \{w_1, \cdots, w_n\}$ *(*$n \geq 2$*) and a search intention* $I$ *inferred by PCS, the* ranking score *of* $I$ *is defined as a vector* $(r_1, \cdots, r_{n-1})$*, where* $r_i \geq r_{i+1}$ *(*$1 \leq i < n-1$*), and* $r_i$ *is a largest proximity score between two keywords or between a keyword and a group that are grouped together in PCS (calculated by formulae (5) to (7)), or the largest proximity score between two neighboring groups (calculated by Formula (8)). Given two ranking scores $A$ and $B$, $A = (a_1, \cdots, a_{n-1})$ is smaller than $B = (b_1, \cdots, b_n)$ iff the first $a_i$ which is different from $b_i$ is smaller than $b_i$.*

*Example 1.* Suppose the keywords in the query $\{w_1 \ w_2 \ w_3 \ w_4\}$ are grouped as $\{(w_1, w_2), (w_3, w_4)\}$ and the inferred search intention is $I = (wt_1, wt_2, wt_3, wt_4)$. The ranking score of $I$ is a sequence of proximity scores $\text{HP}(w_1, w_2)$, $\text{HP}(w_3, w_4)$ and $P(wt_2, wt_3)$, which will be sorted in descending order. The ranking score $(2.5, 2, 1.2)$ is larger than $(2.5, 1.8, 1)$.

Note that we use $(r_1, \cdots, r_{n-1})$ rather than $\sum_i r_i$ as the ranking score. The reason for this will become clear if one considers the data tree in Fig. 1 and the query {issue 16 database}. The correct search intention $(1, 3, 5)$ (where 1,3, 5 refer to the word types in Table 3.1) will be ranked higher than the intention $(2,4, 5)$ using our ranking scheme. However, if we use $\sum_i r_i$ as the ranking score, then the intention $(1, 3, 5)$ will be ranked lower than $(2,4,5)$.

---

[3] Note that there may exist duplicates in the inferred likely search intentions.

---

**Algorithm 1.** PCS (K,WT)

---

**Input:** Query $K = \{w_1, \cdots, w_n\}$, the corresponding word type sets $\{WT_1, \cdots, WT_n\}$
**Output:** a search intention: list $l = (l_1, \ldots, l_n)$ and it ranking vector $r$

1: let $l = (null, \ldots, null)$; $r = \emptyset$
2: **while** ExistUngroupedKeyword() **do**
3:     let $j = 1$; $k = 1 + w_1$; $largestScore = 0$
4:     **while** $k \leq n$ **do**
5:         **if** $not(w_j.num > 1 \wedge w_k.num > 1)$ **then**
6:             $(score_{j,k}, wt_j, wt_k) = $ computeHPscore$(w_j, w_k)$
7:             **if** $score_{j,k} > largestScore$ **then**
8:                 let $largestScore = score_{j,k}$; $s = j$; $t = k$
9:         let $j = k$; $k = j + w_j.num$
10:    group$(w_s, w_t)$
11:    add $largestScore$ to $r$
12:    let $l_s = wt_s$ if $w_s.num = 1$; let $l_t = wt_t$ if $w_t.num = 1$
13: add to $r$ the HP scores between neighboring groups, then sort $r$

---

**Algorithms.** We implemented PCS in Algorithm 1. Before explaining this algorithm, we first present some notation and functions used in the algorithm.

*Notation.* For each keyword $w_i$, we use an attribute $w_i.num$ to record the number of keywords in the group that $w_i$ sits. If $w_i$ is not grouped with others, $w_i.num = 1$. The function group$(w_j, w_k)$ puts $w_j$ and $w_k$ into the same group if they are not grouped yet, or puts one into the same group as the other (if only one of them is grouped), and while doing this it also sets $w_i.num$ to $w_j.num + w_k.num$ for every keyword $w_i$ in the group. The function computeHPscore$(w_j, w_k)$ returns a score, which is the HP value calculated using formula 5 if neither $w_j$ nor $w_k$ is grouped with others, or calculated using formula 7 if only one of $w_j, w_k$ is grouped with others. The corresponding word types of $w_j, w_k$ that achieve the HP value are also returned as $wt_j$ and $wt_k$ respectively. The function ExistUngroupedKeyword() returns **true** iff there is a keyword in $K$ that has not been grouped with others (A linear scan of each keyword in $K$, checking whether $w_i.num = 1$, will do).

Now we explain Algorithm 1. Line 1 initializes $l$ and $r$. If there exists a keyword which is not grouped with others, lines 3 to 9 will compute the HP value between two keywords or between a keyword and a group, and find a pair of keywords (or a keyword and a group), represented by $w_s$ and $w_t$, that achieves the largest HP value. Line 10 groups the pair together. The largest HP value is then added to $r$ (line 11), and if $w_s$ is not grouped before , we set its word type to $wt_s$, the word type in $WT_s$ that achieves the above largest HP value (line 12). The same is done for $w_t$. Finally, line 13 computes the HP values between neighboring groups and add them to the ranking score $r$, and then sorts $r$.

*Example 2.* Suppose the user submits the query {*volume 11 article Karen*} over the real SigmodRecord data set obtained from [1]. In the first run of lines 3 to 12, HP(*volume*, 11), HP(11, *article*) and HP(*article, Karen*) are computed, and because HP(*volume*, 11) is greater than HP(11, *article*) and HP(*article, Karen*), we group *volume* and *11* together. Since HP(*volume*, 11) is achieved by the word types *SimmodRecord.issue.volume* (for *volume*) and *SimmodRecord.issue.volume* (for *11*), these word types are chosen as the word types of *volume* and *11* (line 12), and HP($volume, 11$) is added into $r$. Since there are still keywords that are not grouped, lines 2 to 12 will execute again. This time, we compute HP(($volume, 11$), $article$) and HP($article, Karen$), and since HP($article, Karen$) is larger and it is achieved by the word types $SigmodRecord.issue.articles.article$ and $SigmodRecord.issue.articles.article.authors.author$. Thus we group $article$ and $Karen$ together and set their word types to the aforementioned word types. We add HP($article, Karen$) to $r$. Now every keyword is in a group (and thus has a word type chosen), the outer loop (lines 2 to 12) stops. Finally we add P($SimmodRecord.issue.volume, SigmodRecord.issue.articles.article$) into $r$ and then sort $r$.

***Time complexity.*** Given a keyword query $K = \{w_1, \cdots, w_n\}$, the worst case time complexity of Algorithm 1 is $O(n^2|WT_1||WT_2|)$, where $WT_1$ and $WT_2$ are the sets of word types of the two keywords which have the most word types. The detailed analysis of time complexity is as follows: Computing the HP score between keywords $w_i$ and $w_{i+1}$ needs $O(|WT_i||WT_{i+1}|)$. Computing the HP score between a keyword $w_i$ and a group $g$ needs $O(|WT_i||g|)$ (note that $|g| < n$). The loop (lines 2 to 12) runs at most $n$ times, and each time, the HP score will be computed for at most $n - 1$ pairs of keywords (and/or groups). The function existUngroupedKeyword() is also in $O(n)$. Computing the HP values between neighboring groups takes $O(n)$, and sorting $r$ takes $O(n^2)$. Thus the algorithm takes $O(n^2|WT_1||WT_2|)$.

**Algorithm for Inferring a Set of Likely Search Intentions.** The algorithm for inferring the set of likely search intentions is rather simple. It simply calls PCS repeatedly, each time it chooses one keyword $w_i$ and uses one word type as the word type set of $w_i$. The detailed steps are omitted in order to save space.

# 5    Generating Results

In this section, we explain how to retrieve result subtrees for a set of likely search intentions. Before defining result subtree, we need to define entity types of a node type.

**Definition 8. (Entity-type of a Node Type)** *If a node type $T$ is the node type of some entity node, its entity-type is itself; otherwise, its entity-type is its ancestor node type $T'$ such that (1) $T'$ is the node type of some entity node, and (2) $T'$ is the longest among all ancestor node types of $T$ satisfying condition (1).*

Note that every node type in the data tree owns one and only one entity-type. For example, in Fig. 1, the entity-type of node type $initPage$ is the node type $article$. Node type $article$'s entity-type is itself.

The result subtree is defined as follows.

**Definition 9. (Result Subtree)** *Given a keyword query $K = \{w_1, \cdots, w_n\}$ and a search intention $\{wt_1, \cdots, wt_n\}$, a subtree of $t$ is a result subtree iff: (1) its root has the node type ENtLCA($wt_1, \cdots, wt_n$), (2) it contains all of the keywords in $K$, and at least one of the occurrences of keyword $w_i$ has the word type $wt_i$.*

---

**Algorithm 2.** GenerateResults($L$)

---

**Input:** A set of likely search intentions $L$
**Output:** result subtrees

1: **for** each $l$ in $L$ **do**
2:     $rt = GetNodeTypeofRoot(l)$ // $rt$: node type of the root
3:     $rl = RetrieveInvertedList(root)$
4:     **for** $1 \leq i \leq |l|$ **do**
5:         $IL_i = RetrieveInvertedList(l_i)$
6:     **while** $rl.isEnd() = false$ **do**
7:         **for** $1 \leq i \leq |l|$ **do**
8:             $IL_i.Moveto(rl.getCurrent())$
9:             **if** $isAncestor(rl.getCurrent(), IL_i.getCurrent()) == false$ **then**
10:                 **break**
11:         **if** $i > |l|$ **then**
12:             $resultList.insert(rl.getCurrent())$
13:         $rl.movetoNext()$
14: Build result subtrees rooted at the nodes in $resultList$ and exclude irrelevant entities.
15: Return result subtrees to the user

---

The purpose of using ENtLCA instead of NtLCA in the definition above is to make the returned result subtrees more informative and meaningful.

**Excluding Irrelevant Entities.** Sometimes, a retrieved result subtree may contain lots of irrelevant information. Consider the query {issue 16 database}. Suppose our approach is to return the result subtrees rooted at *issue* nodes. However, if we return the whole subtrees rooted at *issue* nodes, a number of articles that are not related to database are also returned to the user. Therefore, we should exclude these irrelevant entities. In order to achieve this goal, we first find the entity-type of each keyword's word type, then we know the keywords and their word types that an entity should contain. If an entity does not contain the keywords with the inferred word types, it will be excluded. In the example above, an article entity should contain the keyword "database" with the word type No.5 in Table 1. The entities that do not satisfy this condition will be excluded.

**Algorithm.** The algorithm for generating results is shown in Algorithm 2. We refer the readers to the full version of the paper for a detailed explanation of the algorithm.

## 6   Experiments

In this section, we present the experimental results on the effectiveness of our approach against XReal [2], XBridge [3] and MaxMatch [5]. XReal and XBridge are the most up-to-date XML keyword search system which utilize statistics of data to infer the major search intention of a query. Comparisons of these systems with XSeek can be found in [2] and [3].

### 6.1   Experimental Setup

We implemented XReal, XBridge and our system XInfer in C++. All the experiments were performed on an Intel Pentium-M 1.7G laptop with 1G RAM.

Note that `XBridge` only provides information on how to suggest the promising result types, so we extend `XBridge` with the part of generating result subtrees. Similar to `XReal`, the subtrees that are rooted at the nodes of the suggested result types and contain all of the keywords are considered as the result subtrees. The executable file of `MaxMatch` was kindly provided by its authors. We used the following three data sets that are obtained from [1] for evaluation:

**DBLP**: The structure of this data set is wide and shallow. It has many different types of entities. Many words in this data set have multiple word types.
**SigmodRecord**: This data set has a little more complicated structure than DBLP, but has less entity types. Fig. 1 provides a similar sample.
**WSU**: Similar to DBLP, the structure of this data set is also wide and shallow. However, it has only one entity type and many words have only one word type.

The queries we use for evaluation are listed in Table 2. These queries were chosen by three student users who were given the data sets.

**Table 2.** Queries

| dataset | ID | Query |
|---|---|---|
| DBLP | QD1 | {Automated Software Engineering} |
| | QD2 | {Han data mining} |
| | QD3 | {WISE 2000} |
| | QD4 | {Jeffrey XML} |
| | QD5 | {author Jim Gray} |
| | QD6 | {Relational Database Theory} |
| | QD7 | {article spatial database} |
| | QD8 | {Wise database} |
| | QD9 | {Han VLDB 2000} |
| | QD10 | {twig pattern matching} |
| SigmodRecord | QS1 | {Database Design} |
| | QS2 | {title XML} |
| | QS3 | {article Database} |
| | QS4 | {Karen Ward} |
| | QS5 | {author Karen Ward} |
| | QS6 | {volume 11 database} |
| | QS7 | {issue 11 database} |
| | QS8 | {Anthony 11} |
| | QS9 | {Anthony issue 11} |
| | QS10 | {issue 21 article semantics author Jennifer} |
| WSU | QW1 | {CAC 101} |
| | QW2 | {title ECON} |
| | QW3 | {instructor MCELDOWNEY} |
| | QW4 | {FINITE MATH} |
| | QW5 | {prefix MATH} |
| | QW6 | {place TODD} |
| | QW7 | {COST ACCT enrolled} |
| | QW8 | {CELL BIOLOGY times} |
| | QW9 | {ECON days times place} |
| | QW10 | {prefix ACCTG instructor credit} |

## 6.2 Effectiveness

We conducted a user survey on the search intentions of the queries in Table 2. 19 graduate students participated in the survey. We used the search intentions selected by the majority of people to determine the relevant matches. We evaluate

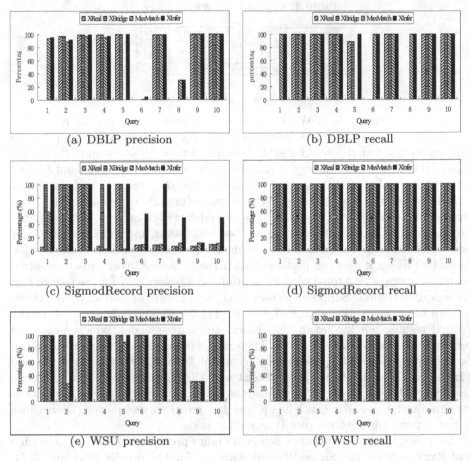

**Fig. 2.** Precision and recall

the effectiveness of XReal, XBridge, MaxMatch and XInfer based on *precision*, *recall* and *F-measure*. Precision is the percentage of retrieved results that are desired by users. Recall is the percentage of relevant results that can be retrieved. F-measure is the weighted harmonic mean of precision and recall.

As shown in Fig. 2(a), XReal and XBridge have a very low precision on the queries QD1, QD6 and QD8. This is mainly because XReal and XBridge infer undesired Search-for Node Types (SNTs). For these queries, XReal and XBridge infer the same SNTs, even though they use different strategies. Suppose the user submits QD1 to retrieve the articles from the journal of Automated Software Engineering. XReal and XBridge just return the *inproceedings* that are related to the automated software engineering, but both MaxMatch and XInfer return the articles from the journal of Automated Software Engineering besides the inproceedings about automated software engineering. Suppose the user submits the query QD8 to retrieve the publications written by Wise. XReal and XBridge

**Table 3.** Comparison on F-Measure

| F-Measure | XReal | XBridge | MaxMatch | XInfer |
|---|---|---|---|---|
| DBLP | 0.69 | 0.69 | 0.79 | 0.90 |
| SigmodRecord | 0.51 | 0.65 | 0.48 | 0.88 |
| WSU | 0.96 | 0.96 | 0.91 | 0.96 |

return the *inproceedings* of WISE conference. `MaxMatch` and `XInfer` return the inproceedings of WISE conference as well as Wise's publications. Suppose the user wants to search the book called Relational Database Theory and submits the query QD6. `XReal` and `XBridge` does not return this book. `MaxMatch` and `XInfer` return this book but have low precisions because they return much irrelevant information at the same time (e.g., the inproceedings about relational database theory, etc). On Query QD2-QD4, `XReal` and `XBridge` achieve a little higher precisions than `MaxMatch` and `XInfer` because `XReal` and `XBridge` just infer one search-for node type which reduces the irrelevant information in the results. Actually this is not a serious problem for `XInfer` because `XInfer` rank the desired search intention as the top-1 search intention, so it is very easy for the user to find their desired results. Suppose the user wants to retrieve the publications written by Jim Gray and submits query QD5. `XReal`, `XBridge` and `XInfer` return the desired results, but `MaxMatch` just returns the author nodes, which means the returned information is too limited. Therefore, `MaxMatch` has a very low precision on this query. Fig. 2(b) presents the recalls of `XReal`, `XBridge`, `MaxMatch` and `XInfer` on the query QD1-QD10. `XReal` and `XBridge` have very low recalls on the query QD1, QD6 and QD8 because they do not return the relevant results as we explained above. `MaxMatch` has a very low recall on the query QD5 because it just returns the subtrees rooted at *author* nodes.

As shown in Fig. 2(c), `XInfer` achieves higher precision than `XReal`, `XBridge` and `MaxMatch` for the SigmodRecord dataset. For the queries QS1 and QS4, `XReal` shows low precision mainly because it infers *issue* as the search-for node type for these two queries, which results in many irrelevant articles being returned to the user. For example, according to the survey, the user intends to retrieve articles about database design with the query QS1. However, lots of articles that are not related to database design are also returned to the user in `XReal`. `XBridge` and `XInfer` returns the articles about database design for QS1 and the articles written by Karen Ward for QS4, which are desired by the user. `XInfer`, `XReal` and `MaxMatch` achieve good precisions on the queries QS2 and QS3. `MaxMatch` gets very low precision on the queries QS4 and QS5. Most participants think the user wants the articles written by Karen Ward with the queries QS4 or QS5, but `MaxMatch` only returns *author* nodes, which do not provide much information desired by the user. In order to retrieve the articles about database from the issues of volume 11, the user submits the query QS6 or QS7. `XReal` and `XBridge` return lots of irrelevant articles (including the articles that are not about database, the database articles whose initPage is 11, etc) to the user. `XInfer` infers five search intentions on QS6 and one search intention on QS7. Compared with QS6, QS7 adds a new keyword "issue" which is used to specify the meaning of "11". XInfer notices this difference, and correctly infers the user's search intention. For the recall, as shown in Fig. 2(d), all of these three approaches present good recalls.

For data set WSU, Fig. 2(e) shows that all of these approaches generally achieve good precision. This is mainly because WSU has a simple and shallow structure compared with the data set SigmodRecord. For the query QW9, they present a relatively low precision. The user intends to retrieve the days, times and place of the courses whose titles contain "ECON", but the systems return the courses whose prefix contain "ECON" as well and give them the highest ranks because most words "ECON" appear in the *prefix* nodes. MaxMatch has a low precision on query QW2 because it also returns the courses whose prefix contain "ECON" even though the user adds a describing word "title". As shown in Fig. 2(f), all of the four approaches present good recalls.

We calculated the average F-measure of the queries over each data set and they are listed in Table 3. It can be seen that XInfer achieves higher F-measure than XReal, XBridge and MaxMatch over all three datasets.

**Other Metrics.** More experimental results are provided in the full version of this paper. These include the categorized efficiency, scalability of efficiency, index structures and index size, the number of returned likely search intentions, ranking effectiveness, and the effect of individual factors (distance, statistics) on the search quality.

## 7  Conclusion

In this paper, we presented a method to improve the effectiveness of XML keyword search by exploiting the relationship between different keywords in a query. We proposed the Pair-wise Comparison Strategy to infer and rank a set of likely search intentions. We developed an XML keyword search system called XInfer which realizes the techniques we propose. The better search quality of XInfer was verified by our experiments.

## References

1. http://www.cs.washington.edu/research/xmldatasets
2. Bao, Z., Ling, T.W., Chen, B., Lu, J.: Effective XML keyword search with relevance oriented ranking. In: ICDE, pp. 517–528 (2009)
3. Li, J., Liu, C., Zhou, R., Wang, W.: Suggestion of promising result types for XML keyword search. In: EDBT, pp. 561–572 (2010)
4. Liu, Z., Chen, Y.: Identifying meaningful return information for XML keyword search. In: SIGMOD Conference, pp. 329–340 (2007)
5. Liu, Z., Chen, Y.: Reasoning and identifying relevant matches for XML keyword search. PVLDB 1(1), 921–932 (2008)
6. Liu, Z., Chen, Y.: Processing keyword search on XML: a survey. World Wide Web 14(5-6), 671–707 (2011)

# A Framework for Grouping and Summarizing Keyword Search Results

Orestis Gkorgkas[1], Kostas Stefanidis[2], and Kjetil Nørvåg[1]

[1] Department of Computer and Information Science, Norwegian University of Science and Technology, Trondheim, Norway
{orestis,Kjetil.Norvag}@idi.ntnu.no
[2] Institute of Computer Science, FORTH, Heraklion, Greece
kstef@ics.forth.gr

**Abstract.** With the rapid growth of the Web, keyword-based searches become extremely ambiguous. To guide users to identify the results of their interest, in this paper, we consider an alternative way for presenting the results of a keyword search. In particular, we propose a framework for organizing the results into groups that contain results with similar content and refer to similar temporal characteristics. Moreover, we provide summaries of results as hints for query refinement. A summary of a result set is expressed as a set of popular keywords in the result set. Finally, we report evaluation results of the effectiveness of our approach.

## 1 Introduction

Keyword-based search is extremely popular as a means for exploring information of interest without using complicated queries or being aware of the underlying structure of the data. Existing approaches for keyword search in relational databases use either the database schema (e.g., [1,12]) or the given database instance (e.g., [5]) to retrieve tuples containing the keywords of a posed query. For example, consider the movie database instance depicted in Fig. 1. For the keyword query $Q = \{comedy,\ J.\ Davis\}$, the results are the *comedy* movies *Deconstructing Harry* and *Celebrity* both with *J. Davis*.

Given the huge volume of available data, keyword-based searches typically return overwhelming number of results. However, users would like to locate only the most relevant results to their information needs. Previous approaches mostly focus on ranking the results of keyword queries to help users retrieve a small piece of them. Such approaches include, among others, adapting IR-style document relevance ranking strategies (e.g., [11]) and exploiting the link structure of the database (e.g., [5]). Still, this flat ranked list of data items could not make it easy for the users to explore and discover important items relevant to their needs.

In this paper, we consider an alternative presentation of the results of the queries expressed through sets of keywords. In particular, we add some structure to the ranked lists of query results. Our goal is to minimize the browsing effort of the users when posing queries, help users receive a broader view of the query results and, possibly, learn about data items that they are not aware of.

B. Catania, G. Guerrini, and J. Pokorný (Eds.): ADBIS 2013, LNCS 8133, pp. 246–259, 2013.

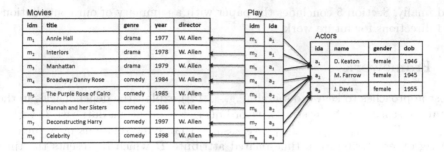

**Fig. 1.** Database instance

Towards this direction, we organize the keyword query results into groups, trying to have groups that exhibit internal cohesion and external isolation. This way, it is easier for the users to scan the results of their queries. Our primary focus is on producing informative, expressive and meaningful groups containing results with similar content that refer to similar temporal characteristics. For example, assume the database instance of Fig. 1 and the keyword query $Q = \{W.\ Allen,\ female\}$. Intuitively, for this query, we can construct three groups of results; the first group refers to the movies *Annie Hall*, *Interiors* and *Manhattan*, the second group refers to the movies *Broadway Danny Rose*, *The Purple Rose of Cairo* and *Hannah and her Sisters* and the third one to the movies *Deconstructing Harry* and *Celebrity*. Each group contains movies with the same actress (*content similarity*) that are produced at the same time period (*temporal similarity*).

To help users refine their queries, we provide them with summaries over the groups of their queries results. The summary of a group presents the most important, in terms of popularity, keywords associated with the specific group of results. Abstractly speaking, for the above constructed groups, we may have the summaries {*drama, D. Keaton*}, {*comedy, M. Farrow*} and {*comedy, J. Davis*}.

Finally, we evaluate the effectiveness of our approach. Our results indicate that users are more satisfied when grouping and summarizing of results are used.

In a nutshell, this paper makes the following contributions:

– It introduces a framework that offers a different way for presenting the results of keyword-based searches.
– It exploits the content of results along with their temporal characteristics to produce groups of results with similar content referring to the same time periods. Summaries for the groups of results are presented to users as hints for query refinement.
– It presents the results of a user study comparing our framework to a standard keyword search technique.

The rest of the paper is organized as follows. In Section 2, we introduce our framework for grouping and summarizing the results of keyword-based searches. In Section 3, we present our evaluation findings. Section 4 describes related work and finally, Section 5 concludes the paper with a summary of our contributions and directions for future work.

## 2 Framework

Most approaches to keyword search (e.g., [1,12]) exploit the dependencies in the database schema for answering keyword queries. Consider a database $\mathcal{D}$ with $n$ relations $\mathcal{R} = \{R_1, R_2, \ldots, R_n\}$. We assume that some relations in $\mathcal{R}$ include, among other attributes, a time-related attribute $B$ which represents the time that the entity described by the tuple was created. The *schema graph* $\mathcal{G}_D$ of a database $\mathcal{D}$ is a directed graph capturing the foreign key relationships in the schema. $\mathcal{G}_D$ has one node for each relation $R_i$ and an edge $R_i \to R_j$, if and only if, $R_i$ has a set of foreign key attributes referring to the primary key attributes of $R_j$. We refer to the undirected version of the schema graph as $\mathcal{G}_U$.

Let $W$ be the potentially infinite set of all keywords. A keyword query $Q$ consists of a set of keywords, i.e., $Q \subseteq W$. Typically, the result of a keyword query is defined with regards to *joining trees of tuples* (JTTs), which are trees of tuples connected through primary to foreign key dependencies [1,5,12].

Our goal in this paper is twofold; first, we focus on organizing into groups the results of a keyword query based on their content similarity and the similarity on the values of their time-related attributes and then, we highlight the important keywords in the produced groups of results.

We start this section with a short introduction to keyword search and then present our approach for organizing the keyword query results in time-dependent groups. Finally, we describe our method for offering the important keywords in the constructed groups.

### 2.1 Keyword Search

This section gives some preliminaries on keyword search, starting by defining the brick of a keyword query result, i.e., the joining tree of tuples.

**Definition 1 (Joining Tree of Tuples (JTT)).** *Given an undirected schema graph* $\mathcal{G}_U$, *a joining tree of tuples (JTT) is a tree of tuples* $T$, *such that, for each pair of adjacent tuples* $t_i$, $t_j$ *in* $T$, $t_i \in R_i$, $t_j \in R_j$, *there is an edge* $(R_i, R_j) \in \mathcal{G}_U$ *and it holds that* $(t_i \bowtie t_j) \in (R_i \bowtie R_j)$.

For example, $(m_7, Deconstructing\ Harry,\ comedy,\ 1997,\ W.\ Allen) - (m_7, a_3) - (a_3, J.\ Davis,\ female,\ 1955)$ represents a JTT for the keyword query $Q = \{comedy,\ J.\ Davis\}$. The size of a JTT is equal to the number of its tuples. In this case the aforementioned JTT has a size equal to 3.

*Total JTT*: A JTT $T$ is *total* for a keyword query $Q$, if and only if, every keyword of $Q$ is contained in at least one tuple of $T$.

*Minimal JTT*: A JTT $T$ that is total for a keyword query $Q$ is also *minimal* for $Q$, if and only if, we cannot remove a tuple from $T$ and get a total JTT for $Q$.

We can now define the result of a keyword query as follows:

**Definition 2 (Query Result).** *Given a keyword query $Q$, the result $Res(Q)$ of $Q$ is the set of all JTTs that are both total and minimal for $Q$.*

We use our movies example (Fig. 1) to briefly describe basic ideas of existing keyword query processing. For instance, the query $Q = \{comedy, J.\ Davis\}$ with result $Q = \{comedy, J.\ Davis\}$ consists of the JTTs: (i) $(m_7, Deconstructing\ Harry, comedy, 1997, W.\ Allen) - (m_7, a_3) - (a_3, J.\ Davis, female, 1955)$ and (ii) $(m_8, Celebrity, comedy, 1998, W.\ Allen) - (m_8, a_3) - (a_3, J.\ Davis, female, 1955)$. Each JTT in the result corresponds to a tree at schema level. That is, both of the above trees correspond to the schema level tree $Movies^{\{comedy\}} - Play^{\{\}} - Actors^{\{J.Davis\}}$, where each $R_i^X$ consists of the tuples of $R_i$ that contain all keywords of $X$ and no other keyword of $Q$. Such sets are called *tuple sets* and the schema level trees are called *joining trees of tuple sets* (JTSs).

Several algorithms in the research literature aim at constructing such trees of tuple sets for a query $Q$ as an intermediate step of the computation of the final results (e.g. [1,12]). We adopt the approach of [12] in which all JTSs with size up to $l$ are constructed. In particular, given a query $Q$, all possible tuple sets $R_i^X$ are computed, where $R_i^X = \{t \mid t \in R_i \wedge \forall a_x \in X, t \text{ contains } a_x \wedge \forall a_y \in Q \backslash X, t \text{ does not contain } a_y\}$. After selecting a random query keyword $a_z$, all tuple sets $R_i^X$ for which $a_z \in X$ are located. These are the initial JTSs with only one node. Then, these trees are expanded either by adding a tuple set that contains at least another query keyword or a tuple set for which $X = \{\}$ (free tuple set). These trees can be further expanded. JTSs that contain all query keywords are returned, while JTSs of the form $R_i^X - R_j^{\{\}} - R_i^Y$, where an edge $R_j \rightarrow R_i$ exists in the schema graph, are pruned, since JTTs produced by them have more than one occurrence of the same tuple for every instance of the database.

## 2.2   Keyword Search Result Vector Representation

Our effort focuses on grouping results based on their content and some temporal information associated with them. Regarding the content of a JTT, we may think of a JTT as the equivalent of a "document". Then, the textual content of a JTT can be represented by a term-vector. For a query $Q$ with result $Res(Q)$, let $\mathcal{A}$ be the set of keywords appearing in the JTTs of $Res(Q)$. The importance score $x_{i,j}$ of a keyword $a_i$ in $\mathcal{A}$ for the JTT $T_j$ of $Res(Q)$, is defined with respect to the TF-IDF model [7]. Specifically, for each $a_i$ in $\mathcal{A}$ for $T_j$, $x_{i,j}$ is equal to: $x_{i,j} = tf_{i,j} * log(N/df_i)$, where $tf_{i,j}$ is the number of occurrences of $a_i$ in the JTT $T_j$ and $df_i$ is the number of tuples in $\mathcal{D}$ that contain $a_i$. $N$ is the maximum $df$ in the database. Then, a JTT-vector for a specific JTT is:

**Definition 3 (JTT-vector).** *Let $Q$ be a keyword query with query result $Res(Q)$ and $\mathcal{A}$ be the set of keywords appearing in the JTTs of $Res(Q)$. The JTT-vector of a JTT $T_j$ in $Res(Q)$ is a vector $u_{T_j} = \{(a_1, x_{1,j}), \ldots, (a_m, x_{m,j})\}$, where $a_i \in \mathcal{A}$, $|\mathcal{A}| = m$, and $x_{i,j}$ is the importance score of $a_i$ for $T_j$, $1 \le i \le m$.*

Many times, two JTTs may contain very similar information. Next, we will exploit similarities between JTTs in order to construct groups of similar results.

## 2.3   Finding Groups of Keyword Search Results

In this work, we consider that each database relation includes in its schema a time-related attribute $B^1$. Then, for a tuple $t_i$ of a relation $R_j$, $1 \leq j \leq n$, we refer to the value $t_i[R_j.B]$ as the *age* of $u$. Naturally time-related attributes of the database relations may vary. For instance, for a relation with *movies* consider the *production year* as a time-related attribute or for a relation with *actors* the *date of birth*. For two tuples $t_i$, $t_x$ of the relations $R_j$, $R_y$, we say that $t_i$ is more recent than $t_x$, if and only if, $t_i[R_j.B] > t_x[R_y.B]$, $1 \leq j, y \leq n$.

Given a joining tree of tuples $T$, we define its *age* with respect to the age of the tuples appearing in the tree. In particular, the age of $T$ is determined by the age of the most recent of its tuples. The motivation behind this, is that before the existence of the entity described in the most recent tuple the tree did not exist. For example, let *Movies.B* be the attribute *year* of the relation *Movies* and *Actors.B* be the attribute *dob* of the relation *Actors*. Furthermore, consider that each tuple $t_i$ of the relation *Play* has value $t_i[Play.B] = 0$. Then, the age of the JTT $(m_7, \text{Deconstructing Harry, comedy, 1997, W. Allen}) - (m_7, a_3) - (a_3, \text{J. Davis, female, 1955})$ is 1997 which is the production year of the movie. Formally:

**Definition 4 (Age of JTT).** *Given a JTT $T$ with tuples $t_1 \in R_{j_1}$, ..., $t_p \in R_{j_p}$, $1 \leq j_1, j_p \leq n$, the age of $T$, $age_T$, is:*

$$age_T = \max_{1 \leq i \leq p} \{t_i[R_{j_i}.B]\}$$

Our goal here is to detect groups of JTTs. Each group contains JTTs that: (i) have similar content and (ii) are continuous in time, which means that their *age* values increase. A straightforward way for quantifying the similarity between two JTTs, is to use a cosine-based definition of similarity, which measures the similarity between their corresponding vectors.

**Definition 5 (Cosine JTT Similarity).** *Given two JTTs $T_1$ and $T_2$ with vectors $u_{T_1} = \{(a_1, x_{1,1}), \ldots, (a_m, x_{m,1})\}$ and $u_{T_2} = \{(a_1, x_{1,2}), \ldots, (a_m, x_{m,2})\}$, respectively, the cosine JTT similarity between $T_1$ and $T_2$ is:*

$$sim_c(T_1, T_2) = \frac{u_{T_1} \cdot u_{T_2}}{||u_{T_1}|| ||u_{T_2}||} = \frac{\sum_{i=1}^{m} x_{i,1} \times x_{i,2}}{\sqrt{\sum_{i=1}^{m} (x_{i,1})^2} \times \sqrt{\sum_{i=1}^{m} (x_{i,2})^2}}$$

Given the similarity between JTTs, we focus on the grouping process. A group of JTTs is expressed as a set of JTTs. The JTTs of a group $G_j$ define a time interval described by two time instances $G_j.s$ and $G_j.e$; $G_j.s$ denotes the starting

---

[1] If a relation does not contain time-related data we consider $B = 0$ for all tuples in the relation.

point of the interval and corresponds to the age of the oldest JTT in the group, while $G_j.e$ denotes the ending point of the interval and corresponds to the age of the most recent JTT. For example, for a group $G_j$ consisting of the JTTs (i) $(m_1,$ Annie Hall, drama, 1977, W. Allen) – $(m_1, a_1)$ – $(a_1,$ D. Keaton, female, 1946), (ii) $(m_2,$ Interiors, drama, 1978, W. Allen) – $(m_2, a_1)$ – $(a_1,$ D. Keaton, female, 1946) and (iii) $(m_3,$ Manhattan, drama, 1979, W. Allen) – $(m_3, a_1)$ – $(a_1,$ D. Keaton, female, 1946), $G_j.s = 1977$ and $G_j.e = 1979$.

Similarly to the JTT-vector, we define the Group-vector which describes the content of a group. In particular, the Group-vector of a group $G_j$ is an aggregation of all vectors of the JTTs belonging to $G_j$. For a query $Q$ with result $Res(Q)$, let $\mathcal{A}$ be the set of keywords appearing in the JTTs of $Res(Q)$ and $G_j$ be a group of JTTs in $Res(Q)$. The importance score $s_{i,j}$ of a keyword $a_i$ in $\mathcal{A}$ for the group $G_j$ is equal to: $s_{i,j} = Aggr_{T_w \in G_j}(x_{i,w})$, where $Aggr$ defines the average, sum, maximum or minimum of the values $x_{i,w}$ of the JTTs of $G_j$.

**Definition 6 (Group-vector).** *Let $G_j$ be a group of JTTs belonging to the query result $Res(Q)$ of a query $Q$ and $\mathcal{A}$ be the set of keywords appearing in the JTTs of $Res(Q)$. The Group-vector of $G_j$ is a vector $u_{G_j} = \{(a_1, s_{1,j}), \ldots, (a_m, s_{m,j})\}$, where $a_i \in \mathcal{A}$, $|\mathcal{A}| = m$, and $s_{i,j}$ is the importance score of $a_i$ for $G_j$, $1 \leq i \leq m$.*

Given the query result $Res(Q)$ of a query $Q$, our aim is to partition the JTTs of $Res(Q)$ into non-overlapping groups. Our definition for non-overlapping groups takes into account both time and content overlaps. Specifically, two groups are: (i) non-overlapping with respect to time, if their time intervals are disjoint and (ii) non-overlapping with respect to content, if they do not contain common JTTs.

**Definition 7 (Non-overlapping Groups).** *Let $G_i$, $G_j$ be two groups of JTTs with time-intervals $[G_i.s, G_i.e]$, $[G_j.s, G_j.e]$. $G_i$, $G_j$ are non-overlapping groups, if and only if: (i) $(G_i.s > G_j.e$ and $G_i.s > G_j.s)$ or $(G_j.s > G_i.e$ and $G_j.s > G_i.s)$, and (ii) $G_i \cap G_j = \emptyset$.*

To partition the joining trees of tuples into non-overlapping groups, we employ a bottom-up hierarchical agglomerative clustering method. Initially, the *JTT Partitioning Algorithm* (Algorithm 1) places each JTT in a cluster of its own. Then, at each iteration, it merges the two most similar clusters. The similarity between two clusters is defined as the minimum similarity between any two JTTs that belong to these clusters (*max linkage*). That is, for two clusters, or groups, $G_1$, $G_2$: $sim(G_1, G_2) = \min_{T_i \in G_1, T_j \in G_2}\{sim_c(T_i, T_j)\}$.

Clearly, two clusters, or groups, $G_1$, $G_2$ can be merged if they are non-overlapping groups. But this is not enough. For constructing groups with JTTs with growing age values there is also a need to ensure that, for the groups $G_1$, $G_2$, there is no other group $G_3$ with time interval between the time intervals of $G_1$ and $G_2$. We refer to such groups as merge-able groups. Formally:

**Definition 8 (Merge-able Groups).** *Let $G_i$, $G_j$ be two groups of JTTs with time-intervals $[G_i.s, G_i.e]$, $[G_j.s, G_j.e]$. $G_i$, $G_j$ are merge-able groups, if and*

---

**Algorithm 1.** JTT Partitioning Algorithm

---

**Input:** A set of JTTs.
**Output:** A set of groups of JTTs.

1: Create a group for each JTT;
2: Repeat
3:    $i = 1$;
4:    Locate the two merge-able groups with the maximum similarity;
5:    **If** there are no merge-able groups or only one group exists**then**
6:        End loop;
7:    **Else**
8:        Merge the two groups;
9:        Compute $K_i$, $C_i$;
10:       $i{+}{+}$;
11: Select the partitioning that constructs $K^*$ groups;

---

only if: (i) $G_i$, $G_j$ are non-overlapping groups, and (ii) $\nexists G_p$ with time interval $[G_p.s, G_p.e]$, such that, the groups $G_i$, $G_p$ and $G_p$, $G_j$ are non-overlapping, and $(G_p.s > G_i.e$ and $G_j.s > G_p.e)$ or $(G_p.s > G_j.e$ and $G_i.s > G_p.e)$

Thus, in overall, we proceed in merging two groups only if the groups are merge-able. The algorithm stops either when a single cluster containing all the JTTs of $Res(Q)$ has already produced or when no more clusters can be merged. As a final step, the algorithm selects to return the clusters of the iteration that present the maximum clustering quality. The clustering quality $C_i$, computed after merging the two clusters of a specific iteration $i$, is:

$$C_i = \sum_{j=1}^{K_i} \sum_{\forall T_p \in G_j} u_{T_p} \cdot u_{G_j} \tag{1}$$

where $K_i$ is the number of clusters after the merging operation of iteration $i$. The selected iteration is the one that constructs $K^*$ clusters, such that:

$$K^* = argmax_i(C_i - \lambda K_i) \tag{2}$$

where $\lambda$ is a penalty for each additional cluster.

Algorithm 1 presents a high level description of the *JTT Partitioning Algorithm*. Although it is possible to pre-specify the number of clusters $K^*$ and directly select to return the clusters of the iteration that produces the $K^*$ ones, we opt for following the above described procedure to ensure high clustering quality, even if the resulting processing cost is high.

We illustrate our approach with the following example. Assume the keyword query $Q = \{ W.\ Allen, female \}$. For the database instance of Fig. 1, the result $Res(Q)$ consists of the JTTs:

(i) $T_1$: ($m_1$, Annie Hall, drama, 1977, W. Allen) - ($m_1$, $a_1$) - ($a_1$, D. Keaton, female, 1946), (ii) $T_2$: ($m_2$, Interiors, drama, 1978, W. Allen) - ($m_2$, $a_1$) - ($a_1$, D. Keaton, female, 1946), (iii) $T_3$: ($m_3$, Manhattan, drama, 1979, W. Allen) - ($m_3$, $a_1$) - ($a_1$, D.

*Keaton, female, 1946*), (iv) $T_4$: ($m_4$, *Broadway Danny Rose, comedy, 1984, W. Allen*) - ($m_4$, $a_2$) - ($a_2$, *M. Farrow, female, 1945*), (v) $T_5$: ($m_5$, *The Purple Rose of Cairo, comedy, 1985, W. Allen*) - ($m_5$, $a_2$) - ($a_2$, *M. Farrow, female, 1945*), (vi) $T_6$: ($m_6$, *Hannah and her Sisters, comedy, 1986, W. Allen*) - ($m_6$, $a_2$) - ($a_2$, *M. Farrow, female, 1945*), (vii) $T_7$: ($m_7$, *Deconstructing Harry, comedy, 1997, W. Allen*) - ($m_7$, $a_3$) - ($a_3$, *J. Davis, female, 1955*) and (viii) $T_8$: ($m_8$, *Celebrity, comedy, 1998, W. Allen*) - ($m_8$, $a_3$) - ($a_3$, *J. Davis, female, 1955*),

with ages 1977, 1978, 1979, 1984, 1985, 1986, 1997 and 1998, respectively. Applying the *JTT Partitioning Algorithm* results in producing three groups $G_1$, $G_2$ and $G_3$ with trees $\{T_1, T_2, T_3\}$, $\{T_4, T_5, T_6\}$ and $\{T_7, T_8\}$.

## 2.4   Summaries of Keyword Query Results

In this section, we describe the notion of group summaries that put in a nutshell the results within groups of keyword searches. In general, group summaries provide hints for query refinement and can lead to discoveries of interesting results that a user may be unaware of.

Let $Res(Q)$ be the query results of a query $Q$ and $\mathcal{A}$ be the set of keywords appearing in the JTTs of $Res(Q)$. Let also $G_1, \ldots, G_z$ be the groups of JTTs produced for $Res(Q)$. Our goal is to compute an importance score $s_{i,j}$ for each keyword $a_i$ in $\mathcal{A}$ for each group $G_j$, $1 \le j \le z$. Then, for each group $G_j$, the top-$k$ keywords, that is, the $k$ keywords with the highest importance scores are used as a summary of the JTTs in $G_j$. Formally:

**Definition 9 (Group Summary).** *Let $G_j$ be a group of JTTs belonging to the query result $Res(Q)$ of a query $Q$ and $\mathcal{A}$ be the set of keywords appearing in the JTTs of $Res(Q)$. The group summary $S_{G_j}$, $S_{G_j} \subseteq \mathcal{A}$, of $G_j$ is a set of $k$ keywords, such that, $s_{i,j} \ge s_{p,j}$, $\forall a_i \in S_{G_j}$, $a_p \in \mathcal{A} \backslash S_{G_j}$.*

For example, for the keyword query $Q = \{W.\ Allen,\ female\}$, the group summaries of the produced groups $G_1$, $G_2$ and $G_3$, for $k = 2$, are $S_{G_1} = \{drama,\ D.\ Keaton\}$, $S_{G_2} = \{comedy,\ M.\ Farrow\}$ and $S_{G_3} = \{comedy,\ J.\ Davis\}$.

To provide users with more detailed summaries that include some information about the schema of the results, we extend the notion of group summaries to take into account the relations that a keyword belongs to. Specifically, for each group $G_j$, instead of reporting the set of the $k$ keywords with the highest importance scores, we report these keywords along with their associated relations. This way, users obtain an overview about the possible origination and meaning of the keywords. We refer to these summaries as enhanced group summaries. Formally:

**Definition 10 (Enhanced Group Summary).** *Let $G_j$ be a group of JTTs and $S_{G_j}$ be the corresponding group summary of $G_j$ with keywords $a_1, \ldots, a_k$. The enhanced group summary $\mathcal{E}_{G_j}$ of $G_j$ is a set of $k$ pairs of the form $(a_i, P_i)$, such that, there is one pair $\forall a_i \in S_{G_j}$ and $P_i$ is the set of relations that contain $a_i$ for the JTTs of $G_j$.*

Returning to our previous example, the enhanced group summaries of $G_1$, $G_2$ and $G_3$ are represented as $\mathcal{E}_{G_1} = \{(drama, \{Movies\}), (D.\ Keaton, \{Actors\})\}$,

$\mathcal{E}_{G_2} = \{(comedy, \{Movies\}), (M.\ Farrow, \{Actors\})\}$ and $\mathcal{E}_{G_3} = \{(comedy, \{Movies\}), (J.\ Davis, \{Actors\})\}$, respectively.

Based on the summaries of the produced groups of results, we define the summary of the query result as a whole, as follows:

**Definition 11 (Query Result Summary).** *Let $G_1, \ldots, G_z$ be the groups of JTTs produced for the query result $Res(Q)$ of a query $Q$. The query result summary $\mathcal{S}_Q$ is a set of z group summaries, $\mathcal{S}_Q = \{\mathcal{S}_{G_1}, \ldots, \mathcal{S}_{G_z}\}$, such that, $\mathcal{S}_{G_j}$ is either the group summary $\mathcal{S}_{G_j}$ or the enhanced group summary $\mathcal{E}_{G_j}$ of $G_j$, $1 \le j \le z$.*

That is, for $Q = \{W.\ Allen,\ female\}$, the query result summary $\mathcal{S}_Q$ taking into account the group summaries is $\{\{drama, D.\ Keaton\}, \{comedy, M.\ Farrow\}, \{comedy, J.\ Davis\}\}$, while for the enhanced group summaries we have the summary $\{(drama, \{Movies\}), (D.\ Keaton, \{Actors\}), (comedy, \{Movies\}), (M.\ Farrow, \{Actors\}), (comedy, \{Movies\}), (J.\ Davis, \{Actors\})\}$.

We could also consider other versions for summaries. For instance, assume that the importance of each keyword is computed separately for each relation. Then, we may report important keywords with respect to their relation-specific scores or keywords for relations of high user interest.

**Summary-Based Exploratory Keyword Queries.** Besides presenting summaries to the users and offering, this way, a side mean for further exploration, we also plan to use the summaries to directly discover interesting pieces of data that are potentially related to the users' information needs. Specifically, to locate such related information, special-purpose queries, called *summary-based exploratory keyword queries*, can be constructed. The focus of these queries is on retrieving results highly correlated with the results of the original users queries.

Our plan is to employ the keywords of summaries to emerge new interesting results. An exploratory keyword query for a query $Q$ will consist of a set of keywords, that is a subset of the keywords in a group summary $G_j$ of $Q$, that frequently appear together in the JTTs of $G_j$. There are also other ways for constructing exploratory queries that qualify different properties. For example, sets of keywords that frequently appear in the result and, at the same time, rarely appear in the database ensure high surprise, or unexpectedness, as a measure of interestingness, as surprise used in the data mining literature (e.g., [17]). Recently, exploratory queries are used for exploration in relational databases through recommendations [8].

## 3   Evaluation

To demonstrate the effectiveness of grouping and summarizing keyword search results, we conducted an empirical evaluation of our approach using a real movie dataset[2] with 30 volunteers with a moderate interest in movies. The schema of our database is shown in Fig. 2 while the size of the database is 1.1 GB.

---

[2] http://www.imdb.com/interfaces

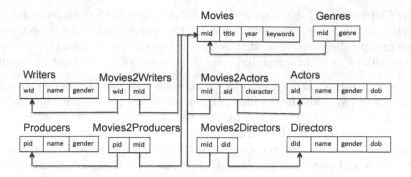

**Fig. 2.** Movies database schema

We run our experiments for queries of different sizes, i.e., number of keywords, and keywords of a different selectivity. We presented the results to the participants using two methods, 1) without any grouping (baseline method) and 2) with groups produced by our approach (grouped method). In the baseline method, for each query, we presented an enhanced group summary of the whole result-set, considering the whole result-set as one group. To help the users understand the context of the significant terms we presented them also the attribute value in which a significant term appeared in. We give also the participants the ability to examine the set of produced JTTs. The results, i.e., the JTTs, are ranked based on their size that corresponds to the relevance of the trees to the query. In the grouped method, the participants are initially presented the groups of JTTs which were formed on the same results that were presented in the baseline. The groups are indicated to the participants by the time period each group covers. When a participant focuses on period he/she is provided with the summary of the group's content and the results belonging to that group.

The participants were asked to evaluate the quality of the results. For characterizing the quality, we use four measures: (i) *group coherence*, which evaluates the similarity of the results content inside a group , (ii) *baseline summary quality*, which evaluates how descriptive is the summary of the baseline method for the whole result set, (iii) *group summary quality*, which evaluates how descriptive is the summary of each group, and (iv) *usefulness evaluation*, that evaluates if the participant found the grouping method more helpful than the baseline method.

For grading the grouping, the participants were asked to evaluate for each group if the movies in the group fit well together. We used 3 values: not coherent (0), quite coherent (1), and very coherent (2). The users were also asked for each summary if it was descriptive of the result and if it was helpful for them to understand the content of the results. The summaries were also graded using three values: not descriptive (0), quite descriptive (1), and very descriptive (2). The degree of overall usefulness was graded with two values: our method is not helpful (0), and our method is helpful (1). Each query was evaluated by at least 3 participants while 95% of the queries were evaluated by at least 4 and 75% by at least 8. On average there were 8 evaluators per query.

**Group Coherence.** Table 1 shows the average values of the coherence measure for each query as they were estimated by the participants. According to the average group coherence value, that is 1.52, the participants found the grouping of the results to be meaningful and helpful for them to understand the results.

**Summary Quality.** Table 2 reports the average values of the quality measures for each query (we omit the detailed per person scores due to space limitations). As it can be seen, in 90% of the queries the quality of group summaries was better than (or equal to) the quality of the baseline summary according to the participants. This comes to complete accordance with the percent of participants (85%) who found our approach helpful. We can also draw the conclusion that while in all queries the majority of participants found the grouped summaries to be quite or very descriptive, the baseline summary was evaluated as quite or very descriptive in only 30% of the queries.

**Time Overhead.** Finally, we study the overall impact of grouping and summarizing keyword search results in terms of time overhead for the above query examples. In particular, we measured the time needed to build the JTTs and the time needed for creating the groups and summaries. The additional computational cost of our approach is small in comparison with the generation of the actual keyword search results. On average, the additional time consumed for creating the summaries and groups was a magnitude smaller than the JTT building time, and in no case was the creation of groups and summaries significantly more expensive in terms of time than the building of the JTTs. For example, the time overhead for the query {Stanley Kubrick, movies} is 3.7% and for the query {Francis Ford Coppola, male actors} is 9.5%.

**Table 1.** Group coherence evaluation for each query

| Query | Average group coherence |
|---|---|
| "Daniel Craig" movies | 1.50 |
| "James Bond" movies | 1.50 |
| "James Bond" male actors | 1.50 |
| "Woody Allen" female actors | 1.27 |
| "Clint Eastwood" movies | 1.50 |
| "Peter Jackson" male actors | 1.75 |
| "Peter Jackson" movies | 1.33 |
| "Denzel Washington" Action | 1.88 |
| "Julia Roberts" Comedy | 1.71 |
| "Julia Roberts" movies | 1.75 |
| "Kevin Spacey" drama | 1.27 |
| "Jack Nicholson" female actors | 1.44 |
| "Al Pacino" movies | 1.45 |
| "Al Pacino" male actors | 1.50 |
| "Al Pacino" directors | 1.50 |
| "Stanley Kubrick" actors | 1.75 |
| "Stanley Kubrick" movies | 1.60 |
| "Lord of The Rings" Tolkien | 1.30 |
| "Robert De Niro" directors | 1.60 |
| "Francis Ford Coppola" male actors | 1.75 |

**Table 2.** Summary quality evaluation for each query

| Query | Baseline summary quality | Group summary quality | Usefulness |
|---|---|---|---|
| "Daniel Craig" movies | 0.75 | 1.50 | 0.75 |
| "James Bond" movies | 0.70 | 1.60 | 0.90 |
| "James Bond" male actors | 1.00 | 1.50 | 0.75 |
| "Woody Allen" female actors | 0.82 | 1.45 | 0.82 |
| "Clint Eastwood" movies | 1.36 | 1.57 | 0.86 |
| "Peter Jackson" male actors | 0.50 | 1.75 | 1.00 |
| "Peter Jackson" movies | 1.67 | 1.67 | 0.67 |
| "Denzel Washington" Action | 1.13 | 1.88 | 1.00 |
| "Julia Roberts" Comedy | 0.88 | 1.43 | 0.86 |
| "Julia Roberts" movies | 1.13 | 1.75 | 0.88 |
| "Kevin Spacey" drama | 0.64 | 1.28 | 0.82 |
| "Jack Nicholson" female actors | 0.22 | 1.56 | 0.89 |
| "Al Pacino" movies | 1.00 | 1.45 | 0.82 |
| "Al Pacino" male actors | 0.50 | 1.63 | 0.88 |
| "Al Pacino" directors | 0.50 | 1.50 | 1.00 |
| "Stanley Kubrick" actors | 1.00 | 2.00 | 1.00 |
| "Stanley Kubrick" movies | 1.50 | 1.30 | 0.80 |
| "Lord of The Rings" Tolkien | 1.36 | 1.20 | 0.60 |
| "Robert De Niro" directors | 0.30 | 1.60 | 0.90 |
| "Francis Ford Coppola" male actors | 1.00 | 2.00 | 1.00 |

## 4   Related Work

Keyword search in relational databases has been the focus of much current research. Schema-based approaches (e.g., [1,12]) use the schema graph to generate join expressions and evaluate them to produce tuple trees. Instance-based approaches (e.g., [5]) represent the database as a graph in which there is a node for each tuple. Results are provided directly by using a Steiner tree algorithm. Based on [5], several more complex approaches have been proposed (e.g., [10,13]). There have also been proposals for providing ranked keyword retrieval, which include incorporating IR-style relevance ranking [11], authority-based ranking [3], automated ranking based on workload and data statistics of query answers [6] and preference-based ranking [18,20].

Our approach is different, in that, we propose grouping keyword search results to help users receive the general picture of the results of their queries. A comparison between a flat ranked list of results and a clustering web search interface shows that the users of the clustering approach view more documents and spend less time per document [21]. However, the relevance of the viewed documents is unknown. [15] presents an approach for clustering keyword search results based on common structure patterns without taking into account the aspect of time. Recently, [2] introduces a prototype and framework for interactive clustering of query results. This technique is applied in document collections, while our work focuses on structured data.

Summaries of keyword queries results resemble the notion of *tag clouds*. A tag cloud is a visual representation for text data. Tags are usually single words, alphabetically listed and in different font size and color to show their importance[3]. Tag clouds have appeared on several Web sites, such as Flickr and del.icio.us. With regard to our approach for summaries, *data clouds* [14] are the most relevant. This work proposes algorithms that try to discover good, not necessarily popular, keywords within the query results. Our approach follows a pure IR technique to locate important, in terms of popularity, keywords. From a database perspective, [9] introduces the notion of *object summary* for summarizing the data in a relational database about a particular *data subject*, or keyword. An object summary is a tree with a tuple containing the keyword as the root node and its neighboring tuples containing additional information as child nodes.

Finally, our work presents some similarities with faceted search (e.g., [4,16]). Faceted search is an exploration technique that provides a form of navigational search. In particular, users are presented with query results classified into multiple categories and can refine the results by selecting different conditions. Our approach is different in that we do not tackle refinement. [8,19] present a different way for database exploration by recommending to users items that are not part of the results of their query but appear to be highly related to them. Such items are computed based on the most interesting sets of attribute values that appear in the results of the original query. The interestingness of a set is defined based on its frequency in the results and the database.

## 5   Conclusions

In this paper we propose a framework for organizing the keyword search results into groups that contain results with similar content that refer to similar temporal characteristics. We employ summaries of results to help users refine their searches. A summary of a result set is expressed as a set of important attribute values in the result set. In addition we evaluate the effectiveness of our approach. Our usability results indicate that users are more satisfied when results are organized with respect to content and time than when results are simply ordered with respect to relevance.

Clearly, there are many directions for future research. Our plans include studying different functions for computing similarities between JTTs, as well as more efficient clustering algorithms. Finally, in this paper, we compute summaries based on popularity; other properties, such as the dependence on the query, should also be sought for.

**Acknowledgments.** The work of the second author is supported by the project "IdeaGarden" funded by the Seventh Framework Programme under grand $n^o$ 318552.

---

[3] http://en.wikipedia.org/wiki/Tag_cloud

# References

1. Agrawal, S., Chaudhuri, S., Das, G.: DBXplorer: A system for keyword-based search over relational databases. In: Proc. of ICDE, pp. 5–16 (2002)
2. Anastasiu, D.C., Gao, B.J., Buttler, D.: A framework for personalized and collaborative clustering of search results. In: Proc. of CIKM, pp. 573–582 (2011)
3. Balmin, A., Hristidis, V., Papakonstantinou, Y.: Objectrank: Authority-based keyword search in databases. In: Proc. of VLDB, pp. 564–575 (2004)
4. Ben-Yitzhak, O., Golbandi, N., Har'El, N., Lempel, R., Neumann, A., Ofek-Koifman, S., Sheinwald, D., Shekita, E.J., Sznajder, B., Yogev, S., Yogev, S.: Beyond basic faceted search. In: Proc. of WSDM, pp. 33–44 (2008)
5. Bhalotia, G., Hulgeri, A., Nakhe, C., Chakrabarti, S., Sudarshan, S.: Keyword searching and browsing in databases using banks. In: Proc. of ICDE, pp. 431–440 (2002)
6. Chaudhuri, S., Das, G., Hristidis, V., Weikum, G.: Probabilistic information retrieval approach for ranking of database query results. ACM Trans. Database Syst. 31(3), 1134–1168 (2006)
7. Dhillon, I.S., Modha, D.S.: Concept decompositions for large sparse text data using clustering. Machine Learning 42(1/2), 143–175 (2001)
8. Drosou, M., Pitoura, E.: ReDRIVE: result-driven database exploration through recommendations. In: Proc. of CIKM, pp. 1547–1552 (2011)
9. Fakas, G.J.: A novel keyword search paradigm in relational databases: Object summaries. Data Knowl. Eng. 70(2), 208–229 (2011)
10. He, H., Wang, H., Yang, J., Yu, P.S.: BLINKS: ranked keyword searches on graphs. In: Proc. of SIGMOD, pp. 305–316 (2007)
11. Hristidis, V., Gravano, L., Papakonstantinou, Y.: Efficient IR-style keyword search over relational databases. In: Proc. of VLDB, pp. 850–861 (2003)
12. Hristidis, V., Papakonstantinou, Y.: DISCOVER: Keyword search in relational databases. In: Proc. of VLDB, pp. 670–681 (2002)
13. Kacholia, V., Pandit, S., Chakrabarti, S., Sudarshan, S., Desai, R., Karambelkar, H.: Bidirectional expansion for keyword search on graph databases. In: Proc. of VLDB, pp. 505–516 (2005)
14. Koutrika, G., Zadeh, Z.M., Garcia-Molina, H.: Data clouds: summarizing keyword search results over structured data. In: Proc. of EDBT, pp. 391–402 (2009)
15. Peng, Z., Zhang, J., Wang, S., Qin, L.: Treecluster: Clustering results of keyword search over databases. In: Yu, J.X., Kitsuregawa, M., Leong, H.-V. (eds.) WAIM 2006. LNCS, vol. 4016, pp. 385–396. Springer, Heidelberg (2006)
16. Roy, S.B., Wang, H., Das, G., Nambiar, U., Mohania, M.K., Mohania, M.K.: Minimum-effort driven dynamic faceted search in structured databases. In: Proc. of CIKM, pp. 13–22 (2008)
17. Silberschatz, A., Tuzhilin, A., Tuzhilin, A.: On subjective measures of interestingness in knowledge discovery. In: Proc. of KDD, pp. 275–281 (1995)
18. Simitsis, A., Koutrika, G., Ioannidis, Y.E.: Précis: from unstructured keywords as queries to structured databases as answers. VLDB J. 17(1), 117–149 (2008)
19. Stefanidis, K., Drosou, M., Pitoura, E.: You May Also Like results in relational databases. In: Proc. of PersDB, pp. 37–42 (2009)
20. Stefanidis, K., Drosou, M., Pitoura, E.: PerK: personalized keyword search in relational databases through preferences. In: Proc. of EDBT, pp. 585–596 (2010)
21. Zamir, O., Etzioni, O.: Grouper: A dynamic clustering interface to web search results. Computer Networks 31(11-16), 1361–1374 (1999)

# QGramProjector: Q-Gram Projection
# for Indexing Highly-Similar Strings

Sebastian Wandelt and Ulf Leser

Knowledge Management in Bioinformatics,
Institute for Computer Science,
Humboldt-Universität zu Berlin, Germany
{wandelt,leser}@informatik.hu-berlin.de

**Abstract.** Q-gram (or n-gram, k-mer) models are used in many research
areas, e.g. in computational linguistics for statistical natural language
processing, in computer science for approximate string searching, and in
computational biology for sequence analysis and data compression. For a
collection of N strings, one usually creates a separate positional q-gram
index structure for each string, or at least an index structure which needs
roughly N times of storage compared to a single string index structure.
For highly-similar strings, redundancies can be identified, which do not
need to be stored repeatedly; for instance two human genomes have more
than 99 percent similarity.

In this work, we propose QGramProjector, a new way of indexing
many highly-similar strings. In order to remove the redundancies caused
by similarities, our proposal is to 1) create all q-grams for a fixed reference, 2) referentially compress all strings in the collection with respect
to the reference, and then 3) project all q-grams from the reference to
the compressed strings.

Experiments show that a complete index can be relatively small compared to the collection of highly-similar strings. For a collection of 1092
human genomes (raw data size is 3 TB), a 16-gram index structure, which
can be used for instance as a basis for multi-genome read alignment, only
needs 100.5 GB (compression ratio of 31:1). We think that our work is an
important step towards analysis of large sets of highly-similar genomes
on commodity hardware.

**Keywords:** positional q-grams, k-mer, large sequences, similarity,
referential compression.

## 1   Introduction

Indexing and searching large collections of *highly-similar* strings became a hot
and challenging topic during the last years [14], for instance in order to perform
population-scale genome analysis[1]. If one can identify the similarities between
strings in the collection, then the amount of storage for saving/indexing the
strings can be greatly reduced. In general, indexing and searching strings has a
long history in computer science research [15]. The literature on string search in

B. Catania, G. Guerrini, and J. Pokorný (Eds.): ADBIS 2013, LNCS 8133, pp. 260–273, 2013.

general is vast and cannot be summarized here; we refer the reader to several excellent surveys [7,13].

Q-grams can be used for indexing large strings following the the seed-and-extend approach [3,19], e.g. to create a disk-based search index [6]. One well-known example from bioinformatics for the use of q-grams (there often called k-mers or seeds) is BLAST[2], which uses q-grams as anchors for finding longer approximate string matches.

In this work, we introduce QGramProjector, where we focus on finding all positions of a given (single) q-gram in a collection of highly-similar strings, for instance, genomes. Two randomly selected human genomes are app. 99% identical [17]. Instead of indexing all q-grams in all genomes, we propose to only explicitly store all q-grams of one chosen reference genome. Furthermore, we propose to compress all other strings in the collection referentially with respect to the reference genome. An index over the referential compressions is used to project all q-grams in the reference string into the other strings of the collection. Since this projection is not complete, because some q-grams do not occur in the reference, we have an additional (smaller) index keeping track of these derivations. In addition, we show a way to further reduce the necessary storage for our q-gram index: We rewrite the reference genome such that longer matches can be encoded.

The most similar work to ours is an index structure for computation of q-gram occurrence frequencies over straight line programs [9][1]. The index's purpose is to retrieve the total occurrence frequencies of all q-grams. For DNA data the index uses half as much size as the original data, achieving a compression ratio of 2:1. However, our projection-based index structure allows for compression (on human genomes) up to 31:1 for $q = 16$. Another related work, string dictionary lookup in a compressed string dictionary with edit distance one is discussed in [4]. Although the authors report nearly optimal complexity results for their matching algorithm (together with a way to trade-off query answering time for index space), there is no practical evaluation of the algorithm. Another approach concerned with searching larger compressed collection of strings is presented in [10]. However, this work has a focus on theoretical results, with a preliminary evaluation on a very small dataset; the program is not publicly available for evaluation on our big datasets.

The remainder of this paper is structured as follows. We introduce the problem of indexing q-grams and the foundations of referential compression in Section 2. Section 3 describes the algorithm used in QGramProjector in detail. Our algorithms are evaluated in Section 4, and Section 5 concludes the paper.

## 2   Referential Compression and Q-Grams

A *string* $s$ is a finite sequence over an alphabet $\Sigma$. The length of a string $s$ is denoted with $|s|$ and the substring starting at position $i$ with length $n$ is denoted

---

[1] Straight line programs can be seen as a generalization of many dictionary/grammar-based compression formats[18].

$s(i, n)$. $s(i)$ is an abbreviation for $s(i, 1)$. All positions in a string are zero-based, i.e. the first character is accessed by $s(0)$. The concatenation of two strings $s$ and $t$ is denoted with $s \circ t$. A string $t$ is a *prefix of* a string $s$, if $s = t \circ u$, for some string $u$. A *q-gram* of string $s$ is an string of length $q$ over $\Sigma$.

**Definition 1.** *A* positional q-gram *of a string $s$ is a tuple $\langle p, g \rangle$, such that $g = s(p, q)$. The set of all positional q-grams of $s$ is denoted with $PQG_q(s)$. The set of positions for a q-gram $g$ in a string $s$ is defined as $s^g = \{ p \mid \langle p, g \rangle \in PQG_q(s) \}$. Given a set of strings $C = \{ s_1, ..., s_n \}$, we define $C^g = \{ < i, p > \mid s_i \in C \wedge p \in s_i{}^g \}$.*

Given a string $s$, there exist several approaches to compute $s^g$. One approach is to explicitly store a map for all q-grams to positions in $s$. The size of this map data structure (in byte) is usually larger than the size of the input. For instance storing the positions as 4-byte integer values for a string of length $n$ , we have to encode $n - q + 1$ positions, which needs $4 * (n - q + 1)$ bytes. Having $m$ highly-similar strings, the data structure grows roughly by a factor of $m$, since we have to encode $4 * (n_1 - q + 1) + ... + 4 * (n_m - q + 1)$ bytes for the positions, and in addition some kind of ID for each string.

*Example 1.* Given the two strings $s_1 = ACGACT$ and $s_2 = ACGAAT$, we have

$$PQG_2(s_1) = \{ \langle 0, AC \rangle, \langle 1, CG \rangle, \langle 2, GA \rangle, \langle 3, AC \rangle, \langle 4, CT \rangle \}$$
$$PQG_2(s_2) = \{ \langle 0, AC \rangle, \langle 1, CG \rangle, \langle 2, GA \rangle, \langle 3, AA \rangle, \langle 4, AT \rangle \}$$

Storing the q-grams for both strings creates redundancies, since the q-grams for positions $0 - 2$ are the same for $s_1$ and $s_2$. For string $s_3 = GACGACT$, we obtain $PQG_2(s_3) = \{ \langle 0, GA \rangle, \langle 1, AC \rangle, \langle 2, CG \rangle, \langle 3, GA \rangle, \langle 4, AC \rangle, \langle 5, CT \rangle \}$. Although strings $s_1$ and $s_3$ are highly similar as well, there is no obvious redundancy in their positional q-grams. The reason is that all the positional q-grams are shifted one position right for $s_3$, compared to $s_1$.

The above examples show that it could be beneficial to identify similarities between strings to reduce the amount of storage for a positional q-gram index of a collection of strings. In the following, we use referential compression to identify and encode these similarities. Based on referentially compressed strings, we devise an index structure to retrieve positional q-grams from a collection of highly-similar strings. We will show that in some cases our q-gram index structure is orders of magnitude smaller than a conventional index. First, we introduce referential compression as it is used in bioinformatics recently [5,11,20]. Referentially compressing a string means to encode the string as a concatenation of substrings from a given reference string. Since there exists no standard format to represent referentially compressed strings, we define a very general notion for encoding referential matches first. The following is taken from [20].

**Definition 2.** *A* referential match entry *is a triple $\langle start, length, mismatch \rangle$, where start is a number indicating the start of a match within the reference, length denotes the match length, and mismatch denotes a symbol. The length of a referential match entry rme, denoted $|rme|$, is $length + 1$.*

Given a reference $ref$ and a to-be-compressed string $s$, the idea of referential compression is to find a small set of rme's of $s$ with respect to $ref$ which is a) sufficient to reconstruct $s$ and b) as small as possible.

**Definition 3.** *Given strings $s$ and $ref$, a referential compression of $s$ with respect to $ref$, is a list of referential match entries,*

$$\downarrow^{ref}(s) = [\langle start_1, length_1, mismatch_1 \rangle, ..., \langle start_n, length_n, mismatch_n \rangle],$$

*such that*

$$(ref(start_1, length_1) \circ mismatch_1) \circ (ref(start_2, length_2) \circ mismatch_2) \circ ... \circ$$
$$(ref(start_n, length_n) \circ mismatch_n) = s.$$

*Sometimes we use $cs$ instead of $\downarrow^{ref}(s)$, if $s$ and $ref$ are known from the context. The offset of a referential match entry $rme_i$ in a referential compression $\downarrow^{ref}(s) = [rme_1, ..., rme_n]$, denoted $offset(\downarrow^{ref}(s), rme_i)$, is defined as $\sum_{j<i} |rme_j|$. Given a referential match entry $\langle start, length, mismatch \rangle$, we write $(start, length, mismatch) \in \downarrow^{ref}(s)$, if and only if $\langle start, length, mismatch \rangle$ is an element in the referential compression $\downarrow^{ref}(s)$.*

The offset of a referential match entry in a referential compression corresponds to the position of the entry in the uncompressed string. The inverse of a referential compression is the decompression of a referential compression with respect to the reference, such that we obtain the original input string. We assume that the reference sequence contains each symbol from $\Sigma$, if it does not, then we just add all symbold of $\Sigma$ to the end of the reference sequence.

*Example 2.* An example compression for $CGGACAAACTGACGTTCGACG$ with respect to the preselected reference $GACGATCGACGACGGACAAACA$ is as folows. The input is compressed into three referential match entries. The first referential match entry is $\langle 12, 9, T \rangle$, which describes a match for the string $CGGACAAACT$ at position 12 of the reference. The mismatch character is $T$ (in the reference an $A$ is found instead of a $T$). The second referential match entry compresses the string $GACGT$. A referential match entry for the string $GACG$ in the reference at position 10 is introduced, together with a mismatch for symbol $T$. The last referential match entry compresses the string $TCGACG$. Although the string can be completely found in the reference, we only encode the first five symbols as a link to the reference and add $G$ as a mismatch symbol. The offset of referential match entry $\langle 5, 5, G \rangle$ is $|\langle 12, 9, T \rangle| + |\langle 10, 4, T \rangle| = 15$.

Clearly, we require the less rme's, the longer the matches, i.e., the shared substrings, are. Therein it does not matter, at which position of the reference these matches lie; in particular, matches need not be in any particular order. To create a referential compression of input string $s$ with respect to $ref$, our algorithm (Algorithm 1) matches prefixes of $s$ with substrings of $ref$ using a compressed suffix tree on $ref$. The longest such prefix is removed from $s$, encoded as a rme and

---

**Algorithm 1.** Referential Compression Algorithm

---

**Input:** to-be-compressed string $s$ and reference string $ref$
**Output:** referential compression $\downarrow^{ref}(s)$ of $s$ with respect to $ref$
1: Let $\downarrow^{ref}(s)$ be an empty list
2: **while** $|s| \neq 0$ **do**
3:     Let $pre$ be the longest prefix of $s$ occurring in $ref$, and let $i$ be a position
       of an occurrence of $pre$ in $ref$
4:     **if** $s \neq pre$ **then**
5:         Add $\langle i, |pre|, s(|pre|) \rangle$ to the end of $\downarrow^{ref}(s)$
6:         Remove the first $|pre| + 1$ symbols from $s$
7:     **else**
8:         Add $\langle i, |pre| - 1, s(|pre| - 1) \rangle$ to the end of $\downarrow^{ref}(s)$
9:         Remove the prefix $pre$ from $s$
10:    **end if**
11: **end while**

---

added to $\downarrow^{ref}(s)$. The algorithm terminates once $s$ contains no more symbols. Note that a referential compression of a string with respect to a reference is not unique. A simple example for a non-unique referential compression with respect to the reference $ref = ATA$ is $\downarrow^{ref}(AA) = [\langle 0, 1, A \rangle]$ and $\downarrow^{ref}(AA) = [\langle 2, 1, A \rangle]$.

Note that a naive storage of rme's does not yield high compression ratios. We store the parsing in a form of delta-encoding, where the position of a rme is stored as a difference to the most recent rme's plus the length of the most recent rme. In addition, we serialize the length value with huffman encoding (the code tree is obtained by precomputation over sequences from different species).

## 3    Retrieval of Positional Q-Grams

In the following, we first describe a naive way to retrieve all positions of a q-gram in a compressed string $\downarrow^{ref}(s)$. Second, we propose an index structure over a collection of compressed strings. The index structure allows more efficient retrieval of q-gram positions than the naive application (see Section 4). In either case, we do not discuss how to compute $ref^g$, we assume that can be precomputed in an arbitrary way.

### 3.1    Retrieving Q-Grams in One String

In Algorithm 2, we describe a simple algorithm to compute the set of positions of a given q-gram in a string s, which is referentially compressed with respect to $ref$. We call this approach *projection*. Initially, the set $result$ is empty (Line 1). The compressed representation of string $s$, $\downarrow^{ref}(s)$, is traversed from left to right (Line 2-3). For each position of $g$ in the reference string it is checked, whether the match is subsumed by the current referential match entry; if yes, a relative match is added to the result (Line 4-7). For each q-gram which overlaps the mismatch character, the algorithm checks whether the q-gram is equal to $g$ (Line 8-9). If it is, then a relative match with respect to the referential match entry is added to

---

**Algorithm 2.** Projecting q-grams in compressed strings

---

**Input:** q-gram $g$, string $s$, reference string $ref$, $ref^g$
**Output:** $s^g$ stored in result
1: Let $result = \emptyset$
2: Let $curpos = 0$
3: **for all** $\langle start, length, mismatch \rangle \in \downarrow^{ref} (s)$ **do**
4:     Let $refmatches = \{p \mid p \in ref^g \wedge (p \geq start) \wedge (p+|g| \leq start+length)\}$
5:     **for all** $p \in refmatches$ **do**
6:         $result = result \cup \{curpos + (p - start)\}$
7:     **end for**
8:     Let $t = s[curpos + length - (|g| - 1), 2 * |g| - 1]$
9:     **for all** $p \in t^g$ **do**
10:         $result = result \cup \{curpos + length - (|g| - 1) + p)\}$
11:     **end for**
12:     $curpos = curpos + |\langle start, length, mismatch \rangle|$
13: **end for**
14: **return** result

---

the result (Line 10). At the end of the loop, the current position is set to the the beginning of the next referential match entry (Line 12). For Algorithm 2, $ref^g$ can be precomputed (and stored), but $t^g$ is computed at runtime.

*Example 3.* Let the reference be $ref = ACGACTAT$, $s_1 = GACGACTAC$. We obtain $\downarrow^{ref} (s_1) = \{\langle 2, 3, G \rangle, \langle 2, 4, C \rangle\}$. Given $g = AC$, we have $ref^g = \{0, 3\}$. The for loop (Line 3-13) of Algorithm 2 iterates two times. In the first iteration for referential match entry $\langle 2, 3, G \rangle$, we have $refmatches = \{3\}$ and add $\{0 + (3-2)\}$ to $result$. We have $t = CGA$ and $t^g = \emptyset$, and thus no additional matches are being added in the first iteration. In the second iteration for referential match entry $\langle 2, 4, C \rangle$, we have $refmatches = \{3\}$ again and add $\{4 + (3-2)\}$ to $result$. We have $t = AC$ and $t^g = \{0\}$. Thus, $\{4 + 4 - (2 - 1) + 0)\}$ is added to result in Line 10. After the execution of Algorithm 2 we obtain $result = \{1, 5, 7\}$.

Algorithm 2 has three starting points for optimization, if a collection of compressed strings is to be searched. First, the whole referentially compressed string needs to be traversed in order to find all q-grams. Second, we need to keep a copy of the uncompressed string (Line 8), or at least decompress parts of the string during the search for each q-gram. Third, we have to run Algorithm 2 on each compressed sequence separately. Partial decompression and repeated interval containment checks make this algorithm not scalable for a large number of strings in the to-be-searched collection. Both issues can be addressed by using appropriate index structures on referential match entries. We introduce these index structures in the following subsection.

## 3.2 Index Structure for Q-Gram Projection

We use two index structures for improving scalability of q-gram projection in collections of highly-similar referentially compressed strings. First, we devise an index structure for managing all the referential match entries. Second, we develop an index structure for the q-grams overlapping mismatch characters, to avoid partial decompression during search.

**Projection of Reference Q-Grams.** Given a q-gram $g$, a reference string $ref$, all positions of $g$ in $ref$, $ref^g$, and a set of referentially compressed strings $C = \{cs_1, ..., cs_n\}$, we want to find each $\langle start, length, mismatch \rangle$ in each compressed string $cs_i \in C$, such that there exists a $p \in ref^g$ with $(p \geq start)$ and $(p + |g| \leq start + length)$ (Line 4-7 of Algorithm 2). Since all referential match entries can be understood as intervals over the whole reference string, we use interval trees [12] for efficiently querying referential match entries. An interval tree for $n$ intervals uses $O(n)$ storage and can be used to answer containment queries, i.e. find all intervals containing a given query point/interval, in time $O(k + \log n)$, where $k$ is the number of matching intervals.

**Definition 4.** *A reference interval for $ref$ is a tuple $\langle start, length \rangle$, such that $0 \leq start$, $0 \leq length$, and $start + length < |ref|$. A occurrence annotation is a tuple of the form $\langle id, pos \rangle$, where id and pos are natural numbers. We use id to store the identifier of a sequence inside the collection. Given a set of referentially compressed strings $C = \{cs_1, ..., cs_n\}$ with respect to $ref$, the annotated reference interval set for a reference $ref$ and $C$, denoted $ARIS_C^{ref}$, is a tuple $\langle intervals, \mathcal{T}, annotations \rangle$, such that intervals is a set of reference intervals for $ref$, $\mathcal{T}$ is a interval tree over these intervals, and annotations is a function from intervals to a set of occurrence annotations, such that*

$$< id, pos > \in annotations(\langle start, length \rangle) \iff$$
$$cs_{id} \in C \land \exists m. \langle start, length, m \rangle \in cs_{id} \land offset(cs_{id}, \langle start, length, m \rangle) = pos).$$

Given $ARIS_C^{ref}$, we can project all q-grams from $ref$ into referential match entries occurring in referentially compressed strings in $C$.

**Searching Q-Grams Overlapping Mismatch Characters.** In order to find all q-grams in compressed strings we need to take into account q-grams overlapping mismatch characters as well. We define a map structure which keeps track of all positions of q-grams overlapping at least one mismatch character in a compressed string in the collection.

**Definition 5.** *Given a set of referentially compressed strings $C = \{cs_1, ..., cs_n\}$, the annotated overlap map for the reference $ref$ and $C$, denoted $AOM_C^{ref}$, is $\langle grams, annotations \rangle$, such that grams is a subset of $\Sigma^q$ and annotations is a function from grams to a set of occurrence annotations, such that*

$$< id, pos > \in annotations(w) \iff (cs_{id} \in C \land \exists rme, d.rme \in cs_{id} \land$$
$$offset(cs_{id}, rme) + |rme| - d = pos) \land 1 \leq d \leq q \land \uparrow^{ref}(cs_{id})(pos, q) = w).$$

---

**Algorithm 3.** Searching q-grams in compressed strings

---

**Input:** $ref$, referential q-gram index $\langle PQG_q(ref), \langle i, \mathcal{T}, annotation_1 \rangle,$ $\langle grams, annotation_2 \rangle \rangle$, q-gram $g$
**Output:** set of retrieved positions in $result$
1: Let $result = \emptyset$
2: **for all** $p \in ref^g$ **do**
3:     Use $\mathcal{T}$ to identify all intervals in $i$ which contain the interval $\langle p, q \rangle$, add all these intervals to $refm$
4:     **for all** $\langle start, length \rangle \in refm$ **do**
5:         **for all** $\langle id, pos \rangle \in annotation_1(\langle start, length \rangle)$ **do**
6:             $result = result \cup \{ \langle id, pos + (p - start) \rangle \}$
7:         **end for**
8:     **end for**
9: **end for**
10: $result = result \cup annotation_2(g)$
11: **return** result

---

**Referential Q-Gram Index.** We combine the positional q-grams of a reference with an annotated reference interval set and an annotated overlap map to obtain a combined index structure for referentially compressed strings.

**Definition 6.** *Given a set of referentially compressed strings $C = \{cs_1, ..., cs_n\}$ with respect to a reference string $ref$, a referential q-gram index is defined as $\langle PQG_q(ref), ARIS_C^{ref}, AOM_C^{ref} \rangle$.*

We show our algorithm for retrieving all occurrences of a given q-gram in a referential q-gram index in Algorithm 3. The algorithm iterates over all matches for $g$ in the reference string (Line 2-9). For each match in the reference, the reference intervals containing $\langle p, q \rangle$, i.e. the to-be-searched q-gram, are collected in $refm$ (Line 3). For all intervals in $refm$, the annotation $annotation_1$ is used to retrieve all projected positions in referentially compressed strings and add these positions to the result (Line 4-6). Finally, all occurrences of $g$ overlapping one mismatch character are being added using the annotation map $annotations_2$ (Line 10).

**Lemma 1.** *For a given set of strings $S = \{s_1, ..., s_n\}$, a q-gram $g$, a reference string $ref$, and a set of referential compressions $C = \{cs_1, ..., cs_n\}$ for $S$ with respect to $ref$, we have that after applying Algorithm 3 with referential q-gram index $\langle PQG_q(ref), ARIS_C^{ref}, AOM_C^{ref} \rangle$, $result = S^q$.*

### 3.3 Example

We conclude this section with a complete example on retrieving all positions of a given q-gram from a set of strings.

*Example 4.* Let $S$ be a collection of strings $\{s_1, s_2, s_3, s_4\}$, $s_1 = ATCAGAATCT$, $s_2 = CATCGATCAGA$, $s_3 = ATCAGACATCGA$, and $s_4 = AGCCAAAATCT$.

Referentially compressing $S$ with respect to $ref = ATCAGCATCG$ yields $C = \{cs_1, cs_2, cs_3, cs_4\}$ with

$$cs_1 = \{\langle 0, 5, A \rangle, \langle 6, 3, T \rangle\}, cs_2 = \{\langle 5, 5, A \rangle, \langle 1, 4, A \rangle\},$$
$$cs_3 = \{\langle 0, 5, A \rangle, \langle 5, 5, A \rangle\}, cs_4 = \{\langle 3, 3, C \rangle, \langle 0, 1, A \rangle, \langle 6, 3, T \rangle\}$$

Let q=2, then we obtain the following index structures:

- $ARIS_C^{ref} = \langle i, \mathcal{T}, ann_1 \rangle$, with $i = \{\langle 0, 5 \rangle, \langle 6, 3 \rangle, \langle 1, 4 \rangle, \langle 5, 5 \rangle, \langle 3, 3 \rangle, \langle 0, 1 \rangle\}$ and

$$ann_1 = \{\langle 0, 5 \rangle \to \{\langle 1, 0 \rangle, \langle 3, 0 \rangle\}, \langle 6, 3 \rangle \to \{\langle 1, 6 \rangle, \langle 4, 6 \rangle\}, \langle 1, 4 \rangle \to \{\langle 2, 6 \rangle\},$$
$$\langle 5, 5 \rangle \to \{\langle 2, 0 \rangle, \langle 3, 6 \rangle\}, \langle 3, 3 \rangle \to \{\langle 4, 0 \rangle\}, \langle 0, 1 \rangle \to \{\langle 4, 4 \rangle\}\}$$

- $AOM_C^{ref} = \langle grams, ann_2 \rangle$, with

$$grams = \{GA, AA, CT, AT, AC, CC, CA\}$$
$$ann_2 = \{GA \to \{\langle 1, 4 \rangle, \langle 2, 4 \rangle, \langle 2, 9 \rangle, \langle 3, 4 \rangle, \langle 3, 10 \rangle\}$$
$$, AA \to \{\langle 1, 5 \rangle, \langle 4, 5 \rangle, \langle 4, 6 \rangle\},$$
$$CT \to \{\langle 1, 8 \rangle, \langle 4, 9 \rangle\}, AT \to \{\langle 2, 5 \rangle\},$$
$$AC \to \{\langle 3, 5 \rangle\}, CC \to \{\langle 4, 2 \rangle\}, CA \to \{\langle 4, 4 \rangle\}\}$$

Without referential compression we would need to store 40 two-grams for the four original strings in $S$. With referential compression only 14 explicit two-grams in the annotated overlap map plus 9 two-grams for the reference string (plus the overhead $ARIS_C^{ref}$) are necessary. Although, in this toy example, we do not save space[2], our evaluation in the next section shows that the approach does save much space in case of many highly-similar strings in practice.

## 4    Evaluation

In the following section, we evaluate our proposed compression scheme. All experiments have been run on a Acer Aspire 5950G with 16 GB RAM and Intel Core i7-2670QM, on Fedora 16 (64-Bit, Linux kernel 3.1). All size measures are in byte, e.g. 1 MB means 1,000,000 bytes. Below, the term compression factor is used to denote the inverse compression ration, e.g. a compression factor of 100 means a compression ratio of 100:1.

We have evaluated our algorithms for referential compression and q-gram projection on three biological datasets: a collection of human genomes, a collection of genomes from Arabidopsis thaliana, and a collection of yeast genomes. The raw size of the datasets is shown in Figure 1(a).

Our first datasets of human genomes was created from 1092 genomes of the 1000 Genome project[1].We use H-# to represent the set of all 1092 sequences

---

[2] Note that in this example the strings in $S$ are quite short, $S$ is rather small, and the referential match entries are very short as well.

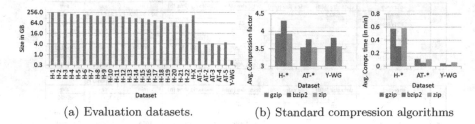

(a) Evaluation datasets.     (b) Standard compression algorithms

**Fig. 1.** Size of evaluation datasets (left) and initial evaluation with standard compression algorithms (right)

for human Chromosome #, e.g. H-1 for human Chromosome 1. The union of all 23 human datasets is denoted with H-*. The largest human dataset is H-1 at 272.1 GB, the smallest dataset is H-22 at 55.9 GB, and the size of H-* is 3.3 TB. Our datasets for Arabidopsis thaliana are taken from the 1001 Genomes project[21] from release GMINordborg2010.[3]. We have extracted 180 genomes with each 5 chromosomes. The Arabidopsis thaliana datasets are prefixed with AT, e.g. AT-1 stands for 180 Chromosome 1 of Arabidopsis thaliana. The union of all 5 Arabidopsis thaliana datasets is denoted with AT-*. The largest Arabidopsis thaliana dataset is AT-1 at 5.4 GB, the smallest dataset is AT-4 at 3.3 GB, and the size of AT-* is 21.4 GB The last dataset, is a collection of yeast genomes[4]. In total, we have downloaded 38 yeast strains, each of them was provided in FASTA format. The yeast dataset is denoted with Y-WG. The size of Y-WG is 0.4 GB.

## 4.1   Existing Standard Compression Algorithms

We have used three standard compression programs to create initial statistics about self-referential compression: gzip, bzip2, and zip. For each species and each chromosome, we have randomly selected five sequences and applied each of the compression algorithms. The results are shown in Figure 1(b). bzip2 is the best compression program among the three tested programs. The best average compression ratio is obtained by bzip2 for all three species and bzip2 is the fastest compression program as well, outperforming the other two programs by a factor of two in average. Using bzip2, it should be possible to compress H-* down to 0.7 TB using bzip2, but the run time is expected to be around 126 hours. AT-* can be compressed down to 5.6 GB in 48 minutes. The compression factor is relatively stable within species for H-*(min: 3.91 for H-3, max: 5.82 for H-22) and AT-*(min: 3.74 for AT-2, max: 3.80 for AT-1).

---

[3] http://1001genomes.org/data/GMI/GMINordborg2010/releases/current/

[4] http://www.yeastgenome.org/download-data/sequence

| | approach | data size (MB) | index size (MB) | total size (MB) | indexing time (h) | search time (s) min | max | avg |
|---|---|---|---|---|---|---|---|---|
| H-22 | grep | 55,932 | | 55,932 | | 427.92357 | 23,821.69608 | 8,826.85440 |
| | decomp/search | 78 | | 78 | | 98.28773 | 251.16323 | 218.43293 |
| | CST | | 139,776 | 139,776 | 5.16 | 0.01420 | 55.69200 | 31.13650 |
| | FMindex 2 | - | | - | - | - | - | - |
| | SQLite | | 1,183,370 | 1,183,370 | 490.27 | 0.00963 | 0.99628 | 0.64103 |
| | QGramProjectionOnline | 78 | | 78 | | 75.64537 | 82.30472 | 77.45636 |
| | QGramProjector | | 1,800 | 1,800 | 0.32 | 0.00005 | 0.31250 | 0.03413 |

| | approach | data size (MB) | index size (MB) | total size (MB) | indexing time (h) | search time (s) min | max | avg |
|---|---|---|---|---|---|---|---|---|
| AT-1 | grep | 5,477 | | 5,477 | | 65.36063 | 1330.07628 | 756.92232 |
| | decomp/search | 48 | | 48 | | 20.73434 | 36.82837 | 32.41139 |
| | CST | | 10,373 | 10,373 | 0.62 | 0.00324 | 4.86677 | 1.87505 |
| | FMindex 2 | | 1,926 | 1,926 | 0.32 | 0.03300 | 0.72100 | 0.03800 |
| | SQLite | | 113,568 | 113,568 | 47.33 | 0.00148 | 0.01294 | 0.00530 |
| | QGramProjectionOnline | 48 | | 48 | | 44.21570 | 60.90107 | 54.55710 |
| | QGramProjector | | 1,200 | 1,200 | 0.20 | 0.00005 | 0.01360 | 0.00129 |

| | approach | data size (MB) | index size (MB) | total size (MB) | indexing time (h) | search time (s) min | max | avg |
|---|---|---|---|---|---|---|---|---|
| Y-WG | grep | 473 | | 473 | | 3.99718 | 109.63334 | 34.42344 |
| | decomp/search | 6 | | 6 | | 3.04898 | 3.42124 | 3.15477 |
| | CST | | 642 | 642 | 0.05 | 0.00046 | 0.20634 | 0.09337 |
| | FMindex 2 | | 156 | 156 | 0.03 | 0.01900 | 0.21200 | 0.02500 |
| | SQLite | | 9,984 | 9,984 | 3.96 | 0.00014 | 0.00370 | 0.00089 |
| | QGramProjectionOnline | 6 | | 6 | | 9.34890 | 9.45812 | 9.36278 |
| | QGramProjector | | 263 | 263 | 0.02 | 0.00005 | 0.00231 | 0.00026 |

**Fig. 2.** Comparison of approaches for finding 16-mers

## 4.2   Indexing Sequences

In order to compare our approach to related work, we have selected the following competitors:

**grep** : Unix grep on the uncompressed raw sequences on the hard disk. We have used grep with options '-o -b' to only report positions of matches.

**decomp/search** : Index-less search with decompression from the hard disk on the fly.

**CST** : One compressed suffix tree[16] per sequence.

**FMindex 2** : a compressed substring index[8] over all sequences based on the Burrows-Wheeler transform.

**SQLite** : A SQLite database with all pre-computed k-mers for each sequence.

**QGramProjectionOnline** : an online version of our approach, where ARIS and AOM are created from the compressed files on-the-fly.

**QGramProjector** : an index-based version with precomputed ARIS and AOM for all sequences.

We have used 300 queries (16-mers), 100 each of the three species in the dataset. The results for datasets H-22, AT-1 and Y-WG are shown in Figure 2. For the other datasets we obtained similar results (within a species and with increasing sequence length, all tested approach show roughly linear index size and runtime behaviour). Note that we have used complete datasets now, e.g. 1092 sequences for the human genomes.

Using grep, the search speed is limited by the throughput of the hard disk. The hard-disk throughput is no longer a limit with naive search/decompress on compressed sequences. The average search time is reduced by a factor of 10-40.

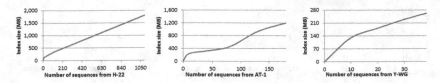

**Fig. 3.** Size of our index structure for different number of sequences

In addition, the amount of storage is clearly reduced, e.g. by a factor of 720:1 for for H-22. Using compressed suffix trees[16], the search times are three and more orders of magnitudes shorter than before. If there are many q-gram positions to be retrieved, then the compressed suffix tree solution slows down recognizably, e.g. down to 55.7 seconds for H-22. We were unable to run FMIndex 2 on the human genome datasets. For H-22 and even only one sequence, FMIndex 2 never returns a after sequences larger than 40 MB (we let it run on one sequence for several hours without any result). For the other datasets FMIndex 2 worked well. It is interesting that the index is actually 3-4 times smaller then the input dataset (as opposed to compressed suffix trees). In average, FMIndex 2 is faster than CST, which is mainly due to the fact that often occuring k-mers can be retrieved more efficiently. Our implementation based on SQLite shows very good average search times. The index structure becomes very large, as expected, since the position of each k-mer for each sequence has to be stored. For H-22 and AT-1 the indexing size and search times were extrapolated from a smaller set of 50 sample sequences.

QGramProjectionOnline (without precomputed index) has similar search times like decomp/search. It seems like the overhead of computing the index structures on the fly for each sequence takes too much time, in general. For highly-similar sequence, e.g. H-22, there can be small advantages, while for less similar sequences the projection algorithm doe snot pay off. Finally, QGramProjector has the best min, max, and avg search times in the test for all species. In average, even 3-4 times faster than using all precomputed k-mers in a SQLite database. The indexing overhead is in between FMIndex 2 and CST for not so similar sequences, but clearly outperforms both for highly-similar sequences.

We analyse the size of QGramProjector's index for a different number of sequences in Figure 3. The index grows linearly with the number of sequences for our human genome dataset H-22, as shown in Figure 3(left). This is because of the high similarity among the sequence. Each sequences adds roughly a similar amount of new ARIS and AOM entries. For AT-1 and Y-WG, Figure 3(middle and right), the picture is different: the size of the index grows not as regular as H-22. The reason is that the sequences are less similar (with respect to the reference) and each sequence adds a different amount of new annotations to ARIS and AOM.

In Figure 4, we show that retrieval time for 100.000 randomly created 16-grams over H-*, which corresponds to 3 TB of data. This test was performed on a server system with a TB of main memory. Each point corresponds to one

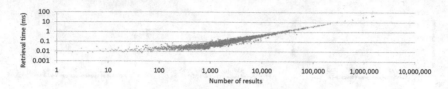

**Fig. 4.** Retrieval of 16-grams

q-gram (only q-grams with at least one occurrence in any of the 1000 genomes are shown). The 16-gram retrieval time is between 0.01 ms and 120 ms (average 0.64 ms; median 0.24 ms). Once a q-gram occurs in many sequences at different positions, the retrieval time increases linearly with the number of results, which is mainly caused by the time spent on decompressing the AOM/ARIS-annotations. We apply a special delta encoding on the sorted positions in occurrence annotations. Without such a compression of the annotations the index structure would increase by a factor of 2-4.

## 5    Conclusions

This work is another step towards analysis of a collection of highly-similar strings based on q-grams. We have shown that our approach allows to manage the q-grams of 1000 genomes using orders of magnitude less memory than the actual data size, and thus can be kept in main memory of a commodity hardware. Clearly, q-gram projection works particularly well for highly-similar sequences with a small alphabet size and large chunks of repeats across strings. Whether it can be directly applied to other areas is still to be evaluated. Another direction for future work is to implement a system for approximate string searching over large sets of genomes, using the q-gram index and following the seed-and-extend approach [3]. Finally, improving average retrieval times for q-grams is an important issue. The main challenge seems to retrieve q-grams with many occurrences. Alternatively, it would be interesting to define a notion of important matches. Since the reference sequence is always searched first, one could try to anticipate the overall number of matches, before retrieving all results. If the reference contains already more matches than a given threshold, one could limit the retrieval over the compressed sequences.

## References

1. 1000 Genomes Project Consortium. A map of human genome variation from population-scale sequencing. Nature, 467(7319), 1061–1073 (October 2010)
2. Altschul, S.F., Gish, W., Miller, W., Myers, E.W., Lipman, D.J.: Basic local alignment search tool. Journal of Molecular Biology 215(3), 403–410 (1990)
3. Baeza-Yates, R.A., Perleberg, C.H.: Fast and practical approximate string matching. In: Apostolico, A., Galil, Z., Manber, U., Crochemore, M. (eds.) CPM 1992. LNCS, vol. 644, pp. 185–192. Springer, Heidelberg (1992)

4. Belazzougui, D., Venturini, R.: Compressed string dictionary look-up with edit distance one. In: Kärkkäinen, J., Stoye, J. (eds.) CPM 2012. LNCS, vol. 7354, pp. 280–292. Springer, Heidelberg (2012)
5. Deorowicz, S., Grabowski, S.: Robust Relative Compression of Genomes with Random Access. Bioinformatics (September 2011)
6. du Mouza, C., Litwin, W., Rigaux, P., Schwarz, T.: As-index: a structure for string search using n-grams and algebraic signatures. In: Proceedings of the 18th ACM Conference on Information and Knowledge Management, CIKM 2009, pp. 295–304. ACM, New York (2009)
7. Ferragina, P.: String algorithms and data structures. CoRR, abs/0801.2378 (2008)
8. Ferragina, P., Manzini, G.: Indexing compressed text. J. ACM 52(4), 552–581 (2005)
9. Goto, K., Bannai, H., Inenaga, S., Takeda, M.: Speeding up $q$-gram mining on grammar-based compressed texts. In: Kärkkäinen, J., Stoye, J. (eds.) CPM 2012. LNCS, vol. 7354, pp. 220–231. Springer, Heidelberg (2012)
10. Kreft, S., Navarro, G.: On compressing and indexing repetitive sequences. Theor. Comput. Sci. 483, 115–133 (2013)
11. Kuruppu, S., Puglisi, S., Zobel, J.: Optimized relative lempel-ziv compression of genomes. In: Australasian Computer Science Conference (2011)
12. McCreight, E.: Efficient algorithms for enumerating intersection intervals and rectangles. Technical report, Xerox Paolo Alte Research Center (1980)
13. Navarro, G.: A guided tour to approximate string matching. ACM Comput. Surv. 33(1), 31–88 (2001)
14. Navarro, G.: Indexing highly repetitive collections. In: Arumugam, S., Smyth, B. (eds.) IWOCA 2012. LNCS, vol. 7643, pp. 274–279. Springer, Heidelberg (2012)
15. Navarro, G., Raffinot, M.: Flexible pattern matching in strings: practical on-line search algorithms for texts and biological sequences. Cambridge University Press, New York (2002)
16. Ohlebusch, E., Fischer, J., Gog, S.: CST++. In: Chavez, E., Lonardi, S. (eds.) SPIRE 2010. LNCS, vol. 6393, pp. 322–333. Springer, Heidelberg (2010)
17. Reich, D.E., Schaffner, S.F., Daly, M.J., McVean, G., Mullikin, J.C., Higgins, J.M., Richter, D.J., Lander, E.S., Altshuler, D.: Human genome sequence variation and the influence of gene history, mutation and recombination. Nature Genetics 32(1), 135–142 (2002)
18. Rytter, W.: Application of lempel–ziv factorization to the approximation of grammar-based compression. Theor. Comput. Sci. 302(1-3), 211–222 (2003)
19. Sutinen, E., Tarhio, J.: On using q-gram locations in approximate string matching. In: Spirakis, P.G. (ed.) ESA 1995. LNCS, vol. 979, pp. 327–340. Springer, Heidelberg (1995)
20. Wandelt, S., Leser, U.: Adaptive efficient compression of genomes. Algorithms for Molecular Biology 7, 30 (2012)
21. Weigel, D., Mott, R.: The 1001 Genomes Project for Arabidopsis thaliana. Genome Biology 10(5), 107+ (2009)

# FlexiDex: Flexible Indexing for Similarity Search with Logic-Based Query Models

Marcel Zierenberg and Maria Bertram

Brandenburg University of Technology Cottbus
Institute of Computer Science, Information and Media Technology
Chair of Database and Information Systems
P.O. Box 10 13 44, 03013 Cottbus, Germany
zieremar@tu-cottbus.de, maria.bertram@berner-mattner.com

**Abstract.** The flexibility of an indexing approach plays an important role for its applicability, especially for logic-based similarity search. A flexible approach allows the use of the same precomputed index structure even if query elements like weights, operators, monotonicity or used features of the aggregation function change in the search process (e.g., when using relevance feedback). While state-of-the-art approaches typically fulfill some of the needed flexibility requirements, none provides all of them. Consequently, this paper present *FlexiDex* , an efficient indexing approach for logic-based similarity search that is more flexible and also more efficient than known techniques.

## 1   Introduction

*Similarity search* is used in domains like image and text retrieval, DNA sequencing, biometric devices, and so forth. It generally seeks to find the most similar objects in a set of database objects, according to a given *query object*. The (dis-)similarity of two objects is determined by a *distance function* operating on *features* extracted from the database objects. Since the number of database objects is usually high and the computation of distances can be expensive in terms of CPU and I/O costs, efficient similarity search is an important research topic.

Early research efforts in similarity search were mainly concerned with search according to a single feature. In recent years, similarity search with multiple features has received more attention. Increasing the number of used features improves the expressiveness of queries and can lead to more effective similarity search. However, in regard of efficiency, adding features significantly increases CPU and I/O costs.

The combination of features is typically achieved through *aggregation functions*, which combine the *partial distances* of multiple features into an *aggregated distance*. While in general arbitrary aggregation functions can be used, the focus of this paper is similarity search with *logic-based query models*. Logic-based similarity search depends on the formulation of queries with the help of Boolean operators. Queries are then converted into arithmetic aggregation functions by means of specific transformation rules. Examples for logic-based query models in similarity search are the *Fuzzy Logic* [1] or the <u>C</u>ommuting <u>Q</u>uantum <u>Q</u>uery <u>L</u>anguage (*CQQL*) [2].

B. Catania, G. Guerrini, and J. Pokorný (Eds.): ADBIS 2013, LNCS 8133, pp. 274–287, 2013.
© Springer-Verlag Berlin Heidelberg 2013

*Indexing* approaches for similarity search try to exclude as many objects as early as possible from the search, consequently decreasing CPU- and I/O costs. In addition to the efficiency of the indexing approach, its *flexibility* plays an important role in logic-based similarity search. A flexible indexing approach allows the use of the same precomputed index structure even if query elements like weights, operators or monotonicity of the aggregation function change in the search process.

The *flexibility requirements* will be denoted by (R1)–(R6) in the following. Flexibility in terms of features (R1) means that the indexing should be independent from the structure of the underlying features. As changes of weights (R2) and operators (R3) in the aggregation function are typical, for example when using *relevance feedback*, the index should be built independently from the used aggregation function(s). Even though the number of features can be high, in most cases only a subset (R4) of all indexed features is used in a query. A flexible index should process such a query without loss of efficiency. Most indexing approaches for multiple features focus on globally monotone aggregation functions (e.g., weighted sum) [3, 4]. However, in logic-based similarity search not all aggregation functions are globally monotone and other types of monotonicity (R5) need to be supported as well. Finally, the index should be dynamic (R6) in the sense that new objects and new features can be added efficiently without rebuilding the whole index.

While a number of specialized indexing approaches for similarity search with multiple features exist [3–6], none of them fulfills all of the flexibility requirements (see Section 2). Consequently, we present *FlexiDex* , an index that is more flexible than state-of-the-art approaches. As a *metric indexing* approach [7], FlexiDex relies solely on distances for indexing and is independent from the structure of features (R1). The index is built without knowledge of the aggregation function and therefore naturally supports changing weights (R2) and operators (R3). Index data is stored on an HDD in the form of one separate signature file per feature, which allows efficient querying for subsets (R4) of the indexed features. All types of logic-based queries are supported by FlexiDex, even queries with aggregation functions that are not globally monotone (R5). Furthermore, FlexiDex is dynamic (R6) as new objects and new features can be simply appended to an existing index.

The paper is structured as follows. A deeper insight into related work is given in Section 2. Section 3 defines the notations and terms used throughout the paper. As the monotonicity of an aggregation function plays an important role in the indexing process, Section 4 presents different types of monotonicity classes. Section 5 details the index creation and search process. The experimental evaluation shown in Section 6 demonstrates that FlexiDex significantly reduces I/O and CPU costs and even outperforms a less flexible state-of-the-art approach. Section 7 summarizes the work and gives an outlook on future research.

## 2   Related Work

This section gives a deeper insight into related work. Indexing approaches are divided into different categories, examples for each category are given and their flexibility is examined in detail.

**Table 1.** Indexing approaches for multiple features and their general flexibility

| Indexing based on | combined to single feature | multiple features | aggregated distances | | partial distances |
|---|---|---|---|---|---|
| **Index type** | R-Tree [8] VA-File [5] (naïve spatial) | GeVAS [9] | M-Tree [10] LAESA [11] (naïve metric) | Multi-Metric [4] | $M^2$-Tree [6] Combiner [3] *FlexiDex* |
| **Flexibility** | low | medium | low | medium | high |

Table 1 shows an overview of indexing approaches for *multiple features* and their general flexibility. The flexibility mainly depends on where the indexing is used in the similarity search process. Indexing based directly on the feature data (single feature, multiple features) is independent of the aggregation function (R2, R3). However, it is typically restricted to vector spaces (*multi-dimensional* or *spatial indexing* [7]), making the index dependent on the structure of the features (R1). On the other side, indexing based on aggregated distances depends on the used aggregation function, disallowing the change of weights (R2) and operators (R3). Only approaches that utilize partial distances, like [3, 6] or FlexiDex, can theoretically support all flexibility requirements. With the exception of the $M^2$-Tree [6], all presented approaches consider only globally monotone increasing aggregation functions (R5).

Naïve approaches for indexing multiple features rely on combining multiple features into a single feature (spatial) or using aggregated distances (metric). This allows the utilization of a plethora of well-known indexing approaches for similarity search with a single feature (e.g., *R-Tree* [8], *VA-File* [5], *M-Tree* [10] or *LAESA* [11]). Unfortunately, it also leads to inflexibility since queries using only subsets of all indexed features (R4) cannot be processed efficiently and adding new features is not possible without rebuilding the whole index (R6). Additionally, in case of naïve metric indexing, weights (R2) or operators (R3) of the aggregation function can not be changed, as the index was created according to a specific (metric) aggregation function.

The spatial indexing approach *GeVAS* [9] is an extension of the VA-File to multiple features. By using a separate signature file for each feature, it allows efficient queries with subsets (R4) and dynamically adding new objects and new features (R6). It is therefore more flexible than the naïve approach. However, the main problem of spatial indexing, the limitation to vector spaces, remains.

*Multi-metric indexing* [4] defines a framework to transform metric indexing approaches for single features into metric indexing approaches for multiple features using partial and aggregated distances at the same time. Even though this technique allows changing weights in the aggregation function (R2), it is restricted to aggregation functions that are metrics (e.g., weighted sum) and the index is still created based on a specific aggregation function, preventing the change of operators (R3). Since aggregated distances are used, querying with subsets (R4) is less efficient and new features can not be added dynamically (R6).

The $M^2$-*Tree* [6] is in general flexible enough to support changing weights (R2) and operators (R3). However, all features are used in the construction of the tree. This means, that with increasing numbers of features, the tree suffers from the curse of dimensionality [7]. Also, the clustering may prove inefficient if only a subset of all indexed features

**Table 2.** Fulfillment of flexibility requirements (R1)–(R6) for indexing approaches

| Index type | Features (R1) | Weights (R2) | Operators (R3) | Subsets (R4) | Monotonic-ity (R5) | Dynamic (R6) |
|---|---|---|---|---|---|---|
| naïve (spatial) | - | ✓ | ✓ | - | - | - |
| GeVAS | - | ✓ | ✓ | ✓ | - | ✓ |
| naïve (metric) | ✓ | - | - | - | - | - |
| Multi-Metric | ✓ | ✓ | - | - | - | - |
| $M^2$-Tree | ✓ | ✓ | ✓ | - | (✓) | - |
| Combiner | ✓ | ✓ | ✓ | ✓ | - | ✓ |
| *FlexiDex* | ✓ | ✓ | ✓ | ✓ | ✓ | ✓ |

is present in a query (R4). Even though the $M^2$-Tree considers globally and locally monotone functions, it still lacks support for some logic-based aggregation functions (R5) (e.g., $x_1 \oplus x_2$ (XOR), see Section 4). New objects can be added dynamically to the $M^2$-Tree, but adding new features requires rebuilding the whole tree (R6).

*Combiner algorithms* [3] are inherently independent of the underlying indexing approach and provide almost all flexibility requirements. Their drawback is that they only consider aggregation functions that are globally monotone increasing (R5), which, as stated before, is not always the case for logic-based queries.

Table 2 summarizes the flexibility of the presented indexing approaches and shows that FlexiDex constitutes the most flexible approach.

While *approximate similarity search* [7] can significantly decrease search time, it comes at the cost of effectiveness, since the results only have a certain probability to be the most similar objects. The focus of our research is exact similarity search and therefore approximate indexing approaches like [12] are not applicable.

## 3 Preliminaries

This section defines the notations and terms used throughout this paper.

Similarity search can be performed by means of a $\underline{k}$-$\underline{N}$earest $\underline{N}$eighbor search. A $k$NN($q$) in the universe of objects $\mathbb{U}$ returns $k$ objects out of a database of objects $D = \{o^1, o^2, \ldots, o^n\} \subseteq \mathbb{U}$ that are closest (most similar) to the query object $q \in \mathbb{U}$. The distance between objects is computed by a distance function $\delta : \mathbb{U} \times \mathbb{U} \mapsto \mathbb{R}_{\geq 0}$ that operates on the features $q'$ and $o'$ extracted from the objects. The result is a (non-deterministic) set $K$ with $|K| = k$ and $\forall o^i \in K, o^j \in D \setminus K : \delta(q, o^i) \leq \delta(q, o^j)$.

A *multi-feature kNN query* consists of $m$ features for each object $q' = (q_1, q_2, \ldots, q_m)$ and $o' = (o_1, o_2, \ldots, o_m)$. A distance function $\delta_j$ is assigned to each single feature to compute *partial distances* $\delta_j(q_j, o_j)$. An *aggregation function* agg $: \mathbb{R}_{\geq 0}^m \mapsto \mathbb{R}_{\geq 0}$ combines all partial distances to an *aggregated distance* $d_{agg}$ and the $k$ nearest neighbors are then determined according to the aggregated distance.

*Logic-based queries* combine multiple features by Boolean operators. Query models like the *Fuzzy Logic* [1] or *CQQL* [2] transform the resulting Boolean expressions into arithmetic formulas. Table 3 shows examples of transformation rules. CQQL uses algebraic transformation rules and, contrary to Fuzzy Logic, normalizes the expressions

**Table 3.** Transformation rules for logic-based queries

| Boolean | Zadeh | Algebraic |
|---------|-------|-----------|
| $\neg a$ | $1-a$ | $1-a$ |
| $a \wedge b$ | $\min(a,b)$ | $a*b$ |
| $a \vee b$ | $\max(a,b)$ | $a+b-a*b$ |
| $(c \wedge a) \vee (\neg c \wedge b)$ | (analog) | $a+b$ |

**Table 4.** Embedding of operand weights with CQQL

| Boolean | Embedding |
|---------|-----------|
| $a \wedge_{\theta_1:\theta_2} b$ | $(a \vee \neg\theta_1) \wedge (b \vee \neg\theta_2)$ |
| $a \vee_{\theta_1:\theta_2} b$ | $(a \wedge \theta_1) \vee (b \wedge \theta_2)$ |

before their transformation. This preserves important properties like the distributivity and idempotency. For example, query $(a \wedge b) \vee (a \wedge c)$ results in formula $a*b+a*c-a*b*a*c$ using algebraic Fuzzy Logic. In contrast, the normalized formula used by CQQL is $a \wedge (b \vee c)$, which results in $a*(b+c-b*c)$.

Additionally, CQQL also supports the direct embedding of operand weights $\theta_i \in \mathbb{R}_{\geq 0}$ into the logic (see Table 4). Therefore, only aggregation functions created by CQQL are examined hereinafter. However, note that the stated results are by no means restricted to CQQL and can be adapted to other logic-based query models as well.

Logic-based query models assume similarity values in the interval $[0,1]$, where 1 means most similar (identity) and 0 means least similar. Hence, the aforementioned partial distances have to be transformed into *partial similarities* by *transformation functions* $t_j : \mathbb{R}_{\geq 0} \mapsto [0,1]$ before the aggregation function can be applied. The aggregated similarity $s_{agg}$ is computed by combining all $m$ partial similarities with the (logic-based) aggregation function agg $: [0,1]^m \mapsto [0,1]$.

A *metric* [7] is a distance function with the properties *positivity* ($\forall x \neq y \in \mathbb{U} : \delta(x,y) > 0$), *symmetry* ($\forall x,y \in \mathbb{U} : \delta(x,y) = \delta(y,x)$), *reflexivity* ($\forall x \in \mathbb{U} : \delta(x,x) = 0$) and *triangle inequality* ($\forall x,y,z \in \mathbb{U} : \delta(x,z) \leq \delta(x,y) + \delta(y,z)$).

Metric indexing approaches exclude objects from search by computing *bounds* on the distance from the query object to database objects. The lower bound $d^{lb}$ and upper bound $d^{ub}$ on the distance $\delta(q,o)$ between query object $q$ and database object $o$ can be computed with the help of the triangle inequality and precomputed distances to a *reference object* (*pivot*) $p$ as follows:

$$d^{lb} = |\delta(q,p) - \delta(p,o)| \leq \delta(q,o) \leq \delta(q,p) + \delta(p,o) = d^{ub}. \qquad (1)$$

For logic-based queries, distance bounds have to be transformed into similarity bounds $s^{lb}$ and $s^{ub}$. Notice that the meaning of lower and upper bound exchanges with this transformation. The upper bound for distances is the maximum dissimilarity, while the upper bound for similarities is the maximum similarity.

Bounds $s_{agg}^{lb}$ and $s_{agg}^{ub}$ on the aggregated similarity $s_{agg}$ can be computed by inserting partial similarity bounds into the aggregation function.

## 4   Monotonicity and Computation of Aggregated Bounds

Most indexing approaches for similarity search with multiple features consider only aggregation functions that are monotone increasing in each of their arguments (globally monotone increasing functions). However, to allow all different kinds of logic-based queries, other classes of monotonicity have to be supported as well (R5). Consequently, this section describes the computation of bounds on aggregated similarities based on the bounds on partial similarities for three different classes of monotonicity: *globally*, *locally* and *flexible monotone aggregation functions*.

Although another naming convention was used, globally and locally monotone functions have already been considered by [13] and [6]. However, queries like $x_1 \oplus x_2$ (XOR) result in aggregation functions that are neither globally nor locally monotone. Therefore, the new class of flexible monotone functions is introduced and we show that the three presented classes of monotonicity are sufficient to support all kinds of (logic-based) aggregation functions created by CQQL.

### 4.1 Globally and Locally Monotone Functions

**Definition 1.** *An aggregation function* agg *is monotone increasing in the i-th argument with* $i \in \{1, 2, \ldots, m\}$, $x = (x_1, x_2, \ldots, x_i, \ldots, x_m)$ *and* $x' = (x_1, x_2, \ldots, x_i', \ldots, x_m)$ *iff:*

$$\forall x, x' \in [0,1]^m : x_i < x_i' \implies \text{agg}(x) \leq \text{agg}(x'). \tag{2}$$

This means, if all arguments except $x_i$ are constant and $x_i$ is increased, the result of the aggregation function will also increase (or be constant).

**Definition 2.** *An aggregation function* agg *is monotone decreasing in the i-th argument iff* agg' $= -$ agg *is monotone increasing in the i-th argument.*

**Definition 3.** *An aggregation function* agg *is globally monotone increasing (decreasing) iff it contains only monotone increasing (decreasing) arguments.*

An example for a globally monotone increasing function is $\text{agg}(x_1, x_2) = x_1 * x_2$, which is the result of the query $x_1 \wedge x_2$.

It can easily be shown that the lower (upper) bound $s_{agg}^{lb}$ ($s_{agg}^{ub}$) on the aggregated similarity of a globally monotone increasing function is computed by inserting all lower (upper) bounds on the partial similarities into the aggregation function:

$$s_{agg}^{lb} = \text{agg}\left(s_1^{lb}, s_2^{lb}, \ldots, s_m^{lb}\right). \tag{3}$$

**Definition 4.** *An aggregation function* agg *is locally monotone iff it contains only monotone increasing and monotone decreasing arguments.*

The aggregation function $\text{agg}(x_1, x_2) = x_1 * (1 - x_2)$ based on the query $x_1 \wedge \neg x_2$ is an example for a locally monotone function.

By definition, every globally monotone function also is a locally monotone function with only monotone increasing/decreasing arguments.

The following equations compute lower bounds for locally monotone aggregation functions. The computation of upper bounds follows the same equations. Only the values $lb$ and $ub$ in function $f_i^{agg}$ have to be exchanged.

$$s_{agg}^{lb} = \text{agg}\left(f_1^{agg}\left(s_1^{lb}, s_1^{ub}\right), f_2^{agg}\left(s_2^{lb}, s_2^{ub}\right), \ldots, f_m^{agg}\left(s_m^{lb}, s_m^{ub}\right)\right) \tag{4}$$

$$f_i^{agg}(lb, ub) = \begin{cases} lb, & \text{if agg is monotone increasing in the } i\text{-th argument} \\ ub, & \text{if agg is monotone decreasing in the } i\text{-th argument} \end{cases} \tag{5}$$

### 4.2   Flexible Monotone Functions

**Definition 5.** *An aggregation function* agg *is flexible monotone in the i-th argument with* $i \in \{1, 2, \ldots, m\}$, $x = (x_1, x_2, \ldots, x_i, \ldots, x_m)$ *and* $x' = (x_1, x_2, \ldots, x_i', \ldots, x_m)$ *iff:*

$$\forall x \in [0, 1]^m :$$
$$\left(\forall x' \in [0, 1]^m : x_i < x_i' \implies \text{agg}(x) \le \text{agg}(x')\right) \tag{6}$$
$$\vee \left(\forall x' \in [0, 1]^m : x_i < x_i' \implies \text{agg}(x) \ge \text{agg}(x')\right).$$

In other words, if all arguments except $x_i$ are constant and $x_i$ is increased, the result of the aggregation function will either always increase (monotone increasing) or always decrease (monotone decreasing) or be constant for the current combination of fixed $m - 1$ argument values. The difference between a monotone increasing/decreasing argument and a flexible monotone argument is that the monotonicity of a flexible monotone argument depends on the values of the other arguments and can therefore change. For monotone increasing/decreasing arguments the monotonicity is always the same, independent of the values of the other arguments.

**Definition 6.** *An aggregation function* agg *is flexible monotone iff it contains only flexible monotone arguments.*

Figure 1 shows a flexible monotone aggregation function based on the query $x_1 \oplus x_2$. As can be seen, the monotonicity of argument $x_2$ changes between monotone increasing and monotone decreasing, depending on the value of $x_1$.

By definition, every monotone increasing/decreasing argument also is a flexible monotone argument. A locally monotone function also is a flexible monotone function where the monotonicity in the flexible monotone arguments never changes.

**Theorem 1.** *The upper and lower bound on the aggregated similarity of a flexible monotone function are always located in the corners of the hyper-rectangle given by the m lower and upper bounds on the partial similarities* $\left\{(s_1^{lb}, s_1^{ub}), (s_2^{lb}, s_2^{lb}), \ldots, (s_m^{lb}, s_m^{ub})\right\}$.

*Proof.* Let aggregation function agg be a flexible monotone function and the partial similarity bounds for $1 \le i \le m$ be $(s_i^{lb}, s_i^{ub}) = (0, 1)$. Without loss of generality now consider a tuple $x \in [0, 1]^m$ with agg$(x)$ to be a global maximum. For an arbitrarily chosen $i$ assume $0 < x_i < 1$, which means $x$ is not located in a corner of the hyper-rectangle $[0, 1]^m$. Now consider two tuples $x', x'' \in [0, 1]^m$ with $x' = (x_1, x_2, \ldots, x_i', \ldots, x_m)$ and $x'' = (x_1, x_2, \ldots, x_i'', \ldots, x_m)$ for which $x_i' < x_i < x_i''$ is true. Because agg$(x)$ is a maximum, agg$(x') \le$ agg$(x)$ and agg$(x'') \le$ agg$(x)$ is true. However, this contradicts

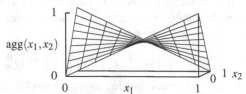

**Fig. 1.** Flexible monotone aggregation function $x_1 * (1 - x_2) + (1 - x_1) * x_2$.

the assumption of a flexible monotone function for which $agg(x') \leq agg(x) \leq agg(x'')$ or $agg(x'') \leq agg(x) \leq agg(x')$ should be true.                                                □

From Theorem 1 immediately follows that lower (upper) bounds $s_{agg}^{lb}$ ($s_{agg}^{ub}$) on the aggregated similarities of a flexible monotone function can be computed by inserting all different combinations of lower and upper bounds for flexible monotone arguments into the aggregation function and subsequently selecting the minimum (maximum) result.

Combined with Equation (4), the computation of bounds for all three classes of monotonicity is therefore defined as follows. Again, the computation of upper bounds $s_{agg}^{ub}$ follows the same principle. Only the values $lb$ and $ub$ in the first two conditions of function $g_i^{agg}$ have to be exchanged.

$$(combinations) \quad C = g_1^{agg}\left(s_1^{lb}, s_1^{ub}\right) \times g_2^{agg}\left(s_2^{lb}, s_2^{ub}\right) \times \ldots \times g_m^{agg}\left(s_m^{lb}, s_m^{ub}\right) \tag{7}$$

$$g_i^{agg}(lb, ub) = \begin{cases} \{lb\}, & \text{if agg is monotone increasing in the } i\text{-th argument} \\ \{ub\}, & \text{if agg is monotone decreasing in the } i\text{-th argument} \\ \{lb, ub\}, & \text{if agg is flexible monotone in the } i\text{-th argument} \end{cases} \tag{8}$$

$$(selection) \quad s_{agg}^{lb} = \min_{c \in C} agg(c) \tag{9}$$

### 4.3   Logic-Based Queries and Monotonicity Classes

As stated in [13], every logic-based aggregation function which contains each argument only once is either globally or locally monotone. For functions that contain an argument twice, [13] proposes to handle those occurrences as independent arguments. Unfortunately, no proof for correctness was given and it can easily be shown that the approach tends to create bounds less tight than Equation (9).

The disadvantage of Equation (9) is that the number of combinations that have to be aggregated increases exponentially with the number of flexible monotone arguments (maximum $2^m$). It is therefore only applicable for aggregation functions with a low number of flexible monotone arguments.

Every aggregation function created by CQQL (*CQQL formula*) is flexible monotone. For space reasons only a sketch of the proof is given[1]: It can be proven that every multivariate linear polynomial is a flexible monotone function by bringing it into the form $f(x) = x_i * c_1 + c_2$, where $c_1$ and $c_2$ are multivariate linear polynomials that do not contain $x_i$. It can also be shown that every CQQL formula is a multivariate linear polynomial. Thus, every CQQL formula is flexible monotone.

## 5   FlexiDex

This section explains the index creation and $k$NN search process of FlexiDex. Figure 2 gives an overview of the process, which is divided into the stages *creation & preparation*, *filtering* and *refinement*. The figure also shows, where the flexibility requirements (R1)–(R6) are involved.

---

[1] Detailed version: http://tiny.cc/8t4wuw (http://dbis.informatik.tu-cottbus.de/down/pdf/CQQL-Monotonicity.pdf)

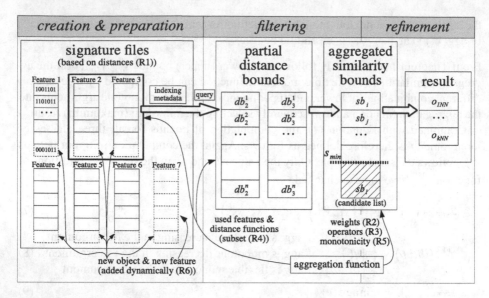

**Fig. 2.** Overview of the creation and search process of FlexiDex

## 5.1   Index Creation

FlexiDex follows the GeVAS [9] approach, creating one *signature file* for each feature. However, to ensure independence from the structure of the features (R1), metric indexing is used instead of spatial indexing. This means, signatures are not based on feature vectors but instead are comprised of precomputed distances to a set of pivots. Since one separate signature file is used for each feature, each signature file can be read from disk independently. Therefore, FlexiDex can efficiently process queries using subsets of all indexed features (R4) and also is dynamic (R6), since new objects can simply be appended to an existing signature file and new signature files can be created at any time.

In the following the index creation process for a single feature is described. The creation process for multiple features works similar. The only constraint is that the order of objects has to be the same in all signature files. The process of signature creation resembles [14], where metric indexing and signature files were used for similarity search with a single feature. Also note that the parameters explained in the following can be set individually for each feature.

Index creation starts with the selection of $P$ pivot objects. Pivots can be selected randomly or by a specific selection strategy like *incremental selection* [15]. Additionally, $2^B$ quantization intervals $i^0, i^1, \ldots, i^{2^B-1}$ are defined which will be used for the compression of the distances. The number of used bits $B$ determines the precision of the intervals. A higher number of bits results in better distance approximations but also increases the time needed for reading the signature file. Selected pivots and indexing parameters are stored on disk as metadata.

When adding a new database object $o$ to the index, the distances $\delta(p^k, o)$ between object $o$ and all pivot objects $p^k$ are computed. Each distance is then replaced with the interval number of the corresponding quantization interval. The *object signature S* is the result of the concatenation of each binary coded interval number and simply appended to the existing signature file.

*Example 1.* Consider a new database object $o$ with two features $o_1$ and $o_2$ ($m = 2$). Indexing parameters are set individually for each feature and distinguished by their subscript. The number of pivots is $P_1 = 4$, $P_2 = 2$ and the interval boundaries are $i_1^0 = [0,4), i_1^1 = [4,8), i_1^2 = [8,12), i_1^3 = [12,16]$ ($B_1 = 2$) and $i_2^0 = [0,2), i_2^1 = [2,4]$ ($B_2 = 1$). Now, let exemplary distances to the pivots for the first feature be given by $\delta_1(p_1^1,o_1) = 5, \delta_1(p_1^2,o_1) = 16, \delta_1(p_1^3,o_1) = 7$ and $\delta_1(p_1^4,o_1) = 1$ and for the second feature by $\delta_2(p_2^1,o_2) = 0$ and $\delta_2(p_2^2,o_2) = 3$. By replacing the distances with their binary coded interval numbers the following object signatures are obtained for object $o$: $S_1 = (01\ 11\ 01\ 00)$ and $S_2 = (0\ 1)$.

## 5.2 Nearest Neighbor Search

Since the creation process is completely independent from the aggregation functions used in $k$NN search, changing weights (R2) and operators (R3) of the aggregation function at query time is supported naturally. With the help of the monotonicity classes defined in Section 4, all types of logic-based queries can be processed (R5).

For $k$NN search the well-known concept of *filter refinement* [5, 14] is utilized. The *filtering phase* reads each signature file sequentially from disk and afterward computes partial distance bounds for each object with the help of the triangle inequality. Equation (1) has to be adapted slightly to incorporate the fact that exact distance $\delta(p,o)$ is replaced by the corresponding interval boundaries $i^{lb}$ and $i^{ub}$:

$$d^{lb} = \max\left\{0, i^{lb} - \delta(q,p), \delta(q,p) - i^{ub}\right\} \le \delta(q,o) \le \delta(q,p) + i^{ub} = d^{ub}. \quad (10)$$

Note that only signature files that are actually present in the current aggregation function are loaded. Distance bounds are then transformed into similarity bounds and combined to aggregated similarity bounds, using Equation (9). Objects having a smaller upper bound $s_{agg}^{ub}$ than the $k$-th greatest lower bound $s_{agg}^{lb}$ ($s_{min}$) can be excluded from the search.

In the *refinement phase* exact aggregated similarities are computed for the objects that could not be excluded in the filtering phase. A priority queue $PQ$, sorted in descending order according to $s_{agg}^{lb}$, acts as a *candidate list* for refinement. Since objects are reinserted into $PQ$ after the computation of the exact aggregated similarity $s_{agg}$, the search can be terminated as soon as $k$ exactly computed objects have reappeared at the top of $PQ$.

# 6 Evaluation

This section compares FlexiDex against linear scan and a combiner algorithm. FlexiDex was implemented as part of the multimedia retrieval system *PythiaSearch*[2] [16].

All experiments were conducted on a 2 x 2.26 GHz Quad-Core Intel Xeon with 8 GB RAM and an HDD with 7,200 rpm. The image collection *Caltech-256 Object Category*

---
[2] http://tiny.cc/mc5wuw (https://saffron.informatik.tu-cottbus.de/livingfeatures)

**Table 5.** Used features and their optimal index parameters

| Feature | $\delta$ | $\rho$ | P | B | D | | T in s | |
|---|---|---|---|---|---|---|---|---|
| ScalableColor (*scal*) [20] | $L_1$ | 2.91 | 8 | 8 | 873 | (2.9 %) | 1.57 | (8.1 %) |
| Tamura (*tam*) [21] | $L_1$ | 2.89 | 8 | 8 | 1,205 | (3.9 %) | 1.93 | (10.6 %) |
| FCTH (*fcth*) [22] | $L_1$ | 5.77 | 32 | 8 | 1,658 | (5.4 %) | 2.57 | (13.0 %) |
| DominantColor (*dom*) [20] | EMD + $L_2$ | 2.16 | 24 | 10 | 1,755 | (5.7 %) | 2.70 | (11.7 %) |
| ColorStructure (*cs*) [20] | $L_2$ | 5.87 | 56 | 8 | 3,143 | (10.3 %) | 3.66 | (21.7 %) |
| ColorHistCenter (*chc*) [20] | $L_1$ | 3.89 | 56 | 8 | 2,999 | (9.8 %) | 3.70 | (20.3 %) |
| ColorLayout (*cl*) [20] | weighted $L_2$ | 5.05 | 48 | 8 | 6,814 | (22.3 %) | 6.42 | (30.5 %) |
| ColorHistBorder (*chb*) [20] | $L_1$ | 5.82 | 64 | 8 | 7,507 | (24.5 %) | 6.77 | (36.2 %) |
| AutoColorCorrel. (*auto*) [20] | $L_2$ | 8.07 | 64 | 8 | 7,789 | (25.4 %) | 7.17 | (36.3 %) |
| ColorHistogram (*ch*) [20] | $L_2$ | 11.48 | 64 | 8 | 10,697 | (34.9 %) | 9.14 | (46.6 %) |
| BIC (*bic*) [23] | $L_1$ | 10.23 | 56 | 8 | 12,908 | (42.2 %) | 9.18 | (49.8 %) |
| EdgeHistogram (*edge*) [20] | weighted $L_1$ | 8.55 | 48 | 10 | 12,056 | (39.4 %) | 10.48 | (50.7 %) |
| CEDD (*cedd*) [24] | $L_2$ | 12.05 | 64 | 8 | 10,863 | (35.5 %) | 11.03 | (45.2 %) |

*Dataset* [17] was used, which consists of 30,607 pictures. Efficiency was assessed by measuring the average number of distance computations and the average $k$NN search time (wall-clock time) of 100 randomly chosen query objects.

To provide a comparison to the state-of-the-art, a combiner algorithm (*threshold algorithm* [3]) was implemented. Sorted lists for the combiner algorithm are provided based on the same signature files as FlexiDex and the implementation of *getNext* is similar to [18].

Table 5 shows the 13 different color, texture and form features extracted for the collection. While most features utilize some type of *Minkowski ($L_p$) distance* function, the feature *dom* employs the *Earth Mover's distance function (EMD)* [19].

Optimization of indexing parameters was performed separately for each feature. Pivots were chosen randomly. The tested precision values $B$ ranged from 4 to 12 and the number of pivots $P$ from 8 to 64. For both parameters higher values decreased the number of needed distance computations but simultaneously increased the time needed for reading the signature files. Optimal parameters were selected based on the average search time for queries with $k = 10$.

The results of the optimization process are presented in Table 5. Search times ($T$) and number of distance computations ($D$) were significantly reduced in comparison to the *linear scan* (see percentage values). Since FlexiDex and the combiner algorithm work similar when using only a single feature, the determined optimal parameters apply to both approaches.

The optimal precision value $B$ was 8 in almost all cases. For lower precision values the number of distance computations increased rapidly, resulting in significantly longer search times. Higher values than 8 gave only slight decreases in distance computations that could not make up for the increased time needed for reading the signature files. The number of used pivots $P$ showed some correlation to the intrinsic dimensionality[3]. For

---

[3] Intrinsic dimensionality $\rho$ is defined as $\rho = \frac{\mu^2}{2*\sigma^2}$ where $\mu$ is the mean and $\sigma^2$ is the variance of a distance distribution. It is frequently used as an estimator for indexability [25].

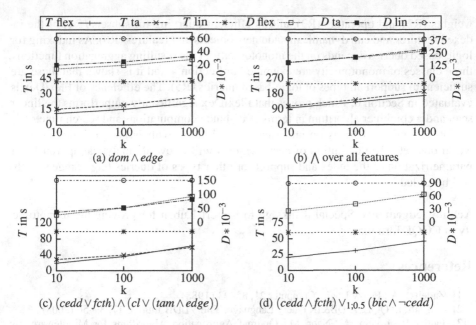

**Fig. 3.** Measured efficiency for selected logic-based $k$NN queries.

low intrinsic dimensionality (features *scal* or *dom*) a small number of pivots excluded more than 90 % of the objects. A greater number of pivots was needed when the intrinsic dimensionality increased.

Figure 3 depicts the average number of distance computations and search time of selected logic-based $k$NN queries for FlexiDex (*flex*), linear scan (*lin*) and the implemented combiner algorithm (*ta*). The selected queries are similar to those used in [26] and represent different types of logic-based queries: only conjunction (3a, 3b), disjunction and conjunction (3c) and weighted disjunction, conjunction and negation (3d).

FlexiDex performed clearly superior to linear scan and combiner algorithm in almost all cases. In the best case FlexiDex was about 33 % faster (query 3c with $k = 10$) than the combiner algorithm and needed about 25 % less distance computations. Only in case of query 3c with $k = 1000$ the combiner algorithm outperformed FlexiDex slightly ($\sim 6\%$ faster). Query 3d resulted in a flexible monotone aggregation function, which was not supported by the combiner algorithm.

## 7 Conclusion and Outlook

The flexibility of an indexing approach plays an important role in logic-based similarity search. It allows the use of the same precomputed index even if query elements like the used features or the monotonicity of the aggregation function change in the search process. Since none of the known approaches presented in Section 2 supports all flexibility requirements (R1)–(R6), we introduce the new index FlexiDex.

The adaption of the GeVAS [9] approach to metric indexing (Section 5) results in a flexible index that is independent from the structure of features (R1), supports changing weights (R2) and operators (R3), allows efficient querying for subsets (R4) of the indexed features and can dynamically add new objects and features (R6). As indexing for logic-based queries depends on the monotonicity of the resulting aggregation functions, three classes of monotonicity are examined in Section 4 and it is shown that these are sufficient to support all types of logic-based queries (R5). The efficiency of FlexiDex is evaluated in Section 6, which proves that FlexiDex significantly outperforms the linear scan and a combiner algorithm in terms of distance computations and search time.

Future work will focus on an extended evaluation with large synthetic and real-world datasets. The definition of cost formulas will allow an analytical approach for parametrization of the index and support for other types of queries (e.g., range search) will be added.

**Acknowledgements.** Special thanks go to Robert Kuban for proving the monotonicity of CQQL formulas.

# References

[1] Zadeh, L.A.: Fuzzy Logic. Computer 21, 83–93 (1988)
[2] Schmitt, I.: QQL: A DB&IR Query Language. The VLDB Journal 17, 39–56 (2008)
[3] Fagin, R., Lotem, A., Naor, M.: Optimal Aggregation Algorithms for Middleware. In: Proceedings of the 20th ACM SIGMOD-SIGACT-SIGART Symposium on Principles of Database Systems, PODS 2001, pp. 102–113. ACM, Santa Barbara (2001)
[4] Bustos, B., Kreft, S., Skopal, T.: Adapting Metric Indexes for Searching in Multi-Metric Spaces. Multimedia Tools Appl. 58(3), 467–496 (2012)
[5] Weber, R., Schek, H.-J., Blott, S.: A Quantitative Analysis and Performance Study for Similarity-Search Methods in High-Dimensional Spaces. In: Proceedings of the 24th International Conference on Very Large Data Bases, VLDB 1998, pp. 194–205. Morgan Kaufmann Publishers Inc., San Francisco (1998)
[6] Ciaccia, P., Patella, M.: The $M^2$-Tree: Processing Complex Multi-Feature Queries with Just One Index. In: DELOS Workshop: Information Seeking, Searching and Querying in Digital Libraries (2000)
[7] Samet, H.: Foundations of Multidimensional and Metric Data Structures. The Morgan Kaufmann Series in Computer Graphics and Geometric Modeling. Morgan Kaufmann Publishers Inc., San Francisco (2005)
[8] Guttman, A.: R-Trees: A Dynamic Index Structure for Spatial Searching. In: Proceedings of the 1984 ACM SIGMOD International Conference on Management of Data, SIGMOD 1984, pp. 47–57. ACM, Boston (1984)
[9] Böhm, K., Mlivoncic, M., Schek, H.-J., Weber, R.: Fast Evaluation Techniques for Complex Similarity Queries. In: Proceedings of the 27th International Conference on Very Large Data Bases, VLDB 2001, pp. 211–220. Morgan Kaufmann Publishers Inc., San Francisco (2001)
[10] Ciaccia, P., Patella, M., Zezula, P.: M-Tree: An Efficient Access Method for Similarity Search in Metric Spaces. In: Proceedings of 23rd International Conference on Very Large Data Bases, VLDB 1997, pp. 426–435. Morgan Kaufmann, Athens (1997)
[11] Micó, M.L., Oncina, J., Vidal, E.: A New Version of the Nearest-Neighbour Approximating and Eliminating Search Algorithm (AESA) with Linear Preprocessing Time and Memory Requirements. Pattern Recogn. Lett. 15, 9–17 (1994)
[12] Lange, D., Naumann, F.: Efficient Similarity Search: Arbitrary Similarity Measures, Arbitrary Composition. In: Proceedings of the 20th ACM International Conference on Information and Knowledge Management, CIKM 2011, pp. 1679–1688. ACM, Glasgow (2011)

[13] Ciaccia, P., Patella, M., Zezula, P.: Processing Complex Similarity Queries with Distance-Based Access Methods. In: Schek, H.-J., Saltor, F., Ramos, I., Alonso, G. (eds.) EDBT 1998. LNCS, vol. 1377, pp. 9–23. Springer, Heidelberg (1998)

[14] Balko, S., Schmitt, I.: Signature Indexing and Self-Refinement in Metric Spaces. Tech. rep. 06/12. Brandenburg University of Technology Cottbus, Institute of Computer Science (September 2012)

[15] Bustos, B., Navarro, G., Chávez, E.: Pivot Selection Techniques for Proximity Searching in Metric Spaces. Pattern Recogn. Lett. 24, 2357–2366 (2003)

[16] Zellhöfer, D., et al.: PythiaSearch: A Multiple Search Strategy-Supportive Multimedia Retrieval System. In: Proceedings of the 2nd ACM International Conference on Multimedia Retrieval, ICMR 2012, pp. 59:1–59:2. ACM, Hong Kong (2012)

[17] Griffin, G., Holub, A., Perona, P.: Caltech-256 Object Category Dataset. Tech. rep. 7694. California Institute of Technology (2007)

[18] Schmitt, I., Balko, S.: Filter Ranking in High-Dimensional Space. Data Knowl. Eng. 56, 245–286 (2006)

[19] Rubner, Y., Tomasi, C., Guibas, L.J.: The Earth Mover's Distance as a Metric for Image Retrieval. Int. J. Comput. Vision 40, 99–121 (2000)

[20] Sikora, T.: The MPEG-7 Visual Standard for Content Description-An Overview. IEEE Transactions on Circuits and Systems for Video Technology 11(6), 696–702 (2001)

[21] Tamura, H., Mori, S., Yamawaki, T.: Texture Features Corresponding to Visual Perception. IEEE Transactions on Systems, Man and Cybernetics 8(6) (1978)

[22] Chatzichristofis, S.A., Boutalis, Y.S.: FCTH: Fuzzy Color and Texture Histogram - A Low Level Feature for Accurate Image Retrieval. In: Proceedings of the 2008 9th International Workshop on Image Analysis for Multimedia Interactive Services, WIAMIS 2008, pp. 191–196. IEEE Computer Society, Washington, DC (2008)

[23] Stehling, R.O., Nascimento, M.A., Falcão, A.X.: A Compact and Efficient Image Retrieval Approach Based on Border/Interior Pixel Classification. In: Proceedings of the 11th International Conference on Information and Knowledge Management, CIKM 2002, pp. 102–109. ACM, McLean (2002)

[24] Chatzichristofis, S.A., Boutalis, Y.S.: CEDD: Color and Edge Directivity Descriptor: A Compact Descriptor for Image Indexing and Retrieval. In: Gasteratos, A., Vincze, M., Tsotsos, J.K. (eds.) ICVS 2008. LNCS, vol. 5008, pp. 312–322. Springer, Heidelberg (2008)

[25] Chávez, E., Navarro, G., Baeza-Yates, R., Marroquín, J.L.: Searching in Metric Spaces. ACM Comput. Surv. 33, 273–321 (2001)

[26] Zellhöfer, D., Schmitt, I.: A User Interaction Model Based on the Principle of Polyrepresentation. In: Proceedings of the 4th Workshop for Ph.D. Students in Information and Knowledge Management, PIKM 2011, pp. 3–10. ACM, Glasgow (2011)

# An Operator-Stream-Based Scheduling Engine for Effective GPU Coprocessing

Sebastian Breß[1], Norbert Siegmund[1], Ladjel Bellatreche[2], and Gunter Saake[1]

[1] University of Magdeburg, Germany
{sebastian.bress,nsiegmun,gunter.saake}@ovgu.de
[2] LIAS/ISAE-ENSMA, Futuroscope, France
bellatreche@ensma.fr

**Abstract.** Since a decade, the database community researches opportunities to exploit graphics processing units to accelerate query processing. While the developed GPU algorithms often outperform their CPU counterparts, it is not beneficial to keep processing devices idle while over utilizing others. Therefore, an approach is needed that effectively distributes a workload on available (co-)processors while providing accurate performance estimations for the query optimizer. In this paper, we extend our hybrid query-processing engine with heuristics that optimize query processing for response time and throughput simultaneously via inter-device parallelism. Our empirical evaluation reveals that the new approach doubles the throughput compared to our previous solution and state-of-the-art approaches, because of nearly equal device utilization while preserving accurate performance estimations.

## 1 Introduction

General Purpose Graphical Processing is a promising approach to use *Graphics Processing Units* (GPUs) for general purpose computations [16]. In this line of research, the database community studied the advantages of this technologies for database systems starting from 2004 by accelerating database operators, such as column scans and aggregations [6]. Other approaches build on top of NVIDIA's Compute Unified Device Architecture to support relational operators [3,5,8].

Mostly, all of the aforementioned approaches aim at improving efficiency of database operations (i.e., performing them as fast as possible). Only few solutions address the challenge to utilize processing devices *effectively*, using the processing device that promises the highest gain w.r.t. an optimization criterion. There are two major classes of solutions in this field: (1) heterogeneous task scheduling approaches and (2) tailor-made coprocessing approaches. With (1), we do not know the specifics of database systems (e.g., fixed set of operations and data representations, access structures, optimizer specifics, concurrency control) and cannot be easily integrated in the optimizer of a *database management system* (DBMS), for example, because the DBMS has its own task abstractions (e.g., Augonnet and others [2] or Ilić and others [10]). With (2), we are tailored for specific operations in a specific system (e.g., He and others [8] or Malik and others [14]).

B. Catania, G. Guerrini, and J. Pokorný (Eds.): ADBIS 2013, LNCS 8133, pp. 288–301, 2013.

*Problem Statement.* We identify a lack of approaches that can exploit DBMS specific optimizations while being independent of an algorithm's implementation details or the hardware in the deployment environment. To close this gap, we present in prior work a decision model that distributes database operations on available processing devices [4]. The model uses a learning-based approach to be independent of implementation details and hardware while providing accurate cost estimations for database operations. We implemented our decision model in a *hybrid query-processing engine* (HyPE), which is designed to be applicable to any hybrid DBMS.[1] However, our decision model has two major drawbacks: (1) it only considers response time of single operations as optimization criterion and (2) it cannot achieve a performance improvement in case a processing device always outperforms the other processing devices.

*Contributions.* In this paper, we make the following contributions:
1. We introduce heuristics that allow us to (1) handle operator streams and (2) effectively utilize inter-device parallelism by adding new optimization heuristics for response time and throughput.
2. We provide an extension to HyPE, which implements the heuristics.
3. We present an exhaustive evaluation of our optimization heuristics w.r.t. varying parameters of the workload using micro benchmarks (e.g., to identify the most suitable heuristic).

The paper is structured as follows. We discuss in Section 2 our preliminary considerations. We present the operator model of our prototype DBMS, which we used for our evaluation, as well as HyPE's extensions in Section 3. We introduce our optimization heuristics in Section 4 and provide an exhaustive evaluation using micro benchmarks in Section 5.

## 2   Preliminary Considerations

In this Section, we provide a brief background over GPUs and our decision model [4], which we extend in this work to support operator-stream-based scheduling.

### 2.1   Graphics Processing Unit

GPUs are specialized processors for graphics applications. A new trend called *General Purpose Computation on Graphics Processing Units* (GPGPU) allows for a broader range of applications to benefit from the processing power of GPUs. Their main properties are: (1) They have higher theoretical processing power compared to CPUs for the same cost. (2) GPUs are optimized for high through-put, because they possess a highly parallel architecture and can efficiently handle thousands of threads concurrently. (3) They can process data only dormant in GPU RAM. Data not available in GPU RAM has to be transfered from the CPU RAM over the PCIe Bus, which is the bottleneck in a CPU/GPU system. Note that data can be concurrently processed on CPU and GPU while other data is copied by the direct memory access controller between CPU and GPU [16].

---

[1] A DBMS that implements for each operation a CPU and a GPU operator, such as *GDB* or *Ocelot*.

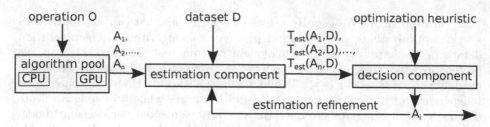

**Fig. 1.** Decision Model Architecture

## 2.2   Decision Model

In prior work, we proposed a decision model, which distributes database operations on CPU and GPU processing devices, so that an operation's response time is minimal [4]. The main idea is to assign each operation $O$ a set of algorithms (e.g., the algorithm pool $AP_O$), where each algorithm utilizes exactly one processing device (e.g., CPU or GPU). Hence, a decision for an algorithm $A$ using processing device $X$ leads to an execution of operation $O$ on $X$. This way, the model does not just decide on a processing device, but on a concrete algorithm on a processing device, thereby removing the need for a separate physical optimization stage without, for most queries and for the chosen hybrid architecture, significant loss of performance in most cases and execution contexts. For an incoming data set $D$, the execution times of all algorithms of operation $O$ are estimated using an *estimation component*, which passes the estimated execution times to a *decision component*. The decision component gets as additional parameter an optimization heuristic, which allows the user to tune for response time or throughput. The decision component returns the algorithm $A_i$ that has to be executed. We explain our heuristics in detail in Section 4. Figure 1 summarizes our decision model.

With the trend of GPGPU, it is very likely that other applications not known to the database system utilize the GPU as well. Depending on the application, this behavior might cause existing approaches to degrade in performance, because they assume that they have an exclusive right on utilizing the GPU resources. In contrast, our approach is able to notice changes in the environment and adjusts its scheduling decisions accordingly. This is a crucial property for use in a query optimizer, because the optimizer relies on cost estimations and decisions of HyPE.

The model is designed to be a stable basis for a query optimizer, because it:

1. Delivers reliable and accurate estimated execution times, which are used to compute the quality of a plan and enables the optimizer for a cost-based optimization and an accurate query-response-time estimation.
2. Refines its estimations at run-time, making them more robust in case of load changes.
3. Decides on the fastest algorithm and therefore, processing device for an optimizer.
4. Requires no a priori knowledge about the deployment environment, for example the hardware in a system, because it learns the hardware characteristics by observing the execution-time behavior of algorithms.

## 2.3    Optimization Criteria

Until now, we considered only response-time optimization [4], which works fairly well for scenarios where the CPU and GPU outperform each other depending on input data size and selectivity. However, for scenarios where the execution times of CPU and GPU algorithms do not have a break even point, meaning that they are equally fast for a given data set, response-time optimization is insufficient, because the faster processing device is over utilized while the slower processing device is idle. This was the main criticism of Pirk and others for operation-based scheduling [15]. Alternatively, they introduced a coprocessing technique, *bitwise distribution*, which utilizes the CPU and the GPU to process one operation.

A different way to approach the problem of *effective* utilization of processing resources in a CPU/GPU system is the purposeful use of slower processing devices to achieve inter-device parallelism. The main challenge is to optimize the response time of single operations, while optimizing throughput for an overall workload.

# 3    Operator Model and Extensions of HyPE

Next, we describe the operator model of CoGaDB, which we used for our evaluation and the extensions to support operator-stream-based scheduling in HyPE.

## 3.1    CoGaDB

We use our prototype CoGaDB as evaluation platform. CoGaDB is an in-memory, column-oriented, and GPU-accelerated DBMS.[2] It processes operators in two phases: First, the specified operator processes the input data and returns a list of *tuple identifiers* (TIDs) representing the result. Second, a materialization operator constructs the final result by applying the TID list on the input data. The first phase can be processed on the CPU or the GPU, respectively.

## 3.2    Extensions of HyPE

We implemented our decision model from prior work [4] and our heuristics, which we discuss in Section 4, in HyPE.[3] To support operation-stream-based scheduling, we refine our decision model as follows. Let $Op(O, D)$ be the application of the operation $O$ to the data set $D$. A workload $W$ is a sequence of operators:

$$W = Op_1(D_1, O_1)Op_2(D_2, O_2) \ldots Op_n(D_n, O_n) \tag{1}$$

Note that a query plan can be linearized into an operator stream by using materialization and chaining of outputs into inputs. Hence, a dataset $D_j$ can be the result of an operator $Op_i$ $(i < j)$, which allows data dependencies between operators. HyPE works in two phases: In the *training phase*, a part of the

---

[2] http://wwwiti.cs.uni-magdeburg.de/iti_db/research/gpu/cogadb
[3] http://wwwiti.cs.uni-magdeburg.de/iti_db/research/gpu/hype/

workload is used to train the model's approximation functions. In the *operational phase*, HyPE provides estimated execution times. On system startup, there are no approximation functions available such that HyPE cannot provide meaningful execution-time estimates, which may lead to poor results. Furthermore, processing-device utilization may change due to other applications running on the same server and also use the GPU. Hence, it is important to schedule not all operations at once, even if meaningful approximation functions exist. Based on these insights, we developed the following scheduling mechanism for HyPE.

We add a ready queue to each processing device. In case a ready queue is full, no more operators may be scheduled to the corresponding processing device. That way, HyPE keeps the processing devices busy, while maintaining the flexibility to react to changing processing device utilization due to other applications. Therefore, selecting the correct operator queue length is an important trade-off.

## 4    Optimization Criteria

In this Section, we discuss our main contribution, the heuristics for response time and throughput optimization for effective processing device utilization.

**Response Time.** Next, we discuss two heuristics for response-time optimization.

*Simple Response Time (SRT).* The decision component gets a set of operators with their estimated execution times as input. The SRT heuristic chooses the algorithm that is likely to have the smallest execution time [4]. The problem with SRT is that using always the fastest algorithm does not consider, when the corresponding algorithm is actually executed. If the model shifts the whole workload to the GPU, operators have to wait, until their predecessors are executed. Therefore, over utilization of a single device slows the processing of a workload down in two ways: (1) individual execution times are likely to increase, and (2) the waiting time until an operator can start its execution increases.

*Waiting-Time-Aware Response Time (WTAR).* We propose an optimization approach WTAR that is aware of the waiting time of operations on all processing devices and schedules an operation to the processing device with the minimal time the operation needs to terminate. WTAR is a modified version of the *heterogeneous earliest finishing time* (HEFT) algorithm [19]. In contrast to HEFT, WTAR is designed to schedule an operator stream, and therefore, does not assume a priori known workload. Furthermore, WTAR uses per operation cost estimations instead of the average algorithm (task) execution cost. This is because HyPE provides accurate performance estimations for algorithms. Let $OQ_X$ be the operator queue of all operators waiting for execution on processing device $X$ and $T_{est}(OQ_X)$ the estimated completion time of all operators in $OQ_X$:

$$T_{est}(OQ_X) = \sum_{Op_i \in OQ_X} T_{est}(Op_i) \qquad (2)$$

Furthermore, let $Op_{run}$ be the operator that is currently executed, $T_{fin}(Op_{run})$ the estimated time until $Op_{run}$ terminates, and $Op_{cur}$ the operator that shall be executed on the processing device that will likely yield the smallest response time, then the model selects the processing device $X$, where $\min(T_{est}(OQ_X) + T_{fin}(Op_{run} + T_{est}(Op_{cur})))$. Note that this approach avoids the overloading of one processing device, because it considers the time, an operator has to wait until it is executed. If this waiting time gets too large on processing device $X$ w.r.t. processing device $Y$, the model will choose $Y$, if possible.

**Throughput.** Next, we discuss heuristics that optimize a workload's throughput.

*Round Robin (RR).* Round robin is a simple and widely used algorithm [18], which assigns tasks alternating to available processing devices without considering task properties. We use it as a reference measure to compare our approaches with throughput-oriented algorithms. RRs simplicity is its major disadvantage: it only achieves good results in case processing devices execute tasks equally fast or else RR over/under utilizes processing devices, which may lead to significant performance penalties. An over utilization of a slow processing device is worse than over utilizing the fastest processing device, as in case of SRT. Hence, we propose a more advanced heuristic for throughput optimization in the following.

*Threshold-based Outsourcing (TBO).* Recall that a decision for an algorithm executes an operation on exactly one processing device (e.g., CPU merge sort on the CPU and GPU merge sort on the GPU). The problem with SRT is that it over utilizes a processing device in case one algorithm always outperforms the others. This violates the basic assumption of our decision model that all algorithms are executed regularly. Therefore, we modify SRT to choose a sub-optimal algorithm (and therefore, a sub-optimal processing device) under the condition that the operation is not significantly slower. Therefore, we need to keep track of (1) the passed time to decide, *when* a different algorithm (processing device) should be used and (2) the estimated slowdown, a sub-optimal algorithm execution may introduce to prevent an outsourcing of operations to unsuited processing devices (e.g., let the GPU process a very small data set).

With (1), we add a timestamp to all operations as well as their respective algorithms. Each operation $O$ is assigned to one logical clock $C_{log}(O)$, which basically provides the number of operation executions. Each time an algorithm $A$ is executed, its timestamp $ts(A)$ is updated to the current time value ($ts(A) = C_{log}(O)$). In case the difference of the timestamps of $A$ and $C_{log}(O)$ exceeds a threshold $\mathcal{T}_{log}$, the respective processing device $X$ of $A$ is considered idle. Therefore, an operation should be outsourced to $X$ in order to balance the system load on the processing devices. With (2), we introduce a new threshold, the maximal relative slowdown $MRS(A_{opt}, A)$ of algorithm $A$ compared to the response time minimal (optimal) algorithm $A_{opt}$.

The operation $O$ may be executed by a sub-optimal algorithm $A$ (using a different processing device) if and only if $A$ was not executed for at least $\mathcal{T}_{log}$ times for $O$ and the relative slowdown does not exceed the threshold $MRS(A_{opt}, A) >$ $\frac{T_{est}(A) - T_{est}(A_{opt})}{T_{est}(A_{opt})}$. For our experiments, we performed a pre-pruning phase to identify a suitable configuration of the parameters. We found that $MRS(A_{opt}, A) = 10\%$ and $\mathcal{T}_{log} = 2$ leads to good and stable performance.

# 5   Evaluation

To judge feasibility of our heuristics, we conducted several experiments that evaluate response time and throughput for two use cases: aggregations and column scans. We selected these use cases, because they are essential stand alone operations during database query processing, but are also sub-operations of complex operations such as the CUBE operator [7]. Although some optimization heuristics have already proven applicable (e.g., response time), we are interested in specific aspects for all optimization heuristics relevant for a database system. The goal of the evaluation is to answer the following research questions:

RQ1: Which of our optimization heuristics perform best under varying workload parameters?

RQ2: How does the optimization criterion impact the quality of estimated execution times?[4]

RQ3: Which optimization heuristic leads to best CPU/GPU utilization ratio and overall performance?

RQ4: How much overhead does the training phase introduce w.r.t. the workload execution time?

RQ5: Are certain optimization heuristics better for aggregations or column scans?

Providing answers for the aforementioned questions is crucial to meet the requirements for a database optimizer and to judge feasibility of our overall approach.

## 5.1   Experiment Overview

In the following, we describe our experiment that we conducted to answer the research questions. First, we present implementation details on our use cases as well as the experimental design (i.e., which benchmarks we used). Second, we discuss the experiment variables. Third, we present the analysis procedure.

*Aggregation.* A data set for an aggregation operation is a table with two integer columns in a key-value form whereas the key refers to a group for which their values (second column) needs to be aggregated. We use as aggregation function SUM, because it is a very common aggregation function in database systems.

*Column Scan.* A data set for a column scan operation is a table with one column. The values in the column are integer values ranging from 0 to 1000. The benchmark generates an operation by computing a random filter value $val \in \{0, \ldots, 1000\}$ and a filter condition $filt_{cond} \in \{=, <, >\}$.

---

[4] This is a major point, because HyPE should be utilizable for reliable execution-time cost estimation in query optimizers.

*Experimental Design.* The used benchmark is a crucial point to conduct a sound experiment. We use a micro benchmark for single operations in CoGaDB. The user has to specify three parameters: a maximal data-set size, the number of data sets in the workload and a data-set generation function, for which we input the first two parameters, and get in return a data set for the respective operation. The specified number of data sets is generated using a use-case-specific data generation function to allow an evaluation of HyPE without restricting generality. We measure the overall runtime, estimation error, device utilization, and training length for a workload. The test machine has a Intel® Core™i5-2500 CPU @3.30 GHz with 4 cores and 8 GB main memory, and a NVIDIA® GeForce® GT 640 GPU (compute capability 2.1) with 2 GB device memory. The operating system is Ubuntu 12.04 (64 bit) with CUDA 4.2. For all experiments, all datasets fit into main memory and swapping of the operating system is disabled by the benchmark tool to avoid side effects due to I/O operations.

*Variables.* We conduct experiments to identify which of our heuristics perform best under certain conditions. We evaluate our approach for the following variables: (1) number of operations in the workload ($\#op$), (2) number of different input data sets in a workload ($\#datasets$), and (3) maximal size of data sets ($size_{max}$).

*Analysis Procedure.* We evaluate our results separate for each use case using boxplots over all related experiments to prove that our optimizations heuristics are stable for the whole parameter space ($\#op$, $\#datasets$, $size_{max}$). We vary the three variables in a ceteris paribus analysis [17] with $(500, 50, 10MB)$ as base configuration and only vary one parameter at a time, leaving the other parameters constant (e.g., $(1000, 50, 10MB), (500, 100, 10MB), (500, 50, 20MB)$):

1. $\#op \in \{500, \dots, 8000\}$
2. $\#datasets \in \{50, \dots, 500\}$
3. $size_{max} \in \{10MB, \dots, 150MB\}$

Note that higher values for $\#datasets$ or $size_{max}$ would result in a database exceeding our main memory and hence, violating our in-memory assumption. As quality measures, we consider (1) the speedup w.r.t. the execution of a workload on the fastest processing device, which can be obtained using static scheduling approaches (e.g., Kerr and others [12]), (2) average estimation errors, which is ideally zero, and (3) device utilization. In case the workload is unevenly distributed, one processing device is over utilized, whereas others are under utilized, increasing execution skew. An ideal device utilization in a scenario of $n$ processing units is that each processing units processes $1/n$ of the workload. For our test environment, a perfect utilization would be to use 50% of workload execution time on CPU and 50% on GPU. (4) Finally, we investigate the relative training times depending on the optimization heuristics.

## 5.2   Results

Now, we present only the results of the experiments. In Section 5.3, we answer the research questions and discuss the achieved speedups, estimation accuracy,

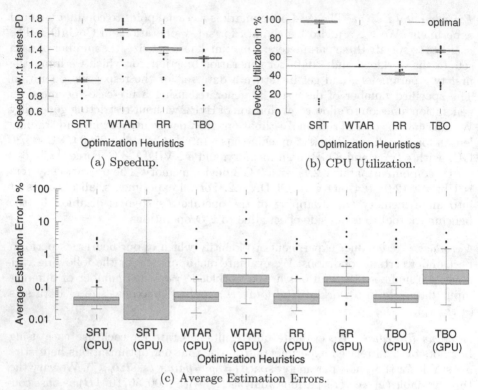

(c) Average Estimation Errors.

PD – Processing Device, SRT – Simple Response Time, WTAR – Waiting Time Aware
Response Time, RR – Round Robin, TBO – Threshold-based Outsourcing

**Fig. 2.** Aggregation Use Case

device utilization and relative training times of the heuristics. The results are
accumulated over all experiments and displayed as box plots to illustrate the typical characteristics (e.g., mean and variance) w.r.t. a quality measure. A box plot
visualizes a data distribution by drawing the median, the interquartile ranges as
box, and extremes as whiskers [1]. Note that 50% of the points are in the box,
and 95% are between the whiskers. Outliers are drawn as individual points.

**Results for Aggregations.** Figures 2(a) illustrates the achieved speedups for
the fastest processing device of our optimization heuristics over all experiments.
To answer RQ1, we observe that (1) SRT has no significant speedup compared
to the fastest processing device, because the box plot is located at 100%, which
means that only a single device (CPU) is utilized. (2) WTAR is significantly faster
than RR, and (3) TBO is inferior to RR and WTAR, but achieves higher performance compared to SRT. Figure 2(b) illustrates the device utilization over all
experiments. The gray horizontal line exemplifies the ideal device utilization. To
answer RQ3, we observe that (1) SRT has a very asymmetric device utilization.
(2) We see that the box plot of WTAR lies on the horizontal line, which represents the best utilization. WTAR and RR are performing best, whereas RR has

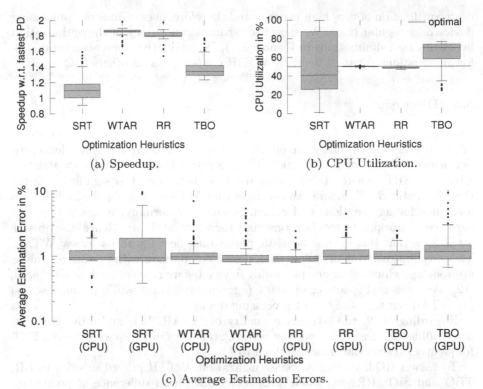

(a) Speedup.                    (b) CPU Utilization.

(c) Average Estimation Errors.

PD – Processing Device, SRT – Simple Response Time, WTAR – Waiting Time Aware
Response Time, RR – Round Robin, TBO – Threshold-based Outsourcing

**Fig. 3.** Column Scan Use Case

slightly worse device utilization compared to WTAR and is slightly better than
TBO. Figure 2(c) shows the estimation accuracy of the optimization heuristics.
To answer RQ2, we observe that (1) the accuracy is typically higher for CPU algorithms compared to GPU algorithms, (2) SRT leads to disappointing accuracy
for GPU algorithms, whereas the other optimization heuristics are acceptable, because the estimation error is smaller than our defined 5% threshold.

**Results for Column Scans.** Figure 3(a) illustrates the achieved speedups
for the fastest processing device of our optimization heuristics over all experiments. We see that WTAR has the lowest response time, which answers RQ1.
Furthermore, we observe that (1) SRT achieves a considerable speedup (up to
40%). This is because for column scans CPU and GPU algorithms are nearly
equally fast, whereas the GPU outperforms the CPU by aggregations. (2) This
property also causes the RR heuristic to be nearly as fast as WTAR. Figure 3(b)
illustrates the device utilization over all experiments. To answer RQ3, we observe
that (1) SRT has a very high variance (column scans), indicating that it is not
suitable for effective task distribution and (2) WTAR and RR achieve nearly
ideal device utilization, whereas TBO tends to over utilize the CPU. SRT has a

device utilization of very high variance and therefore, over utilizes one processing device on a regular basis. We show the estimation accuracy of the optimization heuristics for column scans in Figures 3(c). We make the same observations as for aggregations. That is, heuristic WTAR outperforms all others (RQ2).

## 5.3   Discussion

Overall, WTAR outperforms the other heuristics, especially when relative speed of processing devices differs. To ensure that our results are not coincidence, we performed a t-test with $\alpha = 0.001$ [1]. The result is that WTAR $>>$ RR $>>$ TBO $>>$ SRT, where $A >> B$ means that heuristic $A$ is significant faster than heuristic $B$. Therefore, we conclude that WTAR achieves the highest performance for aggregations and column scans. Furthermore, it has a very low variance in workload execution time and therefore, stability. Hence, the answer for RQ5 is that there is one heuristic performing best for all use cases: WTAR. The speedup experiments allow for a direct comparison with static scheduling approaches, which select one processing device before runtime such as Kerr et al. [12]. We measured speedups of $\approx 60\%$ (aggregations) and $\approx 80\%$ (column scans) for WTAR w.r.t. to the fastest processing device.

Regarding RQ2, the estimation quality of WTAR, RR, and TBO is stable across different use cases over the investigated parameter space, whereas SRT frequently exceeds the error threshold (5%).

To answer RQ3, we discuss device utilization. WTAR proved superior to RR, TBO and SRT. RR gets worse with increasing speed difference of processing devices. In contrast, WTAR delivers nearly ideal device utilization with marginal variance over a large parameter space. SRT has either a very asymmetric device utilization (aggregations) or has a very high variance (column scans) indicating that it is not suitable for effective task distribution. Overall, SRT performs poor in case there is no break-even point of CPU and GPU algorithms execution-time curves, which we observe for our use cases. This is because, SRT over utilizes one processing device, resulting in execution skew and increasing overall workload execution time. However, our prior work clearly shows the benefit of SRT in case a break-even point exists [4].

To answer RQ4, we investigate HyPE's overhead by measuring the training time and compute the relative training time w.r.t. to the workload execution time for aggregations (Figure 4(a)) and columns scans (Figure 4(b)). It is clearly visible that the performance impact of the training is marginal.

*Summary.* For all use cases, WTAR outperformed all other optimization heuristics in terms of performance, estimation accuracy, and equal processing device utilization. In some experiments, RR caused slightly better estimated execution times. Overall, we observe that estimation accuracy strongly depends on the optimization heuristics (RQ2). RR is likely to perform worse than WTAR in case a processing device is significantly faster than the others, because in this case RR leads to an uneven device utilization. We conclude that out of the considered optimization heuristics, WTAR is the most suitable for use in a database optimizer (RQ1–5).

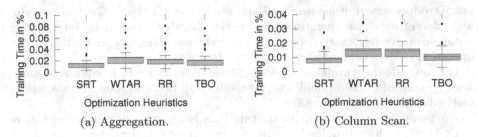

(a) Aggregation.

(b) Column Scan.

Fig. 4. Relative Training Time of Use Cases

## 5.4 Threats to Validity

We now discuss threats to internal and external validity.

*Threats to Internal Validity.* We performed a t-test to ensure that our results are statistically sound and did not occur by pure chance. Furthermore, we have to consider measurement bias when measuring execution times and device utilization. Therefore, we repeated each experiment during our ceteris paribus analysis five times and included all measurements in our box plots, because outliers are plotted as well. This allows a precise evaluation of the reliability of our approach.

*Threats to External Validity.* We are aware that using micro benchmarks does not automatically reflect the performance behavior of real-world DBMS. However, we argue that (1) they are a necessity for an in-depth analysis of our optimization heuristics and (2) we selected a representative set of query types that are very common. Furthermore, the implementation details of database operators in real world DBMS differ. We counter that by using a learning-based approach to allow accurate performance predictions without knowing the algorithms in detail. We address hardware heterogeneity in the same way. We performed experiments on other machines and obtained similar results compared to this paper.

## 6   Related Work

We focus on scheduling, not optimizing algorithms itself. Other work contributes optimized algorithms for aggregation on CPUs (e.g., Zhao and others [21]) as well as GPUs (e.g., Lauer and others [13] or Malik and others [14]). Wu and others develop a GPU-acceleration approach for column scans [20].

He and others developed GDB, a GPU-accelerated DBMS with integrated query processor [8]. In contrast to HyPE, they use an analytical cost model, which needs to be updated for each new generation of GPUs. Furthermore, their model cannot adapt to load changes.

Malik and others proposed a tailor-made scheduling approach for OLAP in hybrid CPU/GPU environments [14]. They introduce an analytical calibration-based-cost model to estimate runtime on CPUs and GPUs. Since the approach is specific to their implementation, it cannot be easily applied on other DBMS in contrast to HyPE. Kerr and others developed a model, which selects CPU and GPU algorithms statically before runtime [12]. Hence, their approach does not introduce any runtime overhead and can utilize CPU and GPU at runtime for different database operations. The major drawback is that no inter-device parallelism can be achieved for a single operation class, because either every operation in the workload is executed on CPU or GPU.

Iverson and others proposed a learning-based approach which requires no hardware specific information similar to our model [11]. However, our used statistical methods and architectures differ.

Augonnet and others introduced StarPU, a heterogeneous scheduling framework that provides a unified execution environment and runtime system [2]. StarPU can distribute parallel tasks in environments with heterogeneous processors such as hybrid CPU/GPU systems and can construct performance models automatically, similar to HyPE. Ilić and others developed CHPS, an execution environment similar to HyPE and StarPU [10]. CHPS main features are (1) support of a flexible task description mechanism, (2) overlapping of processor computation and data transfers and (3) automatic construction of performance models for tasks. Ilić and others applied CHPS on TPC-H queries Q3 and Q6. They observed significant performance gains, but used tailor-made optimizations for the implementation of the queries [9]. A major problem of existing approaches is the high integration effort for DBMS and the fact that the optimizer needs to use the task abstractions of the scheduling frameworks (e.g., CHPS and StarPU). Since optimizer of existing DBMS are extremely complex, an approach is needed that allows for minimal invasive integration in the optimizer, while enabling the optimizer for effective GPU coprocessing. We develop HyPE to close this gap.

## 7    Conclusion

Effective GPU coprocessing is an open challenge yet to overcome in database systems. In this paper, we extended our hybrid query processing engine by the capability to handle operator streams and optimization heuristics for response time and throughput. We validated our extensions on two use cases, namely aggregations and column scans. While these are important stand alone operations, they are also part of complex operations, such as the CUBE operator. We achieved speedups up to 2 compared to our previous solution and static scheduling approaches while delivering accurate performance estimations for CPU and GPU operators without any a priori information on the deployment environment.

**Acknowledgements.** The work of Siegmund is supported by the German ministry of education and science (BMBF), number 01IM10002B. We thank Tobias Lauer from Jedox AG and the anonymous reviewers for their helpful feedback.

## References

1. Anderson, T., Finn, J.D.: The New Statistical Analysis of Data, 1st edn. Springer (1996)
2. Augonnet, C., Thibault, S., Namyst, R., Wacrenier, P.-A.: StarPU: A Unified Platform for Task Scheduling on Heterogeneous Multicore Architectures. Concurrency and Computation: Practice & Experience 23(2), 187–198 (2011)
3. Bakkum, P., Skadron, K.: Accelerating SQL Database Operations on a GPU with CUDA. In: GPGPU, pp. 94–103. ACM (2010)

4. Breß, S., Beier, F., Rauhe, H., Schallehn, E., Sattler, K.-U., Saake, G.: Automatic Selection of Processing Units for Coprocessing in Databases. In: Morzy, T., Härder, T., Wrembel, R. (eds.) ADBIS 2012. LNCS, vol. 7503, pp. 57–70. Springer, Heidelberg (2012)
5. Diamos, G., Wu, H., Lele, A., Wang, J., Yalamanchili, S.: Efficient Relational Algebra Algorithms and Data Structures for GPU. Technical report, Center for Experimental Research in Computer Systems (CERS) (2012)
6. Govindaraju, N.K., Lloyd, B., Wang, W., Lin, M., Manocha, D.: Fast Computation of Database Operations using Graphics Processors. In: SIGMOD, pp. 215–226. ACM (2004)
7. Gray, J., Chaudhuri, S., Bosworth, A., Layman, A., Reichart, D., Venkatrao, M., Pellow, F., Pirahesh, H.: Data Cube: A Relational Aggregation Operator Generalizing Group-By, Cross-Tab, and Sub-Totals. Data Mining and Knowledge Discovery 1(1), 29–53 (1997)
8. He, B., Lu, M., Yang, K., Fang, R., Govindaraju, N.K., Luo, Q., Sander, P.V.: Relational Query Co-Processing on Graphics Processors. ACM Trans. Database Syst. 34, 21:1–21:39 (2009)
9. Ilić, A., Pratas, F., Trancoso, P., Sousa, L.: High-Performance Computing on Heterogeneous Systems: Database Queries on CPU and GPU. In: High Performance Scientific Computing with Special Emphasis on Current Capabilities and Future Perspectives, pp. 202–222. IOS Press (2011)
10. Ilić, A., Sousa, L.: CHPS: An Environment for Collaborative Execution on Heterogeneous Desktop Systems. International Journal of Networking and Computing 1(1), 96–113 (2011)
11. Iverson, M., Ozguner, F., Potter, L.: Statistical Prediction of Task Execution Times Through Analytic Benchmarking for Scheduling in a Heterogeneous Environment. In: HCW, pp. 99–111 (1999)
12. Kerr, A., Diamos, G., Yalamanchili, S.: Modeling GPU-CPU Workloads and Systems. In: GPGPU, pp. 31–42. ACM (2010)
13. Lauer, T., Datta, A., Khadikov, Z., Anselm, C.: Exploring Graphics Processing Units as Parallel Coprocessors for Online Aggregation. In: DOLAP, pp. 77–84. ACM (2010)
14. Malik, M., Riha, L., Shea, C., El-Ghazawi, T.: Task Scheduling for GPU Accelerated Hybrid OLAP Systems with Multi-core Support and Text-to-Integer Translation. In: 26th International Parallel and Distributed Processing Symposium Workshops & PhD Forum (IPDPSW), pp. 1987–1996. IEEE (2012)
15. Pirk, H.: Efficient Cross-Device Query Processing. In: The VLDB PhD Workshop. VLDB Endowment (2012)
16. Sanders, J., Kandrot, E.: CUDA by Example: An Introduction to General-Purpose GPU Programming, 1st edn., vol. 186, pp. 2–6. Addison-Wesley Professional (2010)
17. Schlicht, E.: Isolation and Aggregation in Economics, 1st edn. Springer (1985)
18. Tang, X., Chanson, S.: Optimizing Static Job Scheduling in a Network of Heterogeneous Computers. In: ICPP, pp. 373–382. IEEE (2000)
19. Topcuouglu, H., Hariri, S., Wu, M.-Y.: Performance-Effective and Low-Complexity Task Scheduling for Heterogeneous Computing. IEEE Trans. Parallel Distrib. Syst. 13(3), 260–274 (2002)
20. Wu, R., Zhang, B., Hsu, M., Chen, Q.: GPU-Accelerated Predicate Evaluation on Column Store. In: Chen, L., Tang, C., Yang, J., Gao, Y. (eds.) WAIM 2010. LNCS, vol. 6184, pp. 570–581. Springer, Heidelberg (2010)
21. Zhao, Y., Deshpande, P.M., Naughton, J.F.: An Array-Based Algorithm for Simultaneous Multidimensional Aggregates. In: SIGMOD, pp. 159–170. ACM (1997)

# GPU-Accelerated Collocation Pattern Discovery*

Witold Andrzejewski and Pawel Boinski

Poznan University of Technology, Institute of Computing Science, Piotrowo 2, 60-965 Poznan, Poland

**Abstract.** Collocation Pattern Discovery is a very interesting field of data mining in spatial databases. It consists in searching for types of spatial objects that are frequently located together in a spatial neighborhood. Application domains of such patterns include, but are not limited to, biology, geography, marketing and meteorology. To cope with processing of these huge volumes of data programmable high-performance graphic cards (GPU) can be used. GPUs have been proven recently to be extremely efficient in accelerating many existing algorithms. In this paper we present GPU-CM, a GPU-accelerated version of iCPI-tree based algorithm for the collocation discovery problem. To achieve the best performance we introduce specially designed structures and processing methods for the best utilization of the SIMD execution model. In experimental evaluation we compare our GPU implementation with a parallel implementation of iCPI-tree method for CPU. Collected results show order of magnitude speedups over the CPU version of the algorithm.

## 1 Introduction

The enormous growth of spatial databases limits human abilities to interpret such data and to make useful conclusions. Automatic methods, known as *Knowledge Discovery in Databases* (KDD) are therefore required. KDD has been defined as a non-trivial process of discovering valid, novel, and potentially useful, and ultimately understandable patterns in large data volumes [8]. The most interesting part of this process is called data mining and consists in application of specially designed algorithms to find particular patterns in data.

Popular spatial data mining tasks include spatial clustering, spatial outliers detection, spatial classification and spatial associations. In this work we focus on the problem of detecting classes of spatial objects (the so-called spatial features) that are frequently located together. Each spatial feature can be interpreted as a characteristic of space in a particular location. Typical examples of spatial features include species, business types or points of interest (e.g., hospitals, airports etc.). For example, a mobile company providing multiple services for customers can be interested in relationships between particular factors in the neighborhood of mobile service requests. In ecology and meteorology, the co-occurrence among natural phenomenons can be very interesting for scientists [12].

---

* This paper was funded by the Polish National Science Center (NCN), grant No. 2011/01/B/ST6/05169.

B. Catania, G. Guerrini, and J. Pokorný (Eds.): ADBIS 2013, LNCS 8133, pp. 302–315, 2013.

Shekhar and Huang defined this data mining task as a *collocation pattern discovery* [11]. A spatial collocation pattern (or in short a *collocation*) is a set of spatial features that are frequently located together in a spatial proximity. Identification of such patterns requires computationally demanding step of searching for all instances of these patterns. Many algorithms for collocation pattern discovery problem have been developed [11,12,13,14,15,16]. However, no solutions utilizing hardware support to accelerate collocation pattern discovery have been proposed yet.

In this paper we propose an algorithm which utilizes the power of modern graphics processing units (GPUs) to accelerate the state of the art algorithm for the collocation pattern discovery.

The structure of this paper is as follows. In section 2 we formally define the terms used throughout the rest of the paper. In section 3, we present the state of the art algorithm for collocation pattern discovery and introduce basic concepts of general processing on GPUs. Section 4 presents our contribution - the GPU-accelerated version of the collocation mining algorithm. The results of experimental evaluation are presented in section 5. We summarize our paper and present plans for future work in section 6.

## 2    Definitions

In this section we introduce the basic collocation pattern mining concepts and definition of the collocation pattern mining problem.

**Definition 1.** *Let $f$ be a spatial feature. An object $x$ is an* **instance** *of the feature $f$, if $x$ is a type of $f$ and is described by a location and unique identifier. Let $F$ be a set of spatial features and $S$ be a set of their instances. Given a neighbor relation $R$, we say that the* **collocation pattern** *$C$ is a subset of spatial features $C \subseteq F$ whose instances $I \subseteq S$ form a clique w.r.t. the relation $R$.*

**Definition 2.** *The* **participation ratio** *$Pr(C, f_i)$ of a feature $f_i$ in the collocation $C = \{f_1, f_2, \ldots, f_k\}$ is a fraction of objects representing the feature $f_i$ in the neighborhood of instances of collocation $C - \{f_i\}$. $Pr(C, f_i)$ is equal to the number of distinct objects of $f_i$ in instances of $C$ divided by the number of all instances with feature $f_i$. The* **participation index (prevalence measure)** *$Pi(C)$ of a collocation $C = \{f_1, f_2, \ldots, f_k\}$ is defined as $Pi(C) = \min_{f_i \in C}\{Pr(C, f_i)\}$.*

**Theorem 1.** *The participation ratio and participation index are monotonically non-increasing with increases in the collocation size.*

The collocation pattern mining is defined as follows. Given (1) a set of spatial features $F = \{f_1, f_2, \ldots, f_k\}$ and (2) a set of their instances $S = S_1 \cup S_2 \cup \ldots \cup S_k$ where $S(1 \leq i \leq k)$ is a set of instances of feature $f_i \in F$ and each instance that belongs to $S$ contains information about its feature type, instance id and location, (3) a neighbor relationship $R$ over locations, (4) a minimum prevalence threshold ($min\_prev$), find efficiently a correct and complete set of collocation patterns with a participation index $\geq min\_prev$.

# 3   Related Work

## 3.1   The iCPI-Tree Based Collocation Pattern Discovery

The general approach to collocation mining problem has been proposed in [11] and consists in three major steps. In the first step, a well-known Apriori strategy [1] is used to generate candidate collocations utilizing anti-monotonicity property of the prevalence measure. In the second step instances of such candidates are identified. Finally, in the last step, the prevalence measure is computed for each candidate. Candidates with the prevalence below the given threshold are filtered out.

Although there are researches that do not follow the aforementioned Apriori strategy (e.g., maximal collocation patterns [16], density based collocation patterns [14]), the general approach is the most popular one. Among the most notable general approach methods are *Co-Location Miner* [11], *Joinless* [15] and current state of the art *iCPI-tree* based method [13]. In the next paragraphs we briefly describe the idea behind this algorithm.

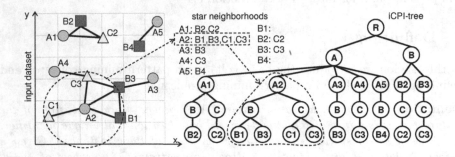

**Fig. 1.** Sample dataset and the corresponding iCPI-tree

At the beginning each spatial feature is denoted as a one element collocation. Next, two element candidates are generated following the Apriori strategy. To compute their prevalences, a list of their instances is required. In the iCPI-tree method, the concept of star neighborhoods (originally introduced in [15]) is used. For each object in space, a list of all neighbors with spatial features greater than the feature of this particular object is called a star neighborhood. Such information is stored in the form of an *iCPI-tree*. Each child of the root node is a subtree that contains neighbors for instances of a specific spatial feature. Sub-trees are composed of nodes representing spatial features of neighbors and leafs corresponding to neighbor instances. A sample dataset and a corresponding iCPI-tree is shown in Fig. 1. For example, a star neighborhood for object $A1$ consists of objects $B1, B3, C1, C3$ and is represented by a subtree below a node $A$ in the iCPI-tree. To identify instances of a candidate, e.g, $B, C$ we can easily read all neighbors with $C$ feature of $B$ instances ($B2, C2$ and $B3, C3$). In $n$-th iteration of the algorithm, $n$ size candidates are processed. For $n = 2$ all instances generated from the tree are cliques (Def. 1). For $n > 2$ the following method for generating collocation instances is used. To identify instances of $n$ size collocation candidates, a set of instances from $n - 1$ iteration is used. For each $n$ size candidate, instances of the prevalent $n - 1$ size collocation with the same first $n - 1$ features as candidate are expanded. Only common neighbors of all collocation instance objects can be used. For example, given the candidate $A, B, C$ we use consecutive instances of the collocation $A, B$, e.g., the instance $A2, B3$. We try to extend it with instances of feature $C$. To get clique instances of $A, B, C$ we look for $C$ instances that are simultaneously neighbors of $A2$ as well as of $B3$. Using the obtained iCPI-tree we can easily find that $A2$ has neighbors $C1$ and $C3$ while $B3$ has only the neighbor $C3$. Therefore, the only common neighbor is $C3$ and a new instance of candidate is $A2,B3,C3$. Such processing is repeated for each candidate. The algorithm stops when there are no new candidates. For details of the iCPI-tree based algorithm please consult the paper [13].

## 3.2   General Processing on Graphics Processing Units

The rapid development of computer graphics cards has led to creating powerful devices performing highly parallel single instruction multiple data computations. This high performance may be utilized not only for graphics applications but also for any general processing applications. Newly developed APIs such as NVIDIA CUDA [10] and OpenCL [9] allow to relatively easily develop programs which utilize *graphics processing units* (GPUs) of graphics cards to accelerate their normal data processing tasks. In our solutions we utilize NVIDIA CUDA, though many of the results presented in this paper are also applicable to OpenCL based applications. NVIDIA CUDA is a low level API which while designed to be universal, is currently only implemented for NVIDIA GPUs. This will probably change in short time as there are new compilation frameworks currently in development, which allow to run CUDA programs on AMD and Intel GPUs as well [7]. Below we give a short description of NVIDIA CUDA API and its capabilities.

Computation tasks are submitted to GPUs in a form of kernels. A kernel is a function which is composed of a sequence of operations which need to be performed concurrently in multiple threads. Threads are divided by the programmer into equally sized *blocks*. A block is a one, two or three dimensional array of at most 1024 threads (or 512 depending on the graphics cards' architecture), where each thread can be uniquely identified by its position in this array. The set of blocks forms a so-called *computation grid*. Threads in a single block may communicate by using the same portion of the so-called shared memory which is physically located on-chip and therefore very fast. Threads running in different blocks may only communicate through the very slow *global memory* of the graphics card. Different memory types have different efficient access pattern requirements. Synchronization capabilities of the threads are limited. Threads in a single block may be synchronized; however, global synchronization of threads is achievable only by means of a costly workaround. Threads in a block are executed in 32 thread SIMD groups called warps (all of these threads should perform the same instruction at the same time). A programmer implementing an algorithm using the NVIDIA CUDA API must take into account all of these low level GPU limitations to obtain the most efficient implementations.

To facilitate creation of programs performing parallel computations (not necessarily GPUs) many parallel primitives have been developed. For the purpose of solutions presented in this paper we utilize the following: *inclusive* and *exclusive scan*, *compact*, *sort*, *unique*, *reduce* and *reduce by key*. Most of these primitives are implemented for GPUs in such libraries as *Thrust* [3]. In our implementation we use this library though we implement our own version of the compact algorithm. Below we give short description of each primitive.

Given any array $a$ an *inclusive scan* finds an array $b$ of the same size such that each $b[i] = \sum_{k=1}^{i} a[i]$. An *exclusive scan* works similarly. For any array $a$ an *exclusive scan* finds and array $b$ of the same size such that each $b[i] = \sum_{k=1}^{i-1} a[i]$. Any associative binary operator may be used instead of sum. A *compact* algorithm given any array $a$ removes all entries that fulfill some condition. Most commonly an additional array of flags storing 0 and 1 (remove/keep) is supplied. A *sort* algorithm sorts any array. Sorting may also be based on some additional key array. A *unique* algorithm is used to find a set of all distinct values stored in a user supplied array. Thrust implementation requires data to be sorted first. A *reduce* algorithm performs reduction of all values within a user specified array $a$ by using any associative binary operator such as sum, i.e., it can be used to find a sum of all values within an array. A *reduce by key* algorithm is a more advanced version of *reduce* algorithm. Given two equally sized arrays $k$ and $v$ where array $k$ stores keys and array $v$ stores values, reduce by key performs reduction of all values belonging to the same key, i.e., for each distinct key value a single reduced result is obtained. Thrust implementation requires data to be sorted by key value.

## 4   The GPU-CM Algorithm

Our GPU-CM algorithm assumes that the iCPI-tree has already been built. Efficient algorithms for constructing this structure from raw input data already exist [13,15]. Moreover this algorithm step takes only a small fraction of the whole mining process time [4].

### 4.1   Data Structures

Following the solutions presented in [4] we represent an iCPI-tree as a hash map. Each entry in the hash map stores a list of instances of a single spatial feature $f_1$ which are neighbors of a single instance $i$ of some other spatial feature $f_2$. This structure is implemented in GPU memory as follows.

Lists of instance neighbors are stored in a single memory block. Due to performance reasons, each list has a constant length $L$ which can be any power of 2 (up to 32). If the number of neighbors is smaller than this value some entries are left empty. The memory block size is therefore equal to the number of lists times $L$. Each entry on a list is an instance representation. Each feature instance is represented as a 32 bit word where the least significant 16 bits store instance number and more significant 8 bits are used for storing the feature number. The most significant 8 bits are set to zero. An empty entry is represented by the value $FFFFFFFFh$. Each list occupies continuous $L \times 4$ bytes of memory. Each list is sorted. We will denote this data structure as an *instance neighbor buffer*.

A hash map is represented by two arrays: *keys* and *values*. Each key stores feature $f_1$ number as well as feature $f_2$ number and instance number $i$. This is encoded on a 32 bit word where the least significant 16 bits encode instance number $i$, more significant 8 bits encode feature $f_2$ identifier and the most significant 8 bits encode the feature $f_1$ number. Each entry in the *keys* array stores either a key value or is empty (equal to $FFFFFFFFh$). We use the open addressing scheme for hashing [2]. The array *values* stores, for each key stored in *keys* array, a pointer to the start of the appropriate list in the memory block described above. We will denote this hash map as an *instance neighbor hash map*.

During the collocation pattern mining process, the algorithm stores for each pattern (or a candidate for one) a list of its instances. To accelerate the search for instances of a specified pattern, another hash map is employed. This structure is more complicated than iCPI-tree representation and is implemented as follows.

Depending on the number of features, each pattern is represented as a sequence of several 32 bit words, where each bit corresponds to a single feature. We denote the number of pattern words as $BL$. All patterns are stored in a single memory block of size equal to the number of patterns times $BL$. Due to the access methods employed during collocation mining each pattern does not occupy continuous regions of memory. Instead, consecutive words in memory store first words of all patterns, then second words and so on. Patterns are sorted in the lexicographic order. We will denote this structure as *a pattern buffer*. Pattern instances are stored similarly in a separate memory block. Each pattern instance is represented as a sequence of 32 bit words where each word is a feature instance encoded as described at the beginning of this section. Consequently, this memory block's size is equal to the number of all pattern instances times the pattern length. Similarly as with patterns, continuous regions of memory store first words of all pattern instances, then second words and so on. Pattern instances of a single pattern occupy neighboring regions of memory, i.e., pattern instances are not interleaved. Moreover, pattern instances are sorted in the lexicographic order (within the groups corresponding to their patterns). We will denote this structure as a *pattern instance buffer*.

A hash map used for finding instances of a pattern is represented by two arrays: *keys* and *values*. Array *keys* stores pointers to starting positions of corresponding patterns in the pattern memory block described above or nulls if entry is empty. Array *values* stores, for each corresponding key, records that consist of: the number of pattern instances, the prevalence of the pattern and the pointer to the first word of a first pattern instance in the pattern instance memory block. We use open addressing scheme for hashing. We will denote this hash map as a *pattern hash map*.

One might notice that the *keys* array of the pattern hash map could store patterns instead of pointers to patterns. The current solution was used to achieve atomic insertions into the hash map. In general we follow a solution presented in [2] where a parallel hash map based on open addressing scheme is introduced (among others). This solution is based on compare and swap scheme. Unfortunately the *atomicCAS* CUDA function (perfoming atomic compare and swap) works only on 32 bit and 64 bit words. Atomic insertion of longer patterns is not possible by means of this function. We have decided to atomically store pointers (which are either 32 or 64 bit) to larger representations of patterns instead.

## 4.2   Initialization of Mining

The mining process starts with reading of the initial iCPI-tree and construction of iCPI-tree hashmap. Instance neighbor buffer is constructed sequentially. Additional data such as numbers of instances of each feature are also retrieved. Moreover, a temporary array of all instance neighbor hash map keys and values (pointers to lists) is sequentially constructed as well. Finally, all of these key - value pairs are inserted into the instance neighbor hash map in parallel.

After the instance neighbor buffer and the instance neighbor hash map are constructed, a pattern buffer, a pattern instance buffer and a pattern hash map for patterns of length 1 are constructed as well. While we have designed a parallel algorithm for this step, we will omit the details as the processing time of this step is negligible.

## 4.3   Generation of Candidates - Pattern Join

Generation of $n$-size candidates utilizes a pattern buffer which stores $n-1$ size patterns. As patterns in the pattern buffer are sorted in the lexicographic order, all $n-1$ size patterns that may be joined into an $n$ size pattern form groups of several consecutive, joinable patterns. We will refer to these groups as *join groups*. At the final stage of the algorithm, the join groups will be converted into *result groups*. A result group is a group of several consecutive patterns that are obtained from joining all patterns within a single join group. The order of groups is retained, i.e., given any two join groups $A$ and $B$, if group $A$ is before group $B$ in the input pattern buffer, the corresponding result groups will be in the same order in the output pattern buffer (unless one of the join groups contains only a single pattern and therefore does not create any join results). If the patterns within a result group are sorted in lexicographic order then this property will guarantee that the resulting pattern buffer is sorted globally.

The pattern join algorithm works in several sequential stages, though each stage can be composed of parallel operations. First stages identify join groups, find their number (denoted $k_{JG}$), compute their size and the size of corresponding result groups (the number of combinations of two). Several important arrays are created:

- *groupSizes* - an array of size $k_{JG}$ which contains for each join group its size,

– *joinCounts* - an array of size $k_{JG}$ which contains for each join group the size of corresponding result group,
– *positions* - an array of size $k_{JG}$ which contains for each join group an index of the last pattern for this group within the input pattern buffer,
– *scannedJoinCounts* - an array of size $k_{JG}$ which is a result of exclusive scan operation performed on *joinCounts* array.

Next stages compute an additional important auxiliary array called *scanned-JoinFlags* of size equal to the number of result patterns ($k_P$). The obtained array contains (for each pattern in each result group) a reference number of the corresponding join group.

The final stage creates the final pattern join. Each *join group* is converted into a *result group*. First, a pattern buffer of size $k_p \times BL$ is allocated (recall that $BL$ is the number of 32 bit words needed to encode a single pattern). Next, $k_p$ threads are started. Each thread, based on the array *scannedJoinFlags* determines the reference number of the corresponding join group. Given this value, the thread retrieves from the *scannedJoinCounts* array the position at which its corresponding result group should start in the result pattern buffer. The thread also computes the difference between its global number and the retrieved position to find its position *pos* within the corresponding result group. This value is then decomposed into numbers of two sequences within the corresponding join block via the following formulas: $p_1 = bs - 1 - \left\lceil 0.5(\sqrt{8(jc - 1 - pos) + 9} - 1) \right\rceil$ and $p_2 = pos - 0.5p_1(2bs - p_1 - 3) + 1$, where: $bs$ is the corresponding join group size retrieved from the array *groupSizes* and $jc$ is the corresponding result group size retrieved from the array *joinCounts*.

These formulas accomplish two tasks: (1) a threads position within the corresponding result group is decomposed into a combination of two patterns within the corresponding join group and (2) the joined patterns will be sorted lexicographically within the result group. The positions $p_1$ and $p_2$ are converted into global positions within the input pattern buffer by adding the appropriate value from the *positions* array (and increasing by 1). Finally, the thread joins the two patterns by performing a binary bitwise OR operation between all words of the two patterns and stores the result in the resulting pattern buffer.

## 4.4   Generation of Candidates - Candidate Pruning

Each pattern obtained through joining must be checked whether its every subpattern is prevalent or not. To perform this check for patterns of length $n$ we utilize: a pattern buffer obtained in the previous algorithm step (see section 4.3) and a pattern hash map for prevalent patterns of length $n-1$. Assume there are $k_p$ patterns obtained through joining. First, an array *flags* of size $k_p$ is allocated. Next, $k_p$ threads are started. Each thread retrieves its corresponding pattern from input pattern buffer and sequentially generates all $n$ subpatterns of length $n - 1$. Each subpattern is checked whether the corresponding entry exists in the pattern hash map. If all subpatterns of a single pattern have corresponding entries in the pattern hash map, the thread stores 1 in the corresponding position in the *flags* array, otherwise 0 is stored.

As a second step a parallel compact algorithm is employed to remove all patterns except for patterns with the corresponding flag equal to 1. As parallel compact does not change the order of patterns, the lexicographic order is retained. The obtained pattern buffer will be referred to as a *candidate pattern buffer*.

## 4.5   Generation of Instances

In this section we describe an algorithm which performs the most time consuming step of the collocation pattern discovery - the generation of instances. Let us introduce several useful terms. Let the number of candidate patterns be equal to $k_C$. Any $n - 1$ size pattern that is a prefix of a $n$ size candidate pattern will be denoted as a *candidate prefix pattern*. Instances of a candidate prefix pattern will be called *candidate prefix pattern instances*. The last feature of a candidate pattern will be called an *extending feature*.

The basic idea for instance generation algorithm is based on the following observations. In the basic iCPI-tree based algorithm finding instances of some candidate pattern $C$ with extending feature $f_e$ involves: (1) retrieving every candidate prefix instance $P_i$, (2) finding for each feature instance of $P_i$ a list of its neighbors with feature $f_e$ by means of iCPI-tree and (3) finding a common part of these lists. As processing of every candidate prefix instance is independent it is a natural candidate for parallelization. Our algorithm processes each candidate prefix instance of every candidate pattern in parallel. Let $k_I$ be the number of processed prefix instances. Notice that each candidate prefix instance might be processed more than once if there are several candidates with the same candidate prefix pattern but different extending feature. Each candidate prefix instance can have two (local and global) numbers. A *local candidate prefix instance number* is a number of the candidate prefix instance within the group of candidate prefix instances of a single candidate prefix pattern. A *global candidate prefix instance number* is a number of candidate prefix instance within the set of all candidate prefix instances used in generation of candidate pattern instances. As an input the algorithm utilizes: a candidate pattern buffer containing $n$ size patterns obtained in the previous stage, a pattern buffer, a pattern instance buffer, a pattern hash map of prevalent $n - 1$ size patterns as well as an instance neighbor buffer and an instance neighbor hash map. The main instance generation algorithm requires also several auxiliary arrays which can be computed in parallel as well:

- *listPointers* - an array of size $k_C$ which for every candidate pattern stores a pointer to the first candidate prefix instance in the pattern instance buffer,
- *instanceCounts* and *scannedInstanceCount* - instanceCounts is an array of size $k_C$ which for every candidate pattern stores the number of corresponding candidate prefix instances, scannedInstanceCount is a result of perfoming a parallel inclusive scan on the instanceCounts array,
- *correspondingPatterns* and *extendingFeatures* - arrays of size $k_I$ which map global candidate prefix instance number to the number of the corresponding

candidate pattern and its extending feature respectively, entries in both of these arrays are sorted by the corresponding pattern.

In the first step, for each candidate prefix instance a list of common $f_e$ feature instances is generated. Two arrays are created: (1) *listSizes* of size $k_I$ which will store lengths of each neighbor list and (2) *newNeighbors* of size $k_I \times L$ (recall that $L$ is the length of neighbor list in instance neighbor buffer structure) which will store the neighbor lists. The memory alignment and the structure of the *newNeighbors* array is the same as that of the instance neighbor buffer. To perform this step $k_I \times L$ threads are started. Each group of consecutive $L$ threads cooperates to generate one neighbor list in the *newNeighbors* array. Each thread at start determines the following information: (1) the global number candidate prefix instance $c \in 0, \ldots, k_I - 1$ (the same for each of $L$ consecutive threads), (2) the corresponding position within the neighbor list $l \in 0, \ldots, L - 1$, (3) the corresponding candidate pattern number $p$ from the *correspondingPatterns* array and (4) the extending feature $f_e$ from the *extendingFeatures* array. Next, based on the global candidate prefix instance number $c$ each thread at start determines the local candidate prefix instance number. This is done by substracting *scannedInstanceCounts[p-1]* from $c$. If $p = 0$ the candidate pattern instance number is equal to $c$. Each thread also retrieves from the *listPointers* array a pointer to the first candidate prefix instance of their corresponding candidate pattern $p$. Based on this pointer and the local candidate prefix instance number each thread computes the address of the first feature instance of the corresponding candidate prefix instance that it is going to process. Now each of the L threads work synchronously to find an intersection of the lists of neighbors with $f_e$ feature of every feature instance being a part of the processed candidate prefix instance. The whole process is performed almost solely in the fast shared memory of GPU. When the threads finish their work they copy their results into the *newNeighbors* array and store the lengths of the obtained lists into the *listSizes* array.

After the neighbor lists are found, the algorithm performs a modified parallel compact algorithm which removes empty entries from the obtained lists, but every remaining entry is materialized in the resulting array as a complete candidate pattern instance, i.e., the corresponding candidate prefix instance with the appended entry. The resulting array is built in such a way that its structure and properties are the same as that of the pattern instance buffer.

## 4.6   Computation of Prevalence

The computation of prevalence is based on several classic parallel algorithms such as sort, reduce by key and unique (see section 2 for details). The process starts with computing, for every feature instance of each candidate pattern instance, a value composed of: its position within the candidate instance, a candidate pattern number, and an instance identifier constructed as described in section 4.1. An array of such values is then sorted in the lexicographic order. Next, non unique values are removed via the unique algorithm. The obtained array now stores for every candidate pattern a list of unique feature instances at each position of its instances. Next, in parallel a feature instance number is removed from every value in the obtained array (although feature identifier remains). Finally, the reduce by key algorithm is used to count the number of distinct values in the array obtained in the previous step (the array is treated as key array and value array is composed of ones). The obtained results store for each position of every candidate pattern a number of unique feature instances appearing in candidate pattern instances. Based on these values, the participation ratios and prevalences of all candidate patterns are computed (in parallel).

Non prevalent candidate patterns and their corresponding instances are removed via the parallel compact algorithm. Only the prevalent patterns and their instances remain (in the pattern buffer and the pattern instance buffer respectively). Finally, a new pattern hash map is constructed (in parallel). These structures now form an input of the next iteration of the collocation pattern discovery algorithm.

## 5 Experiments

### 5.1 Implementation and Testing Environment

For the purpose of this paper we have prepared two implementations of the iCPI-tree based collocation pattern discovery algorithm: for GPU and for CPU. The GPU version uses the solutions described in section 4. The CPU version uses similar data structures as the ones described in section 4.1, however the parallelization of computations is done differently. Instead of SIMD approach, multiple instruction multiple data approach is used (MIMD). CPU implementation uses OpenMP [6] to parallelize instance generation and prevalence computation on a multi-core CPU. The implementation starts the number of threads equal to the number of cores of a CPU and the computation tasks are distributed among the started threads. The parameter $L$ (length of lists in instance neighbor buffer) for both implementations was set to 8.

Experiments were run on a computer with Core2 Duo 2,1Ghz CPU (CPU implementation started two threads) and 8GB of RAM and GeForce 580GTX graphics card with 1.5GB of RAM (Fermi architecture) working under Microsoft Windows 7 operating system.

### 5.2 Data Sets and Experiments

To evaluate our GPU version of the algorithm we have prepared 10 synthetic datasets. We have used a synthetic generator similar to the one described in [15]. The number of data objects ranged from 25 K to 120 K, the number of spatial features ranged from 30 to 90 and 20 to 80 percent of total instances were noisy instances. Two kinds of datasets have been prepared: dense and sparse. Dense datasets have been generated by reducing the size of spatial framework ten times in each dimension while preserving the number of objects (sparse datasets were generated over a grid of size of 10000x10000 units).

We have conducted three series of experiments. Each time we measured *speedup*, i.e., the ratio of execution time of CPU version of the algorithm to the execution time of GPU implementation. In the first series we have examined how increasing of the minimum prevalence threshold affects speedup in both dense and sparse datasets. In the second series we investigated the influence of the distance threshold between neighbors on speedup for two opposite levels of minimum prevalence. Finally we examined how speedup changes with increasing size of the input dataset.

## 5.3   Results and Interpretation

Figure 2(a) presents results of the first experiment which tested the influence of the minimum prevalence parameter on the algorithm performance. Two interesting observations can be made. First, notice that the density of the dataset does not influence the speedup, i.e., while processing times may change, GPU version is faster than CPU version by approximately the same factor for the same minimum prevalence. Second, the speedup drops monotonically with the increase of minimum prevalence. In our testing environment for minimum prevalence equal to 0.2 the achieved speedup is roughly twice the speedup achieved for the minimum prevalence 0.6. This is due to the fact that the increased minimum prevalence parameter causes less candidate patterns to be generated and (indirectly) less candidate prefix instances to be processed. This in turn causes less threads to be started. Less threads mean that: (1) memory transfers may not be hidden, (2) GPU multiprocessors might not have recieved a full load and (3) instance generation step takes less time in comparison to other algorithm steps (as we have primarily focused on optimizing instance generation step this could cause the observed speedup to drop).

Figure 2(b) presents results of the second experiment which tested the influence of the distance threshold between neighbors on the algorithm performance. One can notice that in general, the larger the distance threshold, the greater the speedup. The explanation of this observation is very similar to the one made in the previous experiment. For large distance thresholds, each feature instance has more neighbors. Consequently neighbor lists in the instance neighbor buffer are longer and more instances are generated in the instance generation step of the algorithm. As we have shown in the previous paragraph, the larger the number of instances, the higher speedups can be achieved. One can also notice that we have performed this experiment for two values of minimum prevalence parameter: 0.2 and 0.6. Plots corresponding to these two values confirm the observations made in the previous experiment: the lower the minimum prevalence the higher the observed speedup. Moreover, similarly as before, one can notice that for minimum prevalence 0.2 observed speedup is twice the speedup observed for minimum prevalence 0.6.

Figure 2(c) presents results of the third experiment which tested the influence of the number of feature instances (objects) on the observed speedup. What is surprising is that the observed speedup grows linearly (for the tested datasets) with respect to the number of objects. It is of course obvious that the speedup

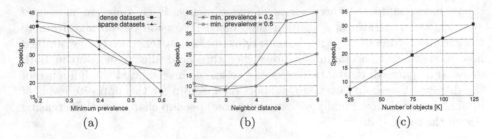

**Fig. 2.** Experiment results

cannot grow forever. The same curve for larger instances would asymptotically approach the maximum achievable speedup dependant on relative performances of GPU and CPU. The monotonic dependency can be explained similarly as before. Greater number of objects leads to greater number of pattern instances and therefore candidate prefix instances. The larger the number of such instances, the larger the number of threads started and the better the utilization of GPU.

Unfortunately, due to limited GPU memory we were not able to run instances large enough to achieve maximal speedups. Algorithm's memory requirements depend on data and query characteristics such as the number of objects, the number of spatial features, data density and the minimal prevalence. In our case, 1.5GB of GPU memory was enough to run dense datasets of size up to 120K objects, 90 spatial features, for the minimal prevalence of 0.2.

## 6    Summary and Future Work

In this paper we have presented an algorithm performing most operations of the state of the art collocation pattern discovery algorithm in parallel on GPU. We have compared our implementation to the multi-threaded CPU implementation and shown that GPU offers an order of magnitude speedup over the CPU version.

While the results are very promising, there is still a lot of work to do. The main problem is the limited graphics cards memory. This problem can be tackled in two ways. First, we plan on modifying our algorithm to be able to process variable length neighbor lists. Current solution wastes both memory and computing power (a lot of threads process empty positions on these lists). Second, we plan on adapting the solutions introduced in [5] which already try to solve the collocation pattern mining problem in limited memory conditions. We also plan designing an algorithm for efficient construction of iCPI-trees on GPU (currently done on CPU). Finally we also want to approach a more difficult problem of the maximal collocation pattern discovery on GPUs.

# References

1. Agrawal, R., Srikant, R.: Fast Algorithms for Mining Association Rules in Large Databases. In: Proceedings of the 20th International Conference on Very Large Data Bases, pp. 487–499. Morgan Kaufmann Publishers Inc., San Francisco (1994)
2. Alcantara, D.A.F.: Efficient Hash Tables on the GPU. Ph.D. thesis, University of California, Davis (2011)
3. Bell, N., Hoberock, J.: Thrust: A Productivity-Oriented Library for CUDA. In: GPU Comuting Coms. Jade edition, pp. 359–371. Morgan-Kauffman (2011)
4. Boinski, P., Zakrzewicz, M.: Collocation Pattern Mining in a Limited Memory Environment Using Materialized iCPI-Tree. In: Cuzzocrea, A., Dayal, U. (eds.) DaWaK 2012. LNCS, vol. 7448, pp. 279–290. Springer, Heidelberg (2012)
5. Boinski, P., Zakrzewicz, M.: Partitioning Approach to Collocation Pattern Mining in Limited Memory Environment Using Materialized iCPI-Trees. In: Morzy, T., Härder, T., Wrembel, R. (eds.) Advances in Databases and Information Systems. AISC, vol. 186, pp. 19–30. Springer, Heidelberg (2013),
http://dx.doi.org/10.1007/978-3-642-32741-4_3
6. Chapman, B., Jost, G., van der Pas, R.: Using OpenMP: Portable Shared Memory Parallel Programming (Scientific and Engineering Computation). The MIT Press (2007)
7. Farooqui, N., Kerr, A., Diamos, G., Yalamanchili, S., Schwan, K.: A framework for dynamically instrumenting gpu compute applications within gpu ocelot. In: Proceedings of the Fourth Workshop on General Purpose Processing on Graphics Processing Units, GPGPU-4, pp. 9:1–9:9. ACM, New York (2011)
8. Fayyad, U., Piatetsky-Shapiro, G., Smyth, P.: From Data Mining to Knowledge Discovery in Databases. AI Magazine 17, 37–54 (1996)
9. Khronos Group: The OpenCL Specification Version: 1.2 (2012),
http://www.khronos.org/registry/cl/specs/opencl-1.2.pdf/
10. NVIDIA Corporation: Nvidia cuda programming guide (2012),
http://developer.download.nvidia.com/compute/DevZone/docs/
html/C/doc/CUDA_C_Programming_Guide.pdf
11. Shekhar, S., Huang, Y.: Discovering Spatial Co-location Patterns: A Summary of Results. In: Jensen, C.S., Schneider, M., Seeger, B., Tsotras, V.J. (eds.) SSTD 2001. LNCS, vol. 2121, pp. 236–256. Springer, Heidelberg (2001)
12. Shekhar, S., Huang, Y.: The multi-resolution co-location miner: A new algorithm to find co-location patterns in spatial dataset. Tech. Rep. 02-019, University of Minnesota (2002)
13. Wang, L., Bao, Y., Lu, J.: Efficient Discovery of Spatial Co-Location Patterns Using the iCPI-tree. The Open Information Systems Journal 3(2), 69–80 (2009)
14. Xiao, X., Xie, X., Luo, Q., Ma, W.Y.: Density Based Co-location Pattern Discovery. In: Proceedings of the 16th ACM SIGSPATIAL International Conference on Advances in Geographic Information Systems, GIS 2008, pp. 29:1–29:10. ACM, New York (2008), http://doi.acm.org/10.1145/1463434.1463471
15. Yoo, J.S., Shekhar, S., Celik, M.: A Join-Less Approach for Co-Location Pattern Mining: A Summary of Results. In: Proceedings of the IEEE International Conference on Data Mining, Washington, pp. 813–816 (2005)
16. Yoo, J.S., Bow, M.: Mining Maximal Co-located Event Sets. In: Huang, J.Z., Cao, L., Srivastava, J. (eds.) PAKDD 2011, Part I. LNCS, vol. 6634, pp. 351–362. Springer, Heidelberg (2011),
http://dx.doi.org/10.1007/978-3-642-20841-6_29

# Resource Allocation for Query Optimization in Data Grid Systems: Static Load Balancing Strategies

Shaoyi Yin, Igor Epimakhov, Franck Morvan, and Abdelkader Hameurlain

IRIT Laboratory, Paul Sabatier University, France
{yin,epimakhov,morvan,hameurlain}@irit.fr

**Abstract.** Resource allocation is one of the principal stages of relational query processing in data grid systems. *Static* allocation methods allocate nodes to relational operations during query compilation. Existing heuristics did not take into account the multi-queries environment, where some nodes may become overloaded because they are allocated to too many concurrent queries. *Dynamic* resource allocation mechanisms are currently developed to modify the physical plan during query execution. In fact, when a node is detected to be overloaded, some of the operations on it will migrate. However, if the resource contention is too heavy in the initial execution plan, the operation migration cost may be very high. In this paper, we propose two load balancing strategies adopted during the static resource allocation phase, so that the workload is balanced at the beginning, the operation migration cost is decreased during the query execution, and therefore the average response time is reduced.

**Keywords:** Resource Allocation, Data Grid Systems, Query Optimization, Load Balancing.

## 1 Introduction

In recent years, large amount of scientific data have been produced and need to be shared by researchers from different organizations all over the world. Examples include the Large Hadron Collider (LHC) [1] at CERN and the Sloan Digital Sky Survey (SDSS) [2]. Data grid systems are designed to support such applications. With the objective of accessing and analyzing huge volume of data, the data grid systems rely on distributed, heterogeneous and autonomous computing resources [3]. On the one hand, a data grid system needs to perform query processing efficiently. On the other hand, a data grid system is characterized as large scale, heterogeneous and dynamic. Putting these two aspects together, many technical problems such as resource allocation become more challenging.

In a data grid system, there are different types of resources such as CPU, memory, storage and network. A node in the data grid system corresponds to a computer containing some of these resources. A user query is often issued on one of the nodes and is expressed using a declarative language such as SQL or OQL [4]. The query processor then transforms the query statement into an algebraic tree, called a logical query

B. Catania, G. Guerrini, and J. Pokorný (Eds.): ADBIS 2013, LNCS 8133, pp. 316–329, 2013.

plan, where nodes denote relational operations and edges represent data flows. Fig. 1 shows a query statement in SQL and its corresponding logical query plan.

At a given time, a node N of the data grid system which has access to a set of resources $R = \{r_1, r_2, ..., r_n\}$ receives a query Q which consists of a set of operations $\{o_1, o_2, ..., o_l\}$. Assume that R is collected during a preceding stage called resource discovery [5]. The problem of resource allocation is to assign one or more resources in R to each $o_i$ such that the execution time of Q is minimized. Thereafter, we call the node N an *allocator*. Note that, the query type that we deal with is the Scan-Project-Join query. Obviously, the complexity of an exhaustive matching algorithm is exponential, so heuristics [6-9] have been proposed to solve this problem. However, most of these heuristics rely on the same principle: they first rank the operations according to certain criteria, then for each operation, they rank the available nodes and allocate the best ones to it. The criteria for ranking the operations, the criteria for ranking the nodes and the number of nodes to allocate are different in each heuristic method.

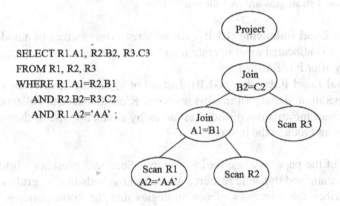

SELECT R1.A1, R2.B2, R3.C3
FROM R1, R2, R3
WHERE R1.A1=R2.B1
AND R2.B2=R3.C2
AND R1.A2='AA' ;

**Fig. 1.** A query statement and its logical query plan

These heuristics did not take into account the following constraints of the data grid system. *Constraint 1: Multi-queries* are treated by the same allocator. For each query, the allocator consumes the same list of candidate nodes, so the most powerful nodes are allocated to too many queries and become overloaded, while the least powerful nodes stay idle. *Constraint 2: Multi-allocators* co-exist in the data grid system (as shown in Fig. 2). The resources discovered for one allocator may also be discovered for other allocators. These resources will become overloaded if they are chosen by too many allocators.

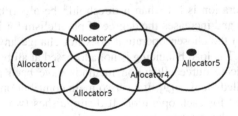

**Fig. 2.** Multi-allocators in a data grid system

The overloading problem caused by the above constraints is currently addressed by adding a resource reallocation phase (which is called *dynamic resource allocation*) during the query execution [10-15], that is, to move part of the work from overloaded resources to less loaded resources. The dynamic phase is very important, not only because the static allocation result may be under optimal, but also because in the data grid system, nodes may leave or enter at any time. However, the efficiency of the dynamic allocation is truly linked with the initial static allocation result. If the workload is already well balanced between nodes after the static resource allocation, much less work needs to be done during the dynamic phase, which is actually the objective of our paper.

In this paper, we propose two load-balancing strategies which aim at improving the physical execution plan generated by the static resource allocation, such that the average response time of the queries is reduced. They cannot replace the dynamic resource allocation phase, but they could make the latter more efficient. The principles of the proposed strategies are as follows:

- **Local Load Balancing (LLB):** virtually reserve resources of a node each time after it is allocated to an operation and virtually release the resources used by a query after it is finished;
- **Global Load Balancing (GLB):** Instead of allocating directly the best node to an operation, the algorithm first proposes K candidates and collects the current workload information of their resources by contacting them, then ranks the K nodes and returns the best one.

The rest of the paper is organized as follows. Section 2 presents a brief survey of the existing static and dynamic resource allocation methods in data grid systems. Section 3 describes the principles of our strategies and the corresponding algorithms. Section 4 evaluates the proposed strategies by combining them with existing resource allocation methods. Section 5 concludes the paper.

## 2     Related Work

The earliest work dealing with the resource allocation problem for query optimization in the data grid environment is the work of Gounaris[6]. It is a static resource allocation method. During the initiation phase, only one node is allocated to each operation. Then the algorithm increases the degree of parallelism for the most costly operation by allocating the most powerful nodes one by one to it. When the benefit of increasing a node to the operation is less than a threshold, the algorithm chooses the next most costly operation and increases the degree of parallelism for it. The iteration continues like this. When the chosen operation does not change any more, the resource allocation is finished. The load balancing is not addressed at all by this work.

Several other static resource allocation methods have been proposed [7-9]. The most recent one is called GeoLoc[9]. It has two main contributions. First, it defines an Allocation Space (AS) for each operation. It distinguishes two kinds of nodes: *data nodes* which contain the relation fragments and *computing nodes* which do not con-tain any relation fragment. For a scan operation, the AS is defined to contain all the nodes storing the used relation fragments. For a join operation, the AS is defined to contain the nodes allocated to its input operations and the nodes geographically close to these nodes. Second, it takes into account the dependency between relational operations. It determines the degree of parallelism for each join operation according to the resource requirements driven by its input operations. For example, if the scan operations send 1000 tuples per second to a join operation, GeoLoc will allocate to that join operation just enough resources which can process 1000 tuples per second. Even though the load balancing problem is not explicitly addressed by the GeoLoc method, the usage of the AS decreases the resource contention. However, it is still possible that a same node is allocated for many queries simultaneously.

In the data grid system, due to the multi-queries and multi-allocators constraints, some nodes may become overloaded during the query execution, thus the average response time is increased. The current solution to this problem is to add a dynamic resource allocation phase which modifies the physical plan during the query execu-tion. There exist two main approaches: centralized and decentralized. In the central-ized approach, the workload status of the nodes is monitored by a dedicated resource broker [10-14]. In the decentralized approach, each node detects if it is over-loaded and makes autonomously the decision of moving operations from it to other nodes [15]. In the work [15], each relational operation is implemented as a mobile agent running on the allocated node, meaning that, it keeps track of its own status and can migrate to another node at any time. Thanks to a two-level cooperation mechanism between the autonom-ous nodes and autonomous operations, the workload is dynamically balanced among the grid nodes during query execution. The dynamic resource allocation phase is very important for query optimization, especially when there are node leavings and enters. However, the total migration cost could be high if the static resource allocation result is too imbalanced. In this paper, we aim at producing a balanced resource allocation re-sult during the static phase, so that the migration cost of the dynamic phase is reduced, and therefore the average query response time is decreased.

## 3     Static Load Balancing Strategies

After the resource discovery phase, the allocator keeps the resource information (for example the size of memory, the CPU speed, the IO throughput and the network bandwidth) of each discovered node in a local table. This information is used as me-tadata during the resource allocation for the arriving queries. When the queries arrive frequently or even in batch, it is not realistic to update the resource information for each query due to the expensive communication cost. It means that, for a group (hun-dreds or thousands) of queries, the allocator consumes the same resource infor-mation table. This may introduce a kind of resource contention: some more powerful nodes may be allocated to too many concurrent queries while some less powerful nodes stay almost idle. Since the nodes are shared by the entire grid system, there is another kind of resource contention: a same node may be discovered for several different allocators and then become overloaded easily. In order to relieve these two kinds of resource contention, we propose in this section two load balancing strategies.

## 3.1     Local Load Balancing (LLB) Strategy

This strategy is proposed to solve the resource contention problem caused by the mul-ti-queries constraint. The principle is to balance the workload between nodes by tak-ing into account the local resource allocation history. In the resource information table RIT, the allocator maintains a counter for each resource type. Table 1 is a snapshot of the RIT, where NB_DM_X denotes the number of active demands for a resource type X and RC_X denotes the resource capacity on X. For example, if node N is allocated to a query which needs only I/O and network resources, we increase each of the coun-ters NB_DM_IO$_N$ and NB_DM_NET$_N$ by one.

**Table 1.** Snapshot of the resource information table RIT

| N_ID | RC_ RAM | RC_ CPU | RC_ IO | RC_ NET | NB_DM _RAM | NB_DM _CPU | NB_DM _IO | NB_DM _NET |
|------|---------|---------|--------|---------|------------|------------|-----------|------------|
| 205 | 15000 | 24000 | 19000 | 27000 | 10 | 5 | 8 | 10 |
| 192 | 8000 | 36000 | 12000 | 35000 | 6 | 12 | 3 | 5 |
| ... | | | | | | | | |

When the allocator allocates nodes for an operation, instead of using RC_X of each node to estimate the execution cost, it uses the currently available resource AR_X, which is computed using the following formula (where N is the node ID):

$$AR\_X_N = \frac{RC\_X_N}{NB\_DM\_X_N} \tag{1}$$

**Table 2.** Snapshot of the resource demand table RDT

| Q_ID | N_ID | DM_RAM | DM_CPU | DM_IO | DM_NET |
|------|------|--------|--------|-------|--------|
| 11 | 205 | 0 | 0 | 1 | 1 |
| 11 | 192 | 0 | 1 | 1 | 0 |
| 12 | | | | | |
| ... | | | | | |

The allocator also maintains a table to register the resource demand of each treated operation. In this table, each tuple contains the query ID, node ID and the demand (DM_X, Boolean) for each resource type X. Each time after allocating a node to an operation, the allocator adds a tuple to this table. Each time a query is finished, the allocator decreases the demand counters in the resource information table RIT accord-ing to resource demand history related to the finished query. Finally, the history records related to that query are removed. Table 2 is a snapshot of the resource de-mand table RDT. For example, when query 11 is finished, we decrease each of the counters NB_DM_IO$_{205}$, NB_DM_NET$_{205}$, NB_DM_CPU$_{192}$ and NB_DM_IO$_{192}$ by one respectively, and we remove the first two tuples from table RDT.

Algorithm 1 describes the process of the maintenance of the resource information table and the resource demand table each time after allocating a node N to an operation O of a query Q. Algorithm 2 describes the process of maintenance each time after a query Q is finished.

---

**Algorithm 1. VirtualResourceReservation()**

INPUT: Query Q, Node N, resource demand DM_X for each X
BEGIN
    FOR each resource type X of node N DO
        IF DM_X > 0 THEN
            Increase the counter NB_DM_X$_N$ by one in table RIT;
        END IF
    END FOR
    Insert the tuple (Q, N, DM_RAM, …) into table RDT;
END

---

**Algorithm 2. VirtualResourceRelease()**

INPUT: Query Q
BEGIN
    FOR each node N related to query Q in RDT DO
        FOR each resource type X of node N DO
            IF DM_X$_N$ > 0 THEN
                Decrease the counter NB_DM_X$_N$ by one in table RIT;
            END IF
        END FOR
    END FOR
    Delete the tuples related to Q from table RDT;
END

---

## 3.2     Global Load Balancing (GLB) Strategy

This strategy is proposed to solve the resource contention problem caused by the multi-allocators constraint. When choosing a node to allocate for an operation, the allocator proposes several candidate nodes according to the local resource information, and then it contacts these nodes to collect their workload status at the moment. Using the new resource information, the allocator ranks these candidate nodes and chooses the best one. Table 3 shows an example of candidate nodes list CNL after contacting them, where CNB_DM_X means the current number of demands for resource X registered by the node. The allocator estimates then the execution cost using the available resource capacity AR_X computed by the new formula:

$$AR\_X_N = \frac{RC\_X_N}{CNB\_DM\_X_N} \qquad (2)$$

**Table 3.** Example of the candidate nodes list CNL

| N_ID | CNB_DM _RAM | CNB_DM _CPU | CNB_DM _IO | CNB_DM _NET |
|:---:|:---:|:---:|:---:|:---:|
| 39 | 12 | 6 | 9 | 10 |
| 267 | 8 | 3 | 6 | 7 |
| ... | | | | |

Algorithm 3 illustrates the steps of using GLB to select a node N for an operation O. Note that, the current workload information collected from the nodes is used only to choose the best node for an operation, but not to update the RIT table. The RIT table could be only modified by the LLB strategy so that the information is kept consistent.

---

**Algorithm 3. NodeSelection()**

---

INPUT: the resource information table RIT, an operation O
OUTPUT: the node N to be allocated to O
BEGIN
    Choose K candidate nodes using RIT;
    FOR each chosen node N DO
        Contact N and get CNB_DM_$X_N$ for each resource X;
        Insert the collected information into CNL;
    END FOR
    FOR each node N in CNL DO
        Estimate the execution cost of O by adding N to it;
    END FOR
    Return the node N with the minimal estimated cost;
END

---

## 4    Performance Evaluation

We first measure the impact of LLB and GLB strategies on two existing static resource allocation methods: method of Gounaris [6] and GeoLoc method [9]. We then add the dynamic resource allocation phase to each method and measure again the impact of LLB and GLB. The dynamic allocation algorithm that we use is the work published in [15]. Our proposed load balancing strategies can be combined with other static or dynamic resource allocation methods, even though we choose only the above representatives for the performance evaluation.

For the experimentation, we use the grid simulator presented in [9], [15] with some extensions. The simulated data grid contains 2000 nodes that store 2000 relations. Each relation contains 5 equal fragments, each of which in turn is duplicated on 4 nodes. So in average we have 20 copies of different fragments on each node. To simulate the dynamicity of the grid system, 5% of the nodes quit the system at the 600th millisecond and reenter at the 1200th millisecond. The main parameters that we used for the simulation are listed in Table 4.

**Table 4.** System configuration and database parameters

|          | Parameter                     | Value             |
|----------|-------------------------------|-------------------|
| Node     | CPU performance               | 100 - 10 000 MIPS |
|          | I/O throughput                | 10 – 1000 Mb/s    |
|          | Memory amount                 | 10 – 1000 MB      |
|          | Network connection bandwidth  | 10 – 1000 Mb/s    |
|          | Network connection latency    | 0.05s             |
| Relation | Number of attributes          | 10                |
|          | Size of attribute             | 1 – 50 Bytes      |
|          | Cardinality of attributes     | 0.01 – 1          |
|          | Number of tuples              | 1000 – 11000      |
|          | Number of fragments           | 5                 |
|          | Number of duplicates per fragment | 4             |

Section 4.1 shows the impact of the proposed strategies on the static resource allocation, in terms of average query response time and optimization cost. Section 4.2 shows the impact of the proposed strategies on the static resource allocation combined with a dynamic resource allocation phase, in terms of average query response time and number of operation migrations. We also measured the effect of increasing the number of concurrent queries.

## 4.1    Impact of LLB and GLB on Static Resource Allocation

We take Gounaris' method and the GeoLoc method as two examples to examine our load balancing strategies. LLB and GLB strategies are evaluated in a gradual way, meaning that, we first measure the initial allocation method (named *Gouna/GeoLoc*), then the initial method combined with LLB (named *GouLLB/GeoLLB*), and finally the initial method combined with LLB and GLB (named *GouLLB+GLB /GeoLLB+GLB*).

**Impact on Gounaris' Method.** The average response time varying with the number of queries is shown Fig. 3. The speedup factor of the two strategies is illustrated in Fig. 4. Not surprisingly, we find that the speedup factor is more significant when there are more concurrent queries.

We measured also the average optimization cost of each query by varying the number of queries. The result can be found in Fig. 5. Interestingly, we see that the optimization time is not always longer when adding our load balancing strategies. This is because the query optimization contains several steps including mainly 1) generation of the logical plan, 2) selection of nodes for each operation and 3) generation of physical operators. When adding LLB or GLB, the first step does not change, the second step is slowed down, but the third step could be sped up or slowed down depending on the result of the second step. For example, when adding LLB strategy, fewer physical operators are generated during the third step, so the optimization time is shorter.

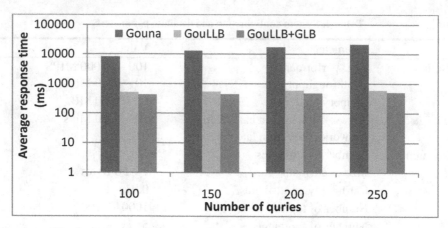

**Fig. 3.** Average query response time varying with the number of queries

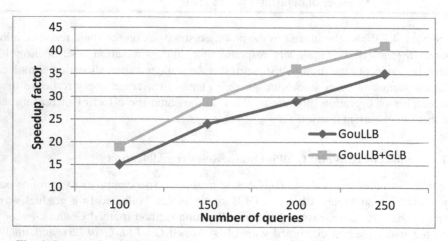

**Fig. 4.** Speedup factor of load balancing strategies varying with the number of queries

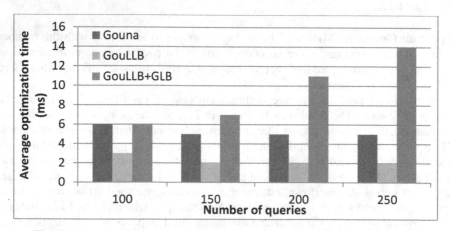

**Fig. 5.** Average optimization time per query varying with the number of queries

**Impact on GeoLoc Method.** The average response time and the speedup factor are shown in Fig. 6 and Fig. 7 respectively. The impact of the load balancing strategies is less significant than for Gounaris' method, because GeoLoc has already balanced the workload to some extent thanks to the using of Allocation Space. However, the same conclusion can be drawn: the speedup factor of LLB and GLB is more important when there are more concurrent queries. The average optimization time of each query is given in Fig. 8. We have three remarks: 1) the optimization time is not always increased by adding the load balancing strategies; 2) even when the optimization time is increased, the discrepancy is not very high; 3) compared to the total response time, the optimization cost is trivial.

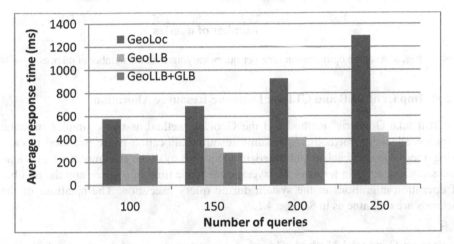

**Fig. 6.** Average query response time varying with the number of queries

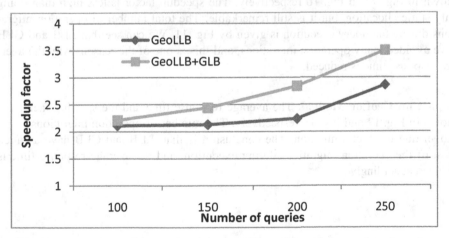

**Fig. 7.** Speedup factor of load balancing strategies varying with the number of queries

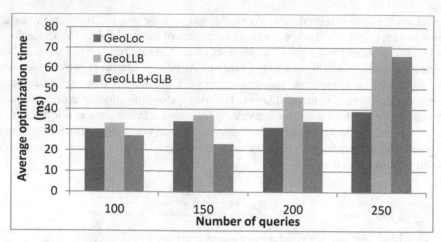

**Fig. 8.** Average optimization time per query varying with the number of queries

## 4.2     Impact of LLB and GLB on Dynamic Resource Allocation

We still take Gounaris' method and the GeoLoc method as two examples of static resource allocation methods. A dynamic resource allocation phase is added to each using the same mechanism published in [15]. We evaluate the impact of our load balancing strategies in terms of the average response time of a query and the number of operation migrations in the system during query execution. The notations of the methods are the same as in Section 4.1.

**Impact on Gounaris' Method.** The average response time and the speedup factor are shown in Fig. 9 and Fig. 10 respectively. The speedup factor is less high than doing only static allocation, but it is still remarkable. The total number of operation migrations during the query execution is given by Fig. 11. We can see that LLB and GLB have avoided many operation migrations, and this is one of the reasons why the average response time is reduced.

**Impact on GeoLoc Method.** The average response time and the speedup factor are shown in Fig. 12 and Fig. 13 respectively. The number of operation migrations is not shown due to space limitation. The conclusion is that, LLB and GLB have avoided most of the operation migrations during execution, and the average response time is reduced accordingly.

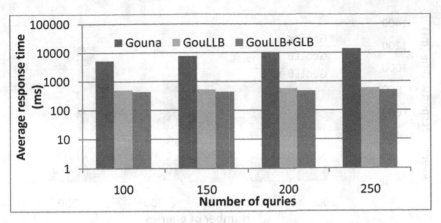

**Fig. 9.** Average query response time varying with the number of queries

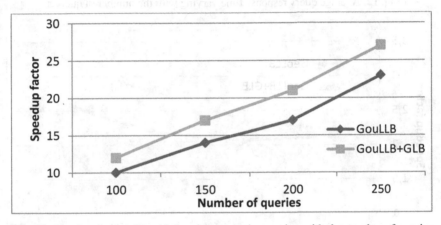

**Fig. 10.** Speedup factor of load balancing strategies varying with the number of queries

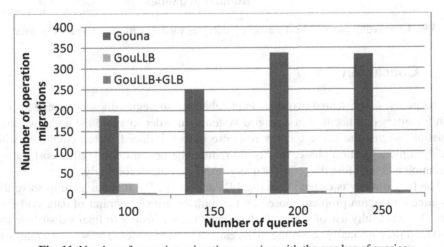

**Fig. 11.** Number of operation migrations varying with the number of queries

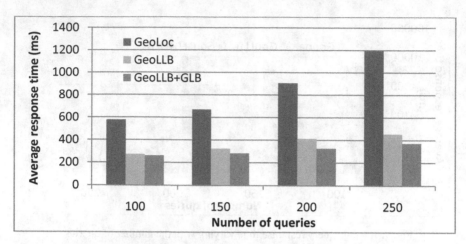

**Fig. 12.** Average query response time varying with the number of queries

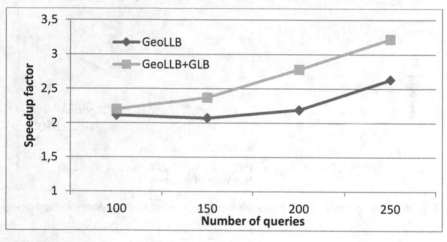

**Fig. 13.** Speedup factor of load balancing strategies varying with the number of queries

## 5     Conclusion

In this paper, we presented two static load balancing strategies during resource alloca-tion for query optimization in data grid systems, in order to avoid the node overload situation, so that the average query response time is reduced. When combined with the dynamic allocation phase, they could reduce the operation migration cost during execution and therefore decrease the query response time accordingly.

The first strategy is called Local Load Balancing (LLB). It is designed to solve the resource contention problem caused by the multi-queries constraint of data grid sys-tems. It makes fully use of the local resource allocation history to finally distribute the workload proportionally according to the capacity of the nodes. The second strategy is called Global Load Balancing (GLB). It is designed to solve the resource contention problem caused by the multi-allocators constraint of data grid systems. It first chooses several candidate nodes for an operation using the local information, then contacts these nodes and collects their current workload status, and finally re-ranks these nodes to select the best one.

The result of the performance evaluation has shown the efficiency of the proposed strategies. For example, by integrating our LLB and GLB strategies with Gounaris' method, the query execution time is reduced by 10 to 40 times; by integrating them with the GeoLoc method, the query execution time is reduced by 2 to 4 times. The proposed strategies could be combined with other existing static and dynamic resource allocation methods in data grid systems. They decrease significantly the query response time, but they don't increase much the allocation cost.

# References

1. http://lhc.web.cern.ch/lhc/
2. http://www.sdss.org/
3. Chervenak, A., et al.: The Data Grid: Towards an Architecture for the Distributed Management and Analysis of Large Scientific Datasets. Journal of Network and Computer Applications 23, 187–200 (1999)
4. Smith, J., Gounaris, A., Watson, P., Paton, N.W., Fernandes, A.A.A., Sakellariou, R.: Distributed Query Processing on the Grid. In: Parashar, M. (ed.) GRID 2002. LNCS, vol. 2536, pp. 279–290. Springer, Heidelberg (2002)
5. Krauter, K., et al.: A taxonomy and survey of grid resource management systems for distributed computing. Journal of Software: Practice and Experience 32, 135–164 (2002)
6. Gounaris, A., et al.: Resource scheduling for parallel query processing on computational grids. In: GRID (2004)
7. Soe, K.M., et al.: Efficient scheduling of resources for parallel query processing on grid-based architecture. In: Information and Telecommunication Technologies (2005)
8. Liu, S., Karimi, H.A.: Grid query optimizer to improve query processing in grids. Future Gener. Comput. Syst. 24, 342–353 (2008)
9. Epimakhov, I., et al.: GeoLoc: Robust Resource Allocation Method for Query Optimization in Data Grid Systems. In: DB&IS (2012)
10. Gounaris, A., et al.: Adaptive query processing and the grid: Opportunities and challenges. In: DEXA Workshops (2004)
11. Gounaris, A., et al.: Practical adaptation to changing resources in grid query processing. In: Proceedings of the 22nd International Conference on Data Engineering, ICDE 2006 (2006)
12. Da Silva, V.F.V., et al.: An adaptive parallel query processing middleware for the grid. Concurrency and Computation: Practice and Experience 18(6), 621–634 (2006)
13. Avnur, R., Hellerstein, J.M.: Eddies: Continuously adaptive query processing. In: Proceedings of the SIGMOD Conference, pp. 261–272 (2000)
14. Patni, J., et al.: Load balancing strategies for grid computing. In: Proceedings of the 3rd International Conference on Electronics Computer Technology, ICECT (2011)
15. Epimakhov, I., et al.: Mobile Agent-based Dynamic Resource Allocation Method for Query Optimization in Data Grid Systems. In: International KES Conference on Agents and Multi-agent Systems – Technologies and Applications (2013)

# Multi-level Parallel Query Execution Framework for CPU and GPU

Hannes Rauhe[1,3], Jonathan Dees[2,3], Kai-Uwe Sattler[1], and Franz Faerber[3]

[1] Ilmenau University of Technology
[2] Karlsruhe Institute of Technology
[3] SAP AG

**Abstract.** Recent developments have shown that classic database query execution techniques, such as the iterator model, are no longer optimal to leverage the features of modern hardware architectures. This is especially true for massive parallel architectures, such as many-core processors and GPUs. Here, the processing of single tuples in one step is not enough work to utilize the hardware resources and the cache efficiently and to justify the overhead introduced by iterators. To overcome these problems, we use just-in-time compilation to execute whole OLAP queries on the GPU minimizing the overhead for transfer and synchronization. We describe several patterns, which can be used to build efficient execution plans and achieve the necessary parallelism. Furthermore, we show that we can use similar processing models (and even the same source code) on GPUs and modern CPU architectures, but point out also some differences and limitations for query execution on GPUs. Results from our experimental evaluation using a TPC-H subset show that using these patterns we can achieve a speed-up of up to factor 5 compared to a CPU implementation.

## 1 Introduction

The recent development in modern hardware architectures provides great opportunities for database query processing, but at the same time pose also significant challenge to be able to exploit these hardware features. For example, the classic iterator model for query execution (sometimes also called Open-Next-Close-model) together with a tuple-at-a-time processing strategy is no longer optimal for modern hardware, because it does not utilize the cache architecture and does not produce enough work to pay off the overhead in case of parallel processing.

Some of the most prominent techniques for addressing these challenges are fine-granular parallelization and vectorized processing. The goal of parallelization is to speed up query response times and/or improving throughput by exploiting multiple processors/cores. Though, parallel query processing dates back to the mid eighties, modern multi- and many-core processors including GPUs require a rethinking of query execution techniques: processing models range from SIMD to multi-processor strategies, the overhead of starting and coordinating threads is much smaller than on old multi-processors systems and even the cache and memory architectures have changed.

B. Catania, G. Guerrini, and J. Pokorný (Eds.): ADBIS 2013, LNCS 8133, pp. 330–343, 2013.

Vectorized processing aims at avoiding the "memory wall", i.e., the imbalance between memory latency and CPU clock speed [3], by using cache-conscious algorithms and data structures, vectorized query processing strategies as used for instance in Vectorwise [2,17] or—as an extreme case—columnwise processing of in-memory database systems like SAP HANA DB.

A promising approach to combine these techniques for making efficient use of the new hardware features is Just-In-Time (JIT) compilation where OLAP queries are translated into a single function that pipelines the data through the operators specified by the query. While lightweight query compilation has been already proposed in other works [12,14], in our work we particularly focus on compilation for parallelization. This approach requires a different handling of parallelism, because we cannot differentiate between inter- and intra-operator parallelism any more. Instead the parallelism must be an integral part of the code generation.

In this paper, we show that we can use the concept of JIT compilation to build a bulk-synchronous model for parallel query processing that works both on multi-core CPUs and on general purpose GPUs. Although the architectures look very different at first, the techniques we have to use to leverage their full potential are very similar.

Compiling queries into executable code should be driven by some target patterns. As we cannot longer rely on the iterator model, we present and discuss some patterns leveraging the different levels of parallelization (threads at the first level and SIMD at the second level). We extend our previous work on query compilation for multi-core CPUs [6] by taking massively parallel GPU architectures into account and discuss as well as show by experimental results similarities and differences of both processing models and their impact on query execution.

## 2   Related Work

Over the last decade there has been significant research on using GPUs for query execution. He et al. built GDB, which is able to offload the execution of single operators to the GPU [8]. They implemented a set of primitives to execute joins efficiently. The main problem is that the data has to be transferred to the GPU, but only a small amount of work is done before copying the results back to the main memory. Peter Bakkum and Kevin Skadron ported SQLite to the GPU [1]. The complete database is stored on the graphics card and only the results are transferred. Since all operations are executed on the GPU, their system is not able to use variable length strings. It can process simple filter operations, but does neither support joins nor aggregations. A similar approach with the focus on transactional processing has been tried in [9]. Their implementation is able to execute a set of stored procedures on the data in the GPU's memory and handles read and write access. To the best of our knowledge there are no publications about running complex analytic queries completely on the GPU.

Compiling a complete query plan (or parts of it) into executable code is a radical and novel approach. Classical database systems generate executable code fragments for single operators, which can be used and combined to perform a SQL query. There are several recent approaches for generating machine code for a complete SQL query. The prototype system HIQUE [12] generates C code that is just-in-time-compiled with gcc, linked as shared library and then executed. The compile times are noticeable (around seconds) and the measured execution time compared to traditional execution varies: TPC-H queries 1 and 3 are faster, query 10 is slower. Note that the data layout is not optimized for the generated C code. HyPer [11,14] also generates code for incoming SQL queries. However, instead of using C, they directly generate LLVM IR code (similar to Java byte code), which is further translated to executable code by an LLVM compiler [13]. Using this light-weight compiler they achieve low compile times in the range of several milliseconds. Execution times for the measured TPC-H queries are also fast, still, they use only a single thread for execution and no parallelization.

One of the fundamental models for parallel query execution is the exchange operator proposed by Graefe [7]. This operator model allows both intra-operator parallelism on partitioned datasets as well as horizontal and vertical inter-operator parallelism. Although this model can also be applied to vectorized processing strategies it still relies on iterator-based execution. Thus, the required synchronization between each operator can induce a noticeable overhead.

# 3    Basics

In this section we explain the technology as well as some basic decisions we had to make before going into the details of our work. We start with explaining the hardware characteristics in Sect. 3.1. Since we have to copy the data to the GPU's memory before we can use it, the transfer over the PCI-Express bus is the main bottleneck for data intensive algorithms. In Sect. 3.2 we address this problem and explain how we cope with it.

## 3.1    Parallel CPU and GPU Architecture

Although modern GPU and CPU architectures look different, we face a lot of similar problems when trying to develop efficient algorithms for both. In order to achieve maximum performance, it is mandatory to use parallel algorithms. Also, we generally want to avoid synchronisation and communication overhead by partitioning the input first and then processing the chunks with independent threads as long as possible. In the literature this is called bulk-synchronous processing [16].

There are some differences in the details of parallelism of both platforms. On the CPU we have to be careful not to oversubscribe the cores with a high amount of threads, while the GPU resp. the programming framework for the GPU handles this on its own. However, on the CPU oversubscription can be easily avoided by using tasks instead of threads: a framework, such as Intel's

TBB[1], creates a fixed number of threads once and assigns work to them in form of tasks. This way the expensive context switches between threads are not necessary. While the usage of tasks is enough to exploit the parallelism of CPUs, another form of parallelism is needed to use the GPU efficiently: *SIMD*. Both the OpenCL platform model[2] and CUDA[3] force the developer to write code in a SIMD fashion.

In addition to the parallel design of algorithms, we face the challenge to organize memory access in a way that fits to the hardware. In both architectures, CPU and GPU, there is a memory hierarchy with large and comparatively slow RAM and small but faster caches. Cache-efficient algorithms work in a way, that the data is in the cache when the algorithm needs it and only data that is really needed is transferred to the cache. On the CPU the hardware decides which data is copied from RAM to cache based on fixed strategies. *Local memory* on the GPU is equivalent to cache on the CPU, but it has to be filled manually.

With all data residing in main memory and many cores available for parallel execution, the main memory bandwidth becomes a new bottleneck for query processing. Therefore, we need to use memory bandwidth as efficiently as possible. On modern server CPUs this requires a careful placement of data structures to support NUMA-aware access, which has been shown in [6].

The GPU's memory is highly optimized for the way graphical algorithms access data: threads of neighboring cores process data in neighboring memory segments. If general purpose algorithms work in a similar way, they can benefit from the GPU architecture. If they load data from memory in a different way, the performance can be orders of magnitude worse. Details can be found in [15].

The challenges we face when writing efficient algorithms are very similar, no matter if they run on the GPU or the CPU. Only in case of the GPU the performance effects of using inadequate patterns and strategies are much worse. So we have to make sure, that our algorithms are able to distribute their work over a high number of threads/tasks and exploit the memory hierarchy on both architectures. On the GPU we have to take special care about using a SIMD approach to distribute small work packages and use coalesced access to the device memory. On the CPU it is important not to oversubscribe it with threads and consider NUMA effects of the parallel architecture. In this paper we suggest a model that supports query execution on both architectures and takes care of their differences.

## 3.2   The Data Transfer Problem

The transfer to the GPU is the main bottleneck for data centric workloads. In general, it takes significantly longer than the execution of the queries, because the transfer of a tuple is usually slower than processing it once. To make matters worse, in general we cannot predict beforehand which values of a column

---

[1] http://www.threadingbuildingblocks.org
[2] http://www.khronos.org/opencl/
[3] http://www.nvidia.com/cuda

are really accessed and therefore need to transfer all values of a column used by the query, which further increases the data volume. One way to cope with this problem is the usage of Universal Virtual Addressing (UVA). UVA is a sophisticated streaming technique for CUDA, which allows the transfer of required data at the moment it is needed be the GPU. The authors of [10] showed how to use it to process joins on the GPU. However, streaming is still rather inefficient for random access to the main memory. Even if the GPU accesses the main memory sequentially we remain limited to the PCI-Express bandwidth of theoretically 8 GB/s (PCI-Express 2).

From our perspective there are three possibilities to cope with this problem:

1. The data is replicated to the GPU memory before query execution and cached there.
2. The data is stored exclusively in the GPU's memory.
3. CPU and GPU share the same memory.

The first option limits the amount of data to the size of the device memory, which is very small at the moment: high end graphic cards for general-purpose computations have about 4–16 GB. Furthermore, updates have to be propagated to the GPU.

Storing the data exclusively in the GPU's memory also limits the size of the database. Furthermore, there are queries that should be executed on the CPU, because they require advanced string operations or have a large result set (see Sect. 4.4). In this case we face the transfer problem again, only this time we have to copy from GPU to CPU.

The third solution, i.e., CPU and GPU sharing the same memory, is not yet available. Integrated GPUs already use the CPU's main memory, but internally it is divided into partitions for GPU and CPU and each can only access its own partition. So we can copy data very fast from GPU to CPU and vice versa as we remain in main memory, but still need to replicate. For the future AMD, for instance, plans to use the same memory for both parts of the chip [4].

We decided to use the first approach for our research, because it makes no sense to compare GPU and CPU performance as long as we need to transfer the data. We also want to be able to execute queries on the CPU and the GPU at the same time. Until real shared memory is available, replication is the best choice for our needs.

## 4 A new Model to Execute Queries in Parallel

In this paper we present a framework that provides simple parallelism for just-in-time query compilation. Section 4.1 gives a short overview about why and how we can use JIT compilation to speed up query execution. There are two ways of applying parallelism to the execution. On the one hand we can distribute the work over independent threads, which are then executed on different computing cores. In our model we call this the first level of parallelism. In Sect. 4.2 we describe the basic model that allows us to execute queries in parallel on the

CPU by using independent threads. On the other hand it is possible to use SIMD to exploit the parallel computing capabilities of one CPU core resp. *Streaming Multi Processor* of the GPU. In Sect. 4.3 we make use of this second level of parallelism to extend our model and execute queries efficiently on the GPU. Our GPU framework has some limitations that we discuss in Sect. 4.4.

## 4.1   JIT-Compilation for SQL Queries

To take full advantage of modern hardware for query processing, we need to use the available processing power carefully. Here, the classical iterator model as well as block-oriented processing have their problems. The iterator model produces a lot of overhead by virtual function calls, while block-oriented processing requires the materialization of intermediate results (see Neuman [14] and Dees et al. [6]). Virtual function calls can cost a lot of processing cycles, similar to materialization, as here we often need to wait for data to be written or loaded. By compiling the whole SQL query into a single function, we can get around virtual function calls and most materialization in order to use processing units as efficiently as possible. For the CPU-only version we used the LLVM framework to compile this function at runtime [6].

Our query execution is fully parallelized since we target modern many-core machines. Neumann [14] already explained that a parallel approach for his code generation can be applied without changing the general patterns. Operations, such as joins and selections, can be executed independently – and therefore in parallel – on different fragments of the data. If other operations are needed we use the patterns described in the next section. Later in Sect. 4.3 we go into details for applying these patterns on GPUs, as here some changes are necessary.

## 4.2   A Model for Parallel Execution

One of the main challenges when designing parallel algorithms is to keep communication between threads low. Every time we synchronize threads to exchange data, some cores are forced to wait for others to finish their work. This problem can be avoided by partitioning the work in equally-sized chunks and let every thread process one or more chunks independently. In the end, the results of each thread are merged into one final result. This approach is called bulk-synchronous model [16] and describes the first level of parallelism we need to distribute work over independent threads.

Our query execution always starts with a certain table (defined by the join order), which is split horizontally. This is performed only logically and does not imply real work. In the first phase of the query execution every partition of the table is processed by one thread in a tight loop, i.e., selections, joins, and arithmetic terms are executed for each tuple independently before the next tuple is processed. We call this the *Compute* phase, since it is the major part of the work. Before we start the second phase of the query execution, called *Accumulate*, all threads are synchronized once. The Accumulate phase merges the intermediate results into one final result.

This model is similar to the MapReduce model, which is used to distribute work over a heterogeneous cluster of machines [5]. However, our model does not have the limitations of using key-value pairs. Additionally, the Accumulate phase does not necessarily reduce the number of results, while in the MapReduce model the Reduce phase always groups values by their keys. In case of a selection, for instance, the Accumulate phase just concatenates the intermediate results.

The Accumulate phase depends on the type of query, which can be one of the following patterns.

**Single Result.** The query produces just a single result line. For example, this is the case SQL queries with a `sum` clause but no `group by`, see TPC-H query 6:

```
select sum(l_ext...*l_discout) from ...
```

This is the simplest case for parallelization: Each thread just holds one result record. Two interim results can be easily combined to one new result by combining the records according to the aggregation function.

**(Shared) Index Hashing.** On SQL queries with a `group by` clause where we can compute and guarantee a maximum cardinality that is not too large, we can use index hashing. This occurs for example at TPC-H query 1:

```
select sum(...) from ... group by l_returnflag, l_linestatus
```

We store the attributes `l_returnflag` and `l_linestatus` in a dictionary-compressed form. The size of each dictionary is equal to the number of distinct values of the attribute. With this we can directly deduce an upper bound for the cardinality of the combination of both attributes. In our concrete example the cardinality of `l_returnflag` is 2 and `l_linestatus` has 3 distinct values for every scale factor of TPC-H. Therefore the combined cardinality is $2 \cdot 3 = 6$. When this upper bound multiplied with the data size of one record (which consists of several sums in query 1) is smaller than the cache size, we can directly allocate a hash table that can hold the complete result. We use index hashing for the hash table, i.e., we compute one distinct index for the combination of all attributes that serves as a hash key. On the CPU, we differentiate whether the results fit into the L3 (shared by all cores of one CPU) or the L2 (single core) cache. If only the L3 cache is sufficient, we use a shared hash table for each CPU where entries are synchronized with latches. We can avoid this synchronization when the results are small enough to fit into L2 cache. In this case we use a separate hash table for each thread while remaining cache efficient.

For combining two hash tables, we just combine all single entries at the same positions in both hash tables, which is performed similar to the single result's handling. There is no difference in combining shared or non-shared hash tables.

**Dynamic Hash Table.** Sometimes we cannot deduce a reasonable upper bound for the maximum `group by` cardinality or we can not compute a small hash index efficiently from the attributes. This can happen if we do not have any estimation of the cardinality of one of the `group by` attributes or the combination of all `group by` attributes just exceeds a reasonable number. In this case we need a dynamic hash table that can grow over time and use a general 64-bit hash key from all attributes. The following is an example query in which it is difficult to give a good estimation for the `group by` cardinality:

```
select sum() ... from ... group by l_extendedprice,
n_nationkey, p_partsupp, c_custkey, s_suppkey
```

Combining two hash tables is performed by rehashing all elements from one table and inserting it into the other table, i.e., combing the corresponding records.

**Array.** For simple `select` statements without a `group by`, we just need to fill an array with all matching lines. The following SQL query is an example for this problem:

```
select p_name from parts where p_name like '%green%'
```

## 4.3   Adjusting the Model for GPU Execution

The first adjustment to execute the *Compute/Accumulate* model on the GPU is a technical one: we replace LLVM with OpenCL, which requires a few changes. Before we execute the query, we allocate space for its result in the graphic card's memory, since the kernel is not able to do that on its own. We then call the kernel with the addresses to the required columns, indexes, and the result memory as parameters. After the actual execution the results are copied to the RAM to give the CPU access to it. The result memory on the device can be freed.

Since the GPU is not able to synchronize the workgroups of one kernel, we need to split the two phases into two kernels. As we can see in Fig. 1(a), the workgroups of the *Compute* kernel work independently and write the intermediate results (1) to global memory. The *Accumulate* kernel works only with one workgroup. It writes the final result (3) to global memory (the GPU's RAM).

To differentiate this first level of parallelism from the second, we call this the *global Compute/Accumulate* process. It is a direct translation of the CPU implementation and shows that workgroups on the GPU are comparable to threads on the CPU. Every workgroup has a number of GPU threads, which work in a SIMD fashion. Therefore, the main challenge and the second adjustment is to extend the model by a second level of parallelism to use these threads.

In contrast to CPU threads, every OpenCL workgroup is capable of using up to 1024 *Stream Processors* for calculations. Threads running on those Stream Processors do not run independently but in a SIMD fashion, i.e., at least 32 threads are executing the same instruction at the same time. These threads form a so called *warp*—the warp size may be different for future architectures.

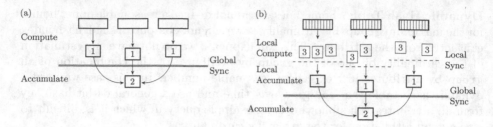

**Fig. 1.** (a) Basic Model, (b) Extended Model
Legend: 1 – intermediate results, 2 – global result, 3 – private result

If there are branches in the code, all threads of one warp step through all branches. Threads that are not used in a certain branch execute NOPs as long as the other threads are calculating. The threads of one warp are always in sync, but all threads of one workgroup must be synchronized manually if necessary. Although this local synchronization is cheap compared to a global synchronization, we have to be careful about using it. Every time it is called, typically all workgroups execute it. Since we use up to 1,000 workgroups (see Sect. 5.2), the costs for one synchronization also multiply.

Therefore we apply a *local Compute/Accumulate* model again in the *global* Compute phase. Since the local results are only accessed by the threads of one workgroup, there is no need to write them to global memory. The whole process can be seen in Fig. 1(b). Every thread computes its result in private memory (3), usually the registers of each Stream Processor. At the end of the local Compute phase, the results are written to local memory, the threads are synchronized, and the local Accumulate phase is started. These local processes work in a SIMD fashion and therefore differ from the global phase in the detail.

The local Compute phase works on batches of tuples. The size of the batch equals the number of threads of one workgroup. The thread ID equals the position of the tuple in the batch, i.e., the first thread of the workgroup works on the first tuple, the second on the second, and so on. This works much better than horizontally partitioning the table as we do in the global Compute phase, because we enforce coalesced memory access. The worst case for this SIMD approach arises if a few tuples of the batch require a lot of work and the rest is filtered out early. As we explained above this leads to a lot of cores executing NOPs, while only a few are doing real work. If all tuples of the batch are filtered—or at least all tuples processed by one warp—the GPU can skip the rest of the execution for this batch and proceed to the next. Fortunately this is often the case in typical OLAP scenarios data, e.g., data of a certain time frame is queried and is stored in the same part of the table.

The aggregation specified by the query is in parts already done in the local Compute phase, where each thread works on private memory. In the local/global Accumulate phase we merge these private/intermediate results. The threads are writing simultaneously to local memory, so we require special algorithms to avoid conflicts. The algorithms (prefix sum, sum, merge, and bitonic sort) are well known and can be found in the example directory that is delivered with NVidia's OpenCL implementation. Therefore we just name them at this point and give a short overview but not a detailed description.

The easiest case is no aggregation at all, where we just copy private results to local memory and later to global memory. However, since all threads copy their result in parallel, the right address has to be determined to guarantee that the global result has no gaps. Therefore, each thread counts the number of its private results and writes the size to local memory. To find the right offset for each threads starting position we calculate the prefix sum in local memory. This is done by pairwise adding elements in $\log n$ steps for $n$ being the number of threads. In the end every position holds the number of private results of all threads with smaller IDs and therefore the position in the global result

We use a very similar approach for the sum-Aggregation. In case of a single result pattern (see Sect. 4.2) we calculate the sum of the private results in parallel by again using pairwise adding of elements in local memory. For index hashing (see Sect. 4.2) this has to be done for each row of the result.

Sorting is the final process of the Accumulate phase. In every case we can use bitonic sort. For the array pattern it is also possible to pre-sort the intermediate results in the Compute phase and use a simple pair-wise merge in the Accumulate phase.

### 4.4    Limitations of the GPU Execution

In this section we discuss the limits of our framework. We categorize these in two classes:

*Hard limits* Due to the nature of our model and the GPU's architecture there are limits that cannot be exceeded with our approach. One of these limits is given by OpenCL: we cannot allocate memory inside kernels, so we have to allocate memory for (intermediate) results in advance. Therefore we have to know the maximum result size and since it is possible that one thread's private result is as big as the final result, we have to allocate this maximum size for every thread. One solution to this problem would be to allocate the maximum result size only once and let every thread write to the next free slot. This method would require a lot of synchronization and/or atomic operations, which is exactly what we want to avoid. With our approach we are limited to a small result size that is known or can be computed in advance, e.g., top-k queries or queries that use group by on keys that have relatively small cardinalities. The concrete number of results we can compute, depends on the temporary memory that is needed for the computation, the hardware and the data types used in the result, but it is in the order of several hundred rows.

*Soft limits* Some use cases are known not to fit to the GPU's architecture. It is possible to handle scenarios of this type on the GPU, but it is unlikely that we can benefit from doing that. The soft limits of our implementation are very common for general-purpose calculations on the GPU. First, the GPU is very bad at handling strings or other variable length data types. They simply do not fit into the processing model and usually require a lot of memory. An indication for this problem is, that even after a decade of research on general-purpose computing on graphics processing units and commercial interest in it, there is no string library available for the GPU, not even simple methods such as `strstr()` for string comparison. In most cases we can avoid handling them on the GPU by using dictionary encoding. Second, the size of the input data must exceed a certain minimum to use the full parallelism of the GPU (see Sect. 5.2).

Furthermore, GPUs cannot synchronize simultaneously running workgroups. Therefore we cannot efficiently use algorithms accessing shared data structures, such as the shared index hashing and the dynamic hash table approach described in Sect. 4.2.

## 5   Evaluation

In this section we present the performance results of our framework. We compare the performance of NVidia's Tesla C2050 to a HP Z600 workstation with two Intel Xeon X5650 CPUs and 24 GB of RAM. The Tesla GPU consists of 14 Streaming Multiprocessors with 32 CUDA cores each and has 4 GB of RAM. The JIT compilation was done by LLVM 2.9 for the CPU and the OpenCL implementation that is shipped with CUDA 5.0 for the GPU.

### 5.1   GPU and CPU Performance

Since the native CPU implementation described in Sect. 4 was already compared to other DBMS [6], we take it as our baseline and compare it to our OpenCL implementation.

In our first experiment we compare the implementations by measuring the raw execution time of each query. As explained in Sect. 3.2 the necessary columns for execution are in memory, we do not consider the compile times, and we include result memory allocation (CPU and GPU) and result transfer (only needed on GPU). We configure the OpenCL framework to use 300 workgroups with 512 threads each on the GPU and 1000 workgroups with 512 threads on the CPU. This configuration gives good results for all queries as we show in Sect. 5.2.

Figure 2 compares the execution times for seven TPC-H queries on different platforms: Z600 is the native CPU implementation on the Z600 machine, Tesla/CL the OpenCL implementation on the Tesla GPU and Z600/CL the OpenCL implementation executed on the CPU of the workstation. As we can see, the OpenCL implementation for the GPU is considerably faster than the native implementation on the workstation for most queries. The only exception is Q1, which takes almost twice as long. The OpenCL implementation of Q1 on the CPU is significantly worse than native implementation as well. The reason for this is hard to find, since we cannot measure time within the OpenCL kernel. The performance results might be worse because in contrast to the other

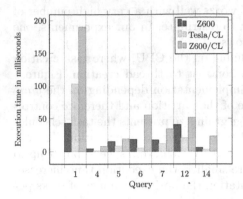

**Fig. 2.** Comparison of execution times

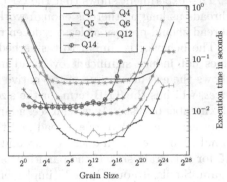

**Fig. 3.** Comparison of different grain sizes

queries the execution of Q1 is dominated by aggregating the results in the local accumulate phase. This indicates that the CPU works better for queries without joins but with aggregations.

Further tests and micro-benchmarks are needed to proof this hypothesis.

It is remarkable that the OpenCL implementation on the CPU achieves almost the same performance as the native CPU implementation for half of the queries, considering that we spent no effort in optimizing it for the actual architecture.

## 5.2   Number of Workgroups and Threads

We illustrate in Fig. 4 how the number of workgroups used for the execution of the query influences the performance. We measure two different configurations: the left figure shows the results when we use 512 threads per workgroup; on the right we use 256 threads per workgroup. Due to implementation details our framework is not able to execute queries 1 and 5 for less than around 140 resp. 200 workgroups. For the other queries we can clearly see in both figures that we need at least around 50,000 threads in total to achieve the best performance. This high number gives the task scheduler on the GPU the ability to execute instructions on all cores (448 on the Tesla C2050) while some threads are waiting for memory access. There is no noticeable overhead if more threads are started with the following exceptions:

Query 4 shows irregular behavior. It is the fastest query in our set, the total execution time is between 1 and 2 ms. Therefore uneven data distribution has a high impact.

Query 5 is getting slower with more threads, because the table chunks are too small and therefore the work done by a thread is not enough. The table, which is distributed over the workgroups, has only 1.5 mio rows, i.e., each thread processes only 3 rows in case of 1,000 workgroups with 512 threads each. This matter is even worse with Query 7, because the table we split has only 100,000 rows. The other queries process tables which have at least 15 mio rows.

This experiment shows that the GPU works well with a very high number of threads as long as there is enough work to distribute. In our experiments, one thread should process at least a dozen rows.

The behavior is similar to task scheduling on the CPU, where task creation does not induce significant overhead in contrast to thread creation. Figure 3 shows the execution time of the native implementation depending on the *grain sizes*, which is the TBB term for the size of the partition and therefore controls the number of tasks. As we can see, a large grain size prevents the task scheduler from using more than one thread and therefore slows down execution. The optimal grain size for all queries is around 1000. We achieve a speed-up of 7 up to 15 for this grain size. If the partitions are smaller, the execution time increases again. Similar to our OpenCL implementation the optimal number of tasks per core is around 100.

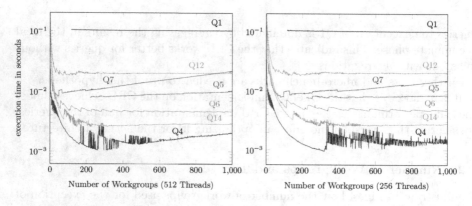

**Fig. 4.** Comparison of different workgroup numbers (left: 512, right: 256 threads per workgroup)

## 6   Conclusion

In this work, we have investigated the parallel execution of OLAP queries on massively parallel processor architectures by leveraging a JIT query compilation approach. Our goal was to identify and evaluate patterns for parallelization that are suitable both for CPUs and GPUs. As a core pattern we have presented a two-phase Compute/Accumulate model similar to the MapReduce paradigm, which requires only one global synchronisation point. For compiling the whole query into executable code we use the OpenCL framework which already provides a JIT compiler for different devices. Surprisingly we were able to execute our GPU-optimized code on the CPU as well. Although, this is still slower than a native CPU optimized implementation, the advantage of having only one implementation for different platforms may be worth the loss in performance. Since OpenCL is an open standard for parallel platforms, we expect that our framework also runs on other coprocessors.

Our experimental results using a subset of TPC-H have shown that GPUs can be used to execute complete OLAP queries and even can outperform modern CPUs for certain queries. However, we have also discussed some limitations and differences between CPU and GPU processing which have to be taken into account for query compilation and execution.

# References

1. Bakkum, P., Skadron, K.: Accelerating SQL database operations on a GPU with CUDA. In: Proc. 3rd Workshop on GPGPU, p. 94 (2010)
2. Boncz, P., Zukowski, M., Nes, N.: MonetDB/X100: Hyper-Pipelining Query Execution. In: Proc. CIDR, vol. 5 (2005)
3. Boncz, P.A., Kersten, M.L., Manegold, S.: Breaking the memory wall in monetdb. Commun. ACM 51(12), 77–85 (2008)
4. Daga, M., Ajl, A., Feng, W.: On the Efficacy of a Fused CPU+GPU Processor (or APU) for Parallel Computing. In: SAAHPC. IEEE (2011)
5. Dean, J., Ghemawat, S.: MapReduce. Comm. of the ACM 51(1), 107 (2008)
6. Dees, J., Sanders, P.: Efficient Many-Core Query Execution in Main Memory Column-Stores. To appear on ICDE 2013 (2013)
7. Graefe, G.: Encapsulation of Parallelism in the Volcano Query Processing System, vol. 19. ACM (1990)
8. He, B., Lu, M., Yang, K., Fang, R., Govindaraju, N.K., Luo, Q., Sander, P.V.: Relational Query Coprocessing on Graphics Processors. ACM Transactions on Database Systems 34 (2009)
9. He, B., Yu, J.X.: High-Throughput Transaction Executions on Graphics Processors. PVLDB 4, 314–325 (2011)
10. Kaldewey, T., Lohman, G., Mueller, R., Volk, P.: GPU Join Processing Revisited. In: Proc. 8th DaMoN (2012)
11. Kemper, A., Neumann, T.: HyPer: A hybrid OLTP&OLAP Main Memory Database System Based on Virtual Memory Snapshots. In: Proc. of ICDE (2011)
12. Krikellas, K., Viglas, S., Cintra, M.: Generating Code for Holistic Query Evaluation. In: Proc. of ICDE, pp. 613–624. IEEE (2010)
13. Lattner, C.: LLVM and Clang: Next Generation Compiler Technology. In: The BSD Conference (2008)
14. Neumann, T.: Efficiently Compiling Efficient Query Plans for Modern Hardware. Proc. of VLDB 4(9), 539–550 (2011)
15. NVidia: CUDA C Best Practices Guide (2012)
16. Valiant, L.G.: A Bridging Model for Parallel Computation. Comm. ACM 33 (1990)
17. Zukowski, M., Boncz, P.A.: From X100 to Vectorwise: Opportunities, Challenges and Things Most Researchers do not Think About. In: SIGMOD Conference (2012)

# Improving the Performance of High-Dimensional $k$NN Retrieval through Localized Dataspace Segmentation and Hybrid Indexing

Michael A. Schuh, Tim Wylie, and Rafal A. Angryk

Montana State University, Bozeman, MT 59717-3880, USA
{michael.schuh,timothy.wylie,angryk}@cs.montana.edu

**Abstract.** Efficient data indexing and nearest neighbor retrieval are challenging tasks in high-dimensional spaces. This work builds upon our previous analyses of iDistance partitioning strategies to develop the backbone of a new indexing method using a heuristic-guided hybrid index that further segments congested areas of the dataspace to improve overall performance for exact $k$-nearest neighbor ($k$NN) queries. We develop data-driven heuristics to intelligently guide the segmentation of distance-based partitions into spatially disjoint sections that can be quickly and efficiently pruned during retrieval. Extensive tests are performed on k-means derived partitions over datasets of varying dimensionality, size, and cluster compactness. Experiments on both real and synthetic high-dimensional data show that our new index performs significantly better on clustered data than the state-of-the-art iDistance indexing method.

**Keywords:** High-dimensional, Partitioning, Indexing, Retrieval, $k$NN.

## 1 Introduction

Modern database-oriented applications are overflowing with rich information composed of an ever-increasing amount of high-dimensional data. While massive data storage is becoming routine, efficiently indexing and retrieving it is still a practical concern. A frequent and costly retrieval task on these databases is $k$-nearest neighbor ($k$NN) search, which returns the $k$ most similar records to any given query record. While a database management system (DBMS) is highly optimized for a few dimensions, most traditional indexing algorithms (*e.g.*, the B-tree and R-tree families) degrade quickly as the number of dimensions increase, and eventually a sequential (linear) scan of every single record in the database becomes the fastest retrieval method.

Many algorithms have been proposed in the past with limited success for true high-dimensional indexing, and this general problem is commonly referred to as *the curse of dimensionality* [4]. These issues are often mitigated by applying dimensionality reduction techniques before using popular multi-dimensional indexing methods, and sometimes even by adding application logic to combine multiple independent indexes and/or requiring user involvement during search. However,

B. Catania, G. Guerrini, and J. Pokorný (Eds.): ADBIS 2013, LNCS 8133, pp. 344–357, 2013.
© Springer-Verlag Berlin Heidelberg 2013

modern applications are increasingly employing highly-dimensional techniques to effectively represent massive data, such as the highly popular 128-dimensional SIFT features [11] in Content-Based Image Retrieval (CBIR). Therefore, it is of great importance to be able to comprehensively index all dimensions for a unified similarity-based retrieval model.

This work builds upon our previous analyses of standard iDistance partitioning strategies [14,19], with the continued goal of increasing overall performance of indexing and retrieval for $k$NN queries in high-dimensional dataspaces. We assess performance efficiency by the total and accessed number of $B^+$-tree nodes, number of candidate data points returned from filtering, and the time taken to build the index and perform queries. In combination, these metrics provide a highly descriptive and unbiased quantitative benchmark for independent comparative evaluations. We compare results against our open-source implementation of the original iDistance algorithm[1].

The rest of the paper is organized as follows. Section 2 highlights background and related work, while Section 3 covers the basics of iDistance and then introduces our hybrid index and heuristics for iDStar. Section 4 presents experiments and results, followed by a brief discussion of key findings in Section 5. We close with our conclusions and future work in Section 6.

## 2 Related Work

The ability to efficiently index and retrieve data has become a silent backbone of modern society, and it defines the capabilities and limitations of practical data usage. While the one-dimensional $B^+$-tree [2] is foundational to the modern relational DBMS, most real-life data has many dimensions (attributes) that would be better indexed together than individually. Mathematics has long-studied the partitioning of multi-dimensional metric spaces, most notably Voronoi Diagrams and the related Delaunay triangulations [1], but these theoretical solutions can often be too complex for practical application. To address this issue, many approximate techniques have been proposed to be used in practice. One of the most popular is the R-tree [7], which was developed with minimum bounding rectangles (MBRs) to build a hierarchical tree of successively smaller MBRs containing objects in a multi-dimensional space. The R*-tree [3] enhanced search efficiency by minimizing MBR overlap. However, these trees (and most derivations) quickly degrade in performance as the dimensions increase [5,12].

Research has more recently focused on creating indexing methods that define a one-way lossy mapping function from a multi-dimensional space to a one-dimensional space that can then be indexed efficiently in a standard $B^+$-tree. These lossy mappings require a filter-and-refine strategy to produce exact query results, where the one-dimensional index is used to quickly retrieve a subset of the data points as candidates (the filter step), and then each of these candidates is verified to be within the specified query region in the original multi-dimensional space (the refine step). Since checking the candidates in the actual dataspace is

---

[1] Publicly available at: `http://code.google.com/p/idistance/`

costly, the goal of the filter step is to return as few candidates as possible while retaining the exact results to satisfy the query.

The Pyramid Technique [5,22] was one of the first prominent methods to effectively use this strategy by dividing up the $d$-dimensional space into $2d$ pyramids with the apexes meeting in the center of the dataspace. For greater simplicity and flexibility, iMinMax($\theta$) [12,16] was developed with a global partitioning line $\theta$ that can be moved based on the data distribution to create more balanced partitions leading to more efficient retrieval. Both the Pyramid Technique and iMinMax($\theta$) were designed for multi-dimensional range queries, and extending to high-dimensional $k$NN queries is not a trivial task.

First published in 2001, iDistance [10,20] specifically addressed exact $k$NN queries in high-dimensional spaces and was proven to be one of the most efficient, state-of-the-art techniques available. In recent years, iDistance has been used in a number of demanding applications, including large-scale image retrieval [21], video indexing [15], mobile computing [8], peer-to-peer systems [6], and video surveillance retrieval [13]. As information retrieval from high-dimensional and large-scale databases becomes more ubiquitous, the motivations for this research will only increase. Many recent works have shifted focus to approximate nearest neighbors [9,18] which can generally be retrieved faster, but these are outside the scope of efficient exact $k$NN retrieval presented in this paper.

# 3   Towards iDStar

Our hybrid index and accompanying heuristics build off the initial concept of iDistance with an approximate Voronoi tessellation of the space using hyperspheres. Therefore we start with a brief overview of the original iDistance algorithm before introducing iDStar.

## 3.1   iDistance

The basic concept of iDistance is to segment the dataspace into disjoint spherical partitions, where all points in a partition are *indexed by their distance* (hence "iDistance") to the reference point of that partition. This results in a set of one-dimensional distance values, each related to one or more data points, for each partition, that are all indexed in a single standard $B^+$-tree. The algorithm was motivated by the ability to use arbitrary reference points to determine the (dis)similarity between any two data points in a metric space, allowing single dimensional ranking and indexing of data points regardless of the dimensionality of the original space [10,20].

**Building the Index.** Here we focus only on *data-based* partitioning strategies, which adjusts the size and location of partitions in the dataspace based on the underlying data distribution, which greatly increases retrieval performance in real-world settings [10,14,20]. For all partitioning strategies, data points are assigned to the single closest partition based on Euclidean distance to each partitions' representative reference point.

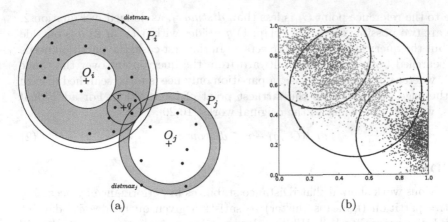

(a)                                   (b)

**Fig. 1.** (a) A query sphere $q$ with radius $r$ and the searched regions (shaded) in the two overlapping partitions $P_i$ and $P_j$ defined by their reference points $O_i$ and $O_j$, and radii $distmax_i$ and $distmax_j$, respectively. (b) An example two dimensional dataset with three cluster-based partitions and their point assignments and radii.

Formally, we have a set of partitions $\mathbb{P} = \langle P_1, \ldots, P_M \rangle$ with respective reference points $\mathbb{O} = \langle O_1, \ldots, O_M \rangle$. We will let the number of points in a partition be denoted $|P_i|$ and the total number of points in the dataset be $N$. After the partitions are defined, a mapping scheme is applied to create separation in the underlying $B^+$-tree between each partition, ensuring that any given index value represents a unique distance for exactly one partition.

Given a partition $P_i$ with reference point $O_i$, the index value $y_p$ for a point $p$ assigned to this partition is defined by Equation 1, where $dist()$ is any metric distance function, $i$ is the partition index, and $c$ is a constant multiplier for creating the partition separation. While constructing the index, each partition $P_i$ records the distance of its farthest point as $distmax_i$. We can safely set $c = 2\sqrt{d}$, which is twice the maximum possible distance of two points in the $d$-dimensional unit space of the data, and therefore no index value in partition $P_i$ will clash with values in any other partition $P_{j \neq i}$.

$$y_p = i \times c + dist(O_i, p) \tag{1}$$

**Querying the Index.** The index should be built in such a way that the filter step returns the fewest possible candidate points without missing the true $k$-nearest neighbors. Fewer candidates reduces the costly refinement step which must verify the true multi-dimensional distance of each candidate from the query point. Performing a query $q$ with radius $r$ consists of three steps: 1) determine the set of partitions to search, 2) calculate the search range for each partition, and 3) retrieve the candidate points and refine by their true distances.

Figure 1(a) shows an example query sphere contained completely within partition $P_i$ and intersecting partition $P_j$, as well as the shaded ranges of each partition that need to be searched. For each partition $P_i$ and its $distmax_i$, the query sphere overlaps the partition if the distance from the edge of the query sphere to the reference point $O_i$ is less than $distmax_i$, as defined in Equation 2. There are two possible cases of overlap: 1) $q$ resides within $P_i$, or 2) $q$ is outside of $P_i$, but the query sphere still intersects it. In the first case, the partition needs to be searched both inward and outward from the query point over the range $(q \pm r)$, whereas in the second case, a partition only needs to be searched inward from the edge $(distmax_i)$ to the farthest point of intersection. For additional details, we refer the reader to the original works [10,20].

$$dist(O_i, q) - r \le distmax_i \tag{2}$$

## 3.2   iDStar

Our previous work showed that iDistance stabilizes in performance by accessing an entire partition (k-means cluster) to satisfy a given query, despite dataset size and dimensionality [14]. While only accessing a single partition is already significantly more efficient than sequential scan, this hurdle was the main motivation to explore further dataspace segmentation to enhance retrieval performance. We achieve this additional segmentation with the creation of intuitive heuristics applied to a novel hybrid index. These extensions are similar to the works of the iMinMax($\theta$) [12] and recently published SIMP [17] algorithms, whereby we can incorporate additional dataspace knowledge at the price of added algorithm complexity and performance overhead. This work proves the feasibility of our approaches and lays the foundations for a new indexing algorithm, which we introduce here as iDStar.

Essentially, we aim to further separate dense areas of the dataspace by *splitting* partitions into disjoint *sections* corresponding to separate *segments* of the $B^+$-tree that can be selectively pruned during retrieval. The previous sentence denotes the technical vocabulary we will use to describe the segmentation process. In other words, we apply a given number of dimensional *splits* in the dataspace which *sections* the partitions into $B^+$-tree *segments*.

We develop two types of segmentation based on the scope of splits: 1) **Global**, which splits the entire dataspace (and consequently any intersecting partitions), and 2) **Local**, which explicitly splits the dataspace of each partition separately. While different concepts, the general indexing and retrieval algorithm modifications and underlying affects on the $B^+$-tree are similar. First, the mapping function is updated to create a constant segment separation within the already separated partitions. Equation 3 describes the new index with the inclusion of the sectional index $j$, and $s$ as the total number of splits applied to the partition. Note that a partition is divided into $2^s$ sections, so we must appropriately bound the number of splits applied. Second, after identifying the partitions to search, we must identify the sections within each partition to actually search, as well as their updated search ranges within the $B^+$-tree. This also requires the overhead of section-supporting data structures in addition to the existing partition-supporting data structures.

$$y_p = i \times c + j \times \frac{c}{2^s} + dist(O_i, p) \tag{3}$$

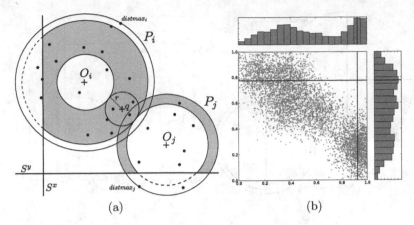

**Fig. 2.** (a) Conceptual global splits $S^x$ and $S^y$ and their effect on query search ranges. (b) The histogram-based splits applied by the G2 heuristic on the example dataset.

**Global.** The global segmentation technique is inspired by the Pyramid and iMinMax($\theta$) methods, whereby we adjust the underlying index based on the data distributions in the space. A global split is defined by a tuple of dimension and value, where the given dimension is split on the specified value. There can be only one split in each dimension, so given the total number of splits $s$, each partition may have up to $2^s$ sections. Because the splits occur across the entire dataspace, they may intersect (and thereby segment) any number of partitions (including zero). An added benefit of global splits is the global determination of which sections require searching, since all partitions have the same global z-curve ordering. We introduce three heuristics to optimally choose global splits.

- **(G1)** Calculate the median of each dimension as the proposed split value and rank the top $s$ dimensions to split in order of split values nearest to 0.5, favoring an even 50/50 dataspace split.
- **(G2)** Calculate an equal-width histogram in each dimension, selecting the center of the highest frequency bin as the proposed split value, and again ranking dimensions to split by values nearest to the dataspace center.
- **(G3)** Calculate an equal-width histogram in each dimension, selecting the center of the bin that intersects the most partitions, ranking the dimensions first by number of partitions intersected and then by their proximity to dataspace center.

Figure 2(a) shows our iDistance example with a split in each dimension $(S^y, S^x)$. We can see that a given split may create empty partition sections, or may not split a partition at all. We explicitly initialize these data structures with a known value representing an "empty" flag. Therefore, we only check for query overlap on non-empty partitions and non-empty sections. For the example dataset and partitions in Figure 1(b), the G2 heuristic applies the splits shown in Figure 2(b).

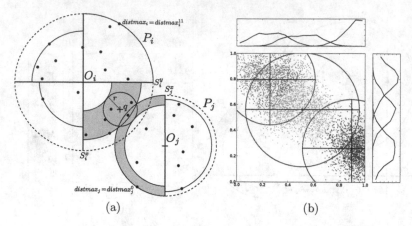

(a)                              (b)

**Fig. 3.** (a) Conceptual local splits $S_i^x$, $S_i^y$, and $S_j^x$ and their effect on query search ranges in overlapping partitions and sections. (b) The local population-based splits applied to each partition by the L3 heuristic on the example dataset.

**Local** We developed the local technique as a purposeful partition-specific segmentation method based only on local information of the given partition rather than the global data characteristics of the entire space. Local splits are defined by a tuple of partition and dimension, where the given partition is split in the specified dimension. Unlike the global technique, we do not need the values of each dimensional split because all splits are made directly on the partition reference point location. Figure 3(a) shows a conceptual example of local splits.

During index creation we maintain a $distmax_i$ of each partition $P_i$, and we now we do this for sections too ($distmax_i^j$ for partition $P_i$ and section $j$). This allows the addition of a section-based query overlap filter which can often prune away entire partitions that would have otherwise been searched. This can be seen in Figure 3(a), where the original partition $distmax$ is now a dotted line and a new $distmax$ is shown for each individual partition section.

We introduce three heuristics to choose local splits. Within each partition, we want to split the data as evenly as possible using the added spatial information. Thus, for all three methods, we rank the dimensions to split by those closest to an equal (50/50) population split. We denote $s$ as the maximum number of splits any partition can have.

- **(L1)** Uniformly apply $s$ splits to each partition using the top ranked.
- **(L2)** Use a population threshold as a cutoff criteria, so each partition could have up to $s$ splits. The intuition here is that if a split only segments a reasonably small portion of the data, it probably is not worth the additional overhead to keep track of it.
- **(L3)** Use $s$ to calculate the maximum number of underlying $B^+$-tree segments created by method L1, but then redistribute them based on partition population. This automatically applies additional splits to dense areas of the dataspace while removing splits where they are not needed and maintain an upper bound on $B^+$-tree segments.

Formally, we have a set of splits $\mathbb{S} = \langle S_1 \ldots, S_M \rangle$ where $S_i$ represents the set of dimensions to split on for partition $P_i$. Given a maximum granularity for the leaves in the $B^+$-tree, we can calculate how many splits each partition should have as a percentage of its population to the total dataset size. Equation 4 gives this assignment used by L3. For example, we can see in Figure 3(b) that the center cluster did not meet the population percentage to have two splits. Thus, the only split is on the dimension that has the closest 50/50 split of the data.

$$|S_i| = \left\lfloor \lg \frac{|P_i|}{N} \cdot M \cdot 2^s \right\rfloor \tag{4}$$

**Hybrid Indexing and Optimization.** The global and local dataspace segmentation generates a hybrid index with one-dimensional distance-based partitions subsequently segmented in a tune-able subset of dimensions. Due to the exponential growth of the tree segments, we must limit the number of dimensions we split on. For the local methods, L3 maintains this automatically. For the spatial subtree segmentation, we use z-curve ordering of each partition's sections, efficiently encoded as a ternary bitstring representation of all overlapping sections that is quickly decomposable to a list of section indices. Figure 3(b) shows an example of this ordering with $distmax_i^{11}$, which also happens to be the overall partition $distmax_i$. Since $P_i$ is split twice, we have a two-digit binary encoding of quadrants, with the upper-right of '11' equating to section 3 of $P_i$.

## 4    Experiments and Results

We methodically determine the effectiveness of our hybrid index and segmentation heuristics over a wide range of dataset characteristics that lead to generalized conclusions about their overall performance.

Every experimental test reports a set of statistics describing the index and query performance of that test. As an attempt to remove machine-dependent statistics, we use the number of $B^+$-tree nodes instead of page accesses when reporting query results and tree size. Tracking nodes accessed is much easier within the algorithm and across heterogeneous systems, and is still directly related to page accesses through the given machine's page size and $B^+$-tree leaf size. We primarily highlight three statistics from tested queries: 1) the number of candidate points returned during the filter step, 2) the number of nodes accessed in the $B^+$-tree, and 3) the time taken (in milliseconds) to perform the query and return the final exact results. Other descriptive statistics included the $B^+$-tree size (total nodes) and the number of dataspace partitions and sections (of partitions) that were checked during the query. We often express the ratio of candidates and nodes over the total number of points in the dataset and the total number of nodes in the $B^+$-tree, respectively, as this eliminates skewed results due to varying the dataset.

The first experiments are on synthetic datasets so we can properly simulate specific dataset characteristics, followed by further tests on real datasets.

All artificial datasets are given a specified number of points and dimensions in the d-dimensional unit space $[0.0, 1.0]^d$. For clustered data, we provide the number of clusters and the standard deviation of the independent Gaussian distributions centered on each cluster (in each dimension). The cluster centers are randomly generated during dataset creation and saved for later use. For each dataset, we randomly select 500 points as $k$NN queries (with $k = 10$) for all experiments, which ensures that our query point distribution follows the dataset distribution.

Sequential scan (SS) is often used as a benchmark comparison for worst-case performance. It must check every data point, and even though it does not use the $B^+$-tree for retrieval, total tree nodes provides the appropriate worst-case comparison. Note that all data fits in main memory, so all experiments are compared without depending on the behaviors of specific hardware-based disk-caching routines. In real-life however, disk-based I/O bottlenecks are a common concern for inefficient retrieval methods. Therefore, unless sequential scan runs significantly faster, there is a greater implied benefit when the indexing method does not have to access every data record, which could potentially be on disk.

## 4.1  First Look: Extensions & Heuristics

The goal of our first experiment is to determine the feasibility and effectiveness of our global and local segmentation techniques and their associated heuristics over dataset size and dimensionality. We create synthetic datasets with 100,000 (100k) data points equally distributed among 12 clusters with a 0.05 standard deviation (*stdev*) in each dimension, ranging from 8 to 512 dimensions. The true cluster centers are used as partition reference points and each heuristic is independently tested with 2, 4, 6, and 8 splits. Here we compare against regular iDistance using the same true center (TC) reference points. Due to space limits, we only present one heuristic from each type of segmentation scope, namely G2 and L1. These methods are ideal because they enforce the total number of splits specified, and do so in a rather intuitive and straightforward manner, exemplifying the over-arching global and local segmentation concepts. We also note that the other heuristics generally perform quite similar to these two, so presentation of all results would probably be superfluous.

Figure 4 shows the performance of G2 compared to TC (regular iDistance) over candidates, nodes accessed, and query time. Unfortunately, above 64 dimensions we do not see any significant performance increase by any global heuristic. Note that we do not show Sequential Scan (SS) results here because they are drastically worse. For example, we see TC returning approximately 8k candidates versus the 100k candidates SS must check.

The same three statistics are shown in Figure 5 for L1 compared to TC. Unlike G2, here we can see that L1 greatly increases performance by returning significantly fewer candidates in upwards of 256 dimensions. Above 64 dimensions, we can see the increased overhead of nodes being accessed by the larger number of splits (6 and 8), but despite these extra nodes, all methods run marginally faster than TC because of the better candidate filtering. It should also be clear that the number of splits is directly correlated with performance, as we can see that

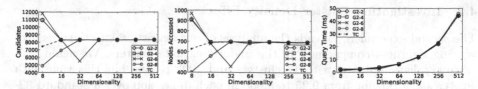

**Fig. 4.** Global heuristic G2 with 2, 4, 6, and 8 splits over dataset dimensionality

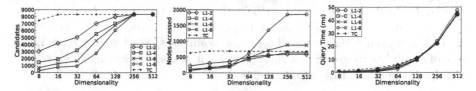

**Fig. 5.** Local heuristic L1 with 2, 4, 6, and 8 splits over dataset dimensionality

8 splits performs better than 6, which performs better than 4, etc. Of course, the increased nodes accessed is a clear indication of the trade-off between the number of splits and the effort to search each partition section to satisfy a given query. Eventually, the overhead of too many splits will outweigh the benefit of enhanced filtering power.

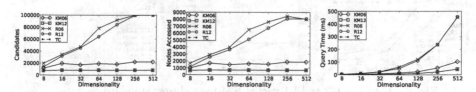

**Fig. 6.** Comparison of TC to k-means (KM) and RAND with 6 (KM6, R6) and 12 (KM12, R12) partitions over dataset dimensionality. Note that the lines for TC and KM12 entirely overlap.

Lastly, Figure 6 revisits a finding from our previous work [14] to show the general performance comparison between TC and k-means (KM), with the random (RAND) space-based partitioning method included as a naïve data-blind approach – all using regular iDistance. Because we establish 12 well-defined clusters in the dataspace, we can use this data knowledge to test KM and RAND, each with 6 and 12 partitions, and compare them to TC (which uses the true 12 cluster centers). Unsurprisingly, we see that RAND (R6 and R12) quickly degrades in performance as dimensionality increases. We also find that KM12 performs almost equivalent to TC throughout the tests. This provides ample justification for the benefits of optimal data clustering for efficient dataspace partitioning, and allows us to focus entirely on k-means for the remainder of our experiments, which is also a much more realistic condition in applications using non-synthetic datasets.

## 4.2   Investigating Cluster Density Effects

The second experiment extends the analysis of our heuristics' effectiveness over varying tightness/compactness of the underlying data clusters. We again generate synthetic datasets with 12 clusters and 100k points, but this time with cluster *stdev* ranging from 0.25 to 0.005 in each dimension over a standard 32-dimensional unit space. Figure 7 reiterates the scope of our general performance, showing SS and RAND as worst-case benchmarks and the rapid convergence of TC and KM as clusters become sufficiently compact and well-defined. Notice however, that both methods stabilize in performance as the *stdev* decreases below 0.15. Essentially, the clusters become so compact that while any given query typically only has to search in a single partition, it ends up having to search the entire partition to find the exact results.

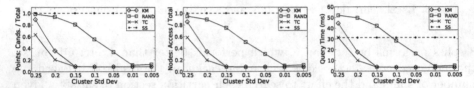

**Fig. 7.** Comparison of iDistance partitioning strategies over cluster standard deviation

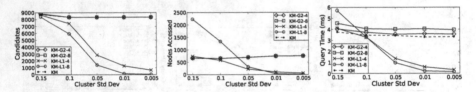

**Fig. 8.** Heuristics G2 and L1 versus KM over cluster standard deviation

Figure 8 compares our heuristics G2 and L1 applied to k-means derived partitions (labeled as KM, using regular iDistance). To simplify the charts, we only look at 4 and 8 splits, which still provide an adequate characterization of performance. We also only look at a cluster *stdev* of 0.15 and less, to highlight the specific niche of poor performance we are addressing. Unfortunately, we again see that G2 performs no better than KM, and it even takes slightly more time due to the increased retrieval overhead and lack of better candidate pruning. However, L1 performance drastically improves over the stalled out KM partitions, which more than proves the effectiveness of localized segmentation heuristics.

## 4.3   A Real-world Comparison

Our last experiment uses a real-world dataset consisting of 90 dimensions and over 500k data points representing recorded songs from a large music archive[2].

---

[2] Publicly available at:
   http://archive.ics.uci.edu/ml/datasets/YearPredictionMSD

First, we methodically determine the optimal number of clusters to use for
k-means, based on our previous discovery of a general iDistance performance
plateau surrounding this optimal number [14]. Using this performance plateau
to choose $k$ represents a best practice for iDistance. Although omitted for brevity,
we tested values of $k$ from 10 to 1000, and found that iDistance performed best
around $k = 120$ clusters (partitions). We also discard global techniques from
discussion given their poor performance on previous synthetic tests.

Figure 9 shows our local L1 and L3 methods, each with 4 and 8 splits, over a
varying number of k-means derived partitions. Since this dataset is unfamiliar,
we include baseline comparisons of KM and SS, representing standard iDistance
performance and worst-case sequential scan performance, respectively. We see
that both local methods significantly outperform KM and SS over all tests. Also
note that L3 generally outperforms L1 due to the more appropriate population-
based distribution of splits, which translates directly to more balanced segments
of the underlying $B^+$-tree.

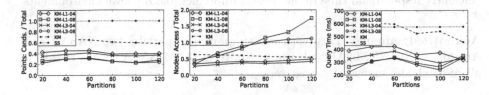

**Fig. 9.** Results of local methods L1 and L3 versus iDistance on real data

# 5   Discussion

One topic worth discussion is the placement of reference points for creating partitions in the dataspace. Building on previous works [10,14,20], we continue to
use the k-means algorithm to cluster the data and obtain a list of cluster centers
to be used directly as reference points. However, Figure 1(b) hints at a possible
shortcoming of using k-means – specifically when building closest-assignment
partitions. Notice how large the center cluster is because of the farthest assigned
data point, and how much additional partition overlap that creates. We hypoth-
esize that just a few outliers in the data can have a tremendously negative effect
on index retrieval due to this property of k-means.

While large overlapping clusters may be somewhat mitigated by increasing
and fine-tuning the total number of clusters in k-means – as we do in the prior
experiment – the more elegant solution tackles the problem of clustering directly.
In a related work, we began investigating the idea of *clustering for the sake
of indexing*, by using a custom Expectation Maximization (EM) algorithm to
learn optimal cluster arrangements designed explicitly for use as reference points
within iDistance [19]. Our current findings further the motivation for this type
of work with possibly new knowledge and insights of how we might define more
optimal clusters for partitions.

# 6   Conclusions and Future Work

This work introduced the foundations of iDStar, a novel hybrid index for efficient and exact kNN retrieval in high-dimensional spaces. We developed global and local segmentation heuristics in the dataspace and underlying B⁺-tree, but only localized partition segmentation proved effective. Results show we can significantly outperform the state-of-the-art iDistance index in clustered data, while performing no worse in unfavorable conditions. This establishes a new performance benchmark for efficient and exact kNN retrieval that can be independently compared to and evaluated against by the community.

Future work surrounds the continued development of iDStar into a fully-fledged indexing algorithm, with specific focus on real-world applications, extended optimizations, and a public repository for the community. Several directions of continued research include testing other heuristics to better guide segmentation and exploring new hybrid components to enable other filtering capabilities. We are also investigating the feasibility of dynamic index updates to efficiently respond and tune to online and unpredictable retrieval environments.

**Acknowledgements.** This work was supported in part by two NASA Grant Awards: 1) No. NNX09AB03G, and 2) No. NNX11AM13A. A special thanks to all research and manuscript reviewers.

# References

1. Aurenhammer, F.: Voronoi diagrams – a survey of a fundamental geometric data structure. ACM Comput. Surv. 23, 345–405 (1991)
2. Bayer, R., McCreight, E.M.: Organization and maintenance of large ordered indices. Acta Informatica 1, 173–189 (1972)
3. Beckmann, N., Kriegel, H.P., Schneider, R., Seeger, B.: The R*-tree: an efficient and robust access method for points and rectangles. In: Proc. of ACM SIGMOD Inter. Conf. on Management of Data, pp. 322–331 (1990)
4. Bellman, R.: Dynamic Programming. Princeton University Press (1957)
5. Berchtold, S., Bhm, C., Kriegal, H.P.: The pyramid-technique: towards breaking the curse of dimensionality. In: Proc. of ACM SIGMOD Inter. Conf. on Management of Data, vol. 27, pp. 142–153 (1998)
6. Doulkeridis, C., Vlachou, A., Kotidis, Y., Vazirgiannis, M.: Peer-to-peer similarity search in metric spaces. In: VLDB 2007 (2007)
7. Guttman, A.: R-trees: a dynamic index structure for spatial searching. In: Proc. of the ACM SIGMOD Inter. Conf. on Management of Data, pp. 47–57 (1984)
8. Ilarri, S., Mena, E., Illarramendi, A.: Location-dependent queries in mobile contexts: Distributed processing using mobile agents. IEEE TMC 5, 1029–1043 (2006)
9. Indyk, P., Motwani, R.: Approximate nearest neighbors: towards removing the curse of dimensionality. In: Proc. of the 30th Annual ACM Sym. on Theory of Computing, STOC 1998, pp. 604–613. ACM (1998)
10. Jagadish, H.V., Ooi, B.C., Tan, K.L., Yu, C., Zhang, R.: iDistance: An adaptive B⁺-tree based indexing method for nearest neighbor search. ACM Trans. Database Syst. 30(2), 364–397 (2005)

11. Lowe, D.: Object recognition from local scale-invariant features. In: The Proc. of the 7th IEEE Inter. Conf. on Computer Vision, vol. 2, pp. 1150–1157 (1999)
12. Ooi, B.C., Tan, K.L., Yu, C., Bressan, S.: Indexing the edges: a simple and yet efficient approach to high-dimensional indexing. In: Proc. of the 19th ACM SIGMOD-SIGACT-SIGART Sym. on Principles of DB Systems, PODS 2000, pp. 166–174 (2000)
13. Qu, L., Chen, Y., Yang, X.: iDistance based interactive visual surveillance retrieval algorithm. In: Intelligent Computation Technology and Automation, ICICTA 2008, vol. 1, pp. 71–75 (October 2008)
14. Schuh, M.A., Wylie, T., Banda, J.M., Angryk, R.A.: A comprehensive study of iDistance partitioning strategies for kNN queries and high-dimensional data indexing. In: Gottlob, G., Grasso, G., Olteanu, D., Schallhart, C. (eds.) BNCOD 2013. LNCS, vol. 7968, pp. 238–252. Springer, Heidelberg (2013)
15. Shen, H.T.: Towards effective indexing for very large video sequence database. In: SIGMOD Conference, pp. 730–741 (2005)
16. Shi, Q., Nickerson, B.: Decreasing Radius K-Nearest Neighbor Search Using Mapping-based Indexing Schemes. Tech. rep., University of New Brunswick (2006)
17. Singh, V., Singh, A.K.: Simp: accurate and efficient near neighbor search in high dimensional spaces. In: Proc. of the 15th Inter. Conf. on Extending Database Technology, EDBT 2012, pp. 492–503. ACM (2012)
18. Tao, Y., Yi, K., Sheng, C., Kalnis, P.: Quality and efficiency in high dimensional nearest neighbor search. In: Proc. of the 2009 ACM SIGMOD Inter. Conf. on Mgmt. of Data, SIGMOD 2009, pp. 563–576. ACM (2009)
19. Wylie, T., Schuh, M.A., Sheppard, J., Angryk, R.A.: Cluster analysis for optimal indexing. In: Proc. of the 26th FLAIRS Conf. (2013)
20. Yu, C., Ooi, B.C., Tan, K.L., Jagadish, H.V.: Indexing the Distance: An Efficient Method to KNN Processing. In: Proc. of the 27th Inter. Conf. on Very Large Data Bases, VLDB 2001, pp. 421–430 (2001)
21. Zhang, J., Zhou, X., Wang, W., Shi, B., Pei, J.: Using high dimensional indexes to support relevance feedback based interactive images retrieval. In: Proc. of the 32nd Inter. Conf. on Very Large Data Bases, VLDB 2006, pp. 1211–1214 (2006)
22. Zhang, R., Ooi, B., Tan, K.L.: Making the pyramid technique robust to query types and workloads. In: Proc. 20th Inter. Conf. on Data Eng., pp. 313–324 (2004)

# A Benchmark for Multidimensional Statistical Data

Philipp Baumgärtel*, Gregor Endler, and Richard Lenz

Institute of Computer Science 6,
University of Erlangen-Nuremberg
{philipp.baumgaertel,gregor.endler,richard.lenz}@fau.de

**Abstract.** ProHTA (Prospective Health Technology Assessment) is a simulation project that aims at estimating the outcome of new medical innovations at an early stage. To this end, hybrid and modular simulations are employed. For this large scale simulation project, efficient management of multidimensional statistical data is important. Therefore, we propose a benchmark to evaluate query processing of this kind of data in relational and non-relational databases. We compare our benchmark with existing approaches and point out differences. This paper presents a mapping to a flexible relational model, JSON documents and RDF. The queries defined for our benchmark are mapped to SQL, SPARQL, the MongoDB query language and MapReduce. Using our benchmark, we evaluate these different systems and discuss differences between them.

## 1 Introduction

ProHTA (Prospective Health Technology Assessment) is a large scale simulation project within the Cluster of Excellence for Medical Technology – Medical Valley European Metropolitan Region Nuremberg (EMN).The objective of this interdisciplinary research project is to study the effects of new innovative medical technologies at a very early stage [7]. At the core of the project is an incrementally growing set of healthcare simulation modules, which are configured and calibrated with data from various sources. Typical data sources are multidimensional statistical data, like cancer registries (e.g. SEER[1]), population statistics or geographical databases. Not all data sources are initially known, though. Moreover, adding new dimensions to a multidimensional classification is common.

All these data are collected and stored in a central ProHTA database, which is required to support uncertainty management, availability and performance [1]. Consequently, the ProHTA database must have a generic general purpose database schema which allows deferred semantic annotations and incremental growth.

---

* On behalf of the ProHTA Research Group
[1] http://seer.cancer.gov/

B. Catania, G. Guerrini, and J. Pokorný (Eds.): ADBIS 2013, LNCS 8133, pp. 358–371, 2013.
© Springer-Verlag Berlin Heidelberg 2013

A straight forward idea to store heterogeneous continuously evolving data sets with varying semantics is to use RDF triplestores. Arbitrary types of information can be expressed through sets of RDF-triples, which simply represent statements of the form subject-predicate-object. In a first prototype we use the RDF triplestore Jena, which is flexible enough to store an arbitrary number of classifications per fact [1]. However, there are many other options to store continuously growing data sets in a generic database. In order to compare the performance of these very different approaches for this problem we developed a benchmark for multidimensional statistical data.

One of the alternatives to RDF is the relational EAV (Entity-Attribute-Value) approach [11], which stores arbitrary attributes for each entity in RDF-like triples. Other possible solutions are document stores like MongoDB and CouchDB. These systems allow for storing arbitrary JSON documents and querying them with system specific query languages or MapReduce [6]. Another alternative are XML databases like BaseX. XML and document stores do not require a schema to be known upfront and entities may contain lists of arbitrary attributes.

We propose a benchmark to compare the alternatives of storing multidimensional data with an arbitrary number of dimensions per fact. To create this benchmark we used Jain's methodology [9]. We present a conceptual model and queries that are based on the requirements of the ProHTA simulation project and map these to different data management systems. We exemplify our benchmark by evaluating PostgreSQL, SQLite, MongoDB and Jena. We chose Jena because our current solution already uses it as data management system. PostgreSQL and SQLite were chosen arbitrarily and we decided to use MongoDB because of it's interesting MongoDB query language and it's MapReduce ability. So far, we have evaluated the MapReduce approach on a single processor only, in order to have a fair comparison with other non-parallel solutions. However, note that the true strength of MapReduce only lies in parallel execution, which is not taken into account in this paper. Commercial systems were excluded to be able to publish our results without restrictions.

In Sect. 2, we discuss related work. Then, we define our benchmark in Sect. 3. Sect. 4 and Sect. 5 present the mapping of the conceptual model and the queries to the specific systems. We evaluate the systems using our benchmark in Sect. 6 and conclude with a short summary and a perspective on future work.

## 2  Related Work

There are many existing approaches to evaluate data warehouse solutions. One of the best known is the TPC-H[2], which measures the performance for a given data warehouse schema. Another approach is the data warehouse benchmark by Darmont et. al. [5], which is able to generate arbitrary data warehouse schemata. Both approaches rely on schemata with a fixed number of dimensions. Therefore, these solutions are not suitable for the data management problem in ProHTA.

---

[2] http://www.tpc.org/tpch/

There are also approaches to evaluate NoSQL systems. The YCSB (Yahoo! Cloud Serving Benchmark) [3] evaluated CRUD (Create, Read, Update, Delete) operations for distributed data management systems. Pavlo et al. [12] compared filter, aggregation and join operations for relational and non-relational systems. They mapped their queries to MapReduce and SQL. Floratou et al. [8] utilized the YCSB to compare a distributed relational DBMS to MongoDB. Additionally, they compared the relational DBMS to Hive[3] using TPC-H. Tudorica and Bucur [14] evaluated the read and write performance of various NoSQL systems and compared the features of these systems. Cudre-Mauroux et al. [4] evaluated SciDB and MySQL for scientific applications by creating a set of 9 scientific queries based on astronomy workloads.

All of these approaches are not suitable for dynamically evolving statistical databases, as they do not evaluate generic multidimensional schemata. Additionally, some of these benchmarks only evaluate read and write operations. For ProHTA, we need to evaluate approaches for managing statistical data cubes with an arbitrary number of dimensions. Stonebraker et al. [13] argue that for each problem domain, a specialized solution performs best. They prove their point by evaluating different systems for scientific applications, data warehousing and data stream processing with application specific workloads. Therefore, we have developed a new benchmark as no existing benchmark covers our specific problem domain.

## 3   Definition of the Benchmark

In this section, we define the conceptual model, exemplary data, and queries of our benchmark. The data and queries are based on the requirements of the ProHTA simulation project.

### 3.1   Conceptual Model

This model is a simplified version of the actual ProHTA data model for heterogeneous multidimensional statistical data [1]. Each fact is a tuple consisting of an identifier, the name of a data cube, a numerical value and a set of classifications.

$$\text{fact}_i = (\text{id}_i, \text{cube}, \text{value}_i, \{\text{classification}_{i,1}, \text{classification}_{i,2}, \dots \}) \qquad (1)$$

We do not support hierarchical classifications for the benchmark, as this is of no importance for performance evaluations. Each classification is a tuple consisting of a number representing the dimension and a numerical value that classifies the fact in this dimension.

$$\text{classification}_{i,j} = (\text{dimension}_{i,j}, \text{value}_{i,j}) \qquad (2)$$

We use data cubes with $d$ dimensions and $n$ possible classification values per dimension as test data. Each data cube is dense and contains $n^d$ facts. Dense

---

[3] http://hive.apache.org/

data cubes are common for multidimensional statistical data in the ProHTA setting. Each fact contains a random value uniformly sampled from the interval $[0, 1]$.

## 3.2 Query Definition

The queries are based on OLAP operators and typical queries from ProHTA simulations. We assume a data cube with $d$ dimensions $0, 1, \ldots, d - 1$ and $n$ classification values $0, 1, \ldots, n - 1$ in each dimension.

**Insert** measures the time to insert $n^d$ facts.

**Dice** queries all facts with a value $\leq \lfloor n/2 \rfloor$ in each classification. That represents a selectivity of approximately 50% for each dimension, which is common for queries in ProHTA.

**Roll Up** groups the facts by the first $\max(\lfloor d/2 \rfloor, 1)$ dimensions and calculates the sum of the facts per group.

With **Add Dimension** the user can extend the classification scheme on demand. This is required for heterogeneous multidimensional data, as we don't know the classification scheme in advance. For the benchmark query, we add a classification with the value 0 and the dimension $d$ to each fact.

**Cube Join** correlates the facts from data cubes $c_1$ and $c_2$ with $n^d$ facts in each cube. For each fact $i_1$ from $c_1$ we search for the fact $i_2$ from $c_2$ with the same classifications. The resulting fact $i'$ has the same classifications as $i_1$ and $i_2$. The value of fact $i_1$ is contained in $i'$ as leftvalue$_{i'}$ and the value of the fact $i_2$ is contained in $i'$ as rightvalue$_{i'}$. The scheme of the resulting facts in the cube $c_{1,2}$ is:

$$\text{joinedfact}_{i'} = (\text{id}_{i'}, c_{1,2}, \text{leftvalue}_{i'}, \text{rightvalue}_{i'}, \{\text{classification}_{i',1}, \ldots \}) \quad (3)$$

The **Cube Join** can correlate facts from different sources to either compare them or to enrich the facts from one source with information from another source.

# 4 Mapping

In this section, we present a mapping from our conceptual model (Sect. 3.1) to the evaluated systems.

## 4.1 Relational

The relational schema is based on the EAV approach:

```
fact (id, value, cube)
classification (id, fact_id[fact], value, dimension)
```

We evaluated multiple alternatives to create indexes for this mapping. We created indexes for the columns `id` (for both relations), `cube`, `factid` and for the combination of the columns (`value`, `dimension`, `factid`). These indexes showed the best performance.

Composite types and arrays are a PostgreSQL-specific alternative to the EAV approach. The classification can be modelled as a composite type and an array of classifications can be stored for each fact. However, this is not suitable for our problem as PostgreSQL does not support searching the content of arrays[4].

Additionally, we evaluated a denormalized version of this mapping to be able to compare the performance to an efficient ROLAP schema. The denormalized solution requires only one table and contains a separate column for each dimension. Despite being not flexible enough for statistical simulation data, we include this solution in our benchmark as baseline.

## 4.2   Document Store

A document store offers two alternatives for mapping our conceptual model to JSON documents. We can either store facts and classifications in separate documents or we can store the classifications as sub-documents. We decided to use the sub-document approach despite it's redundancy. That allows us to take advantage of the MongoDB query language, as this query language does not support joining documents.

As there is no standard schema definition language for JSON documents, we give an example for $d = 2$:

```
{id : 1,
 value: 0.5,
 cube: "Test",
 classifications : [
     {dimension : 0, value : 0},
     {dimension : 1, value : 0}]}
```

As MongoDB provides indexes, we evaluated different alternatives and decided to create indexes on cube and
(classifications.value, classifications.dimension).
For evaluating MapReduce with MongoDB, we did not create any indexes, as MapReduce is not able to utilize them.

## 4.3   RDF

The mapping to RDF is similar to the document store mapping. However, Jena does not support custom indexes. Each fact is linked to an arbitrary number of classifications. We present an exemplary fact for RDF in Turtle[5] notation for $d = 2$.

```
:f1 a               :Fact ;
    :value          0.5 ;
    :cube           "Test" ;
    :classification [a :Classification; :dimension 0; :value 0] ;
    :classification [a :Classification; :dimension 1; :value 0] .
```

---

[4] http://www.postgresql.org/docs/9.1/static/arrays.html
[5] http://www.w3.org/TR/turtle/

# 5     Query Mapping

In this section, we present a translation of exemplary most interesting queries from Sect. 3.2 for each system. Additionally, we discuss how the query complexity depends on the number of dimensions $d$. We assume $n = 10$ and $d = 2$ for all exemplary queries except for **Roll Up** with $d = 4$.

## 5.1     Relational

For example, the query for Cube Join is:

```
SELECT lf.value AS leftvalue, rf.value AS rightvalue,
    lc0.value AS dimvalue0 , lc1.value AS dimvalue1
FROM fact lf, fact rf, classification lc0, classification rc0,
    classification lc1, classification rc1
WHERE lf.cube = 'Test' AND rf.cube = 'Test2'
AND lc0.factid = lf.id AND rc0.factid = rf.id
AND lc0.dimension =  0 AND rc0.dimension =  0
AND lc0.value = rc0.value
AND lc1.factid = lf.id AND rc1.factid = rf.id
AND lc1.dimension =  1 AND rc1.dimension =  1
AND lc1.value = rc1.value;
```

Here, we join the facts `lf` and `rf`. As $d = 2$ for this example, we have to compare the classifications `lc0` und `lc1` of the fact `lf` with the classifications `rc0` and `rc1` of the fact `rf`. Therefore, we need $2d + 1$ joins and $5d + 2$ filter conditions.

**Dice** requires $d$ Joins and $3d + 1$ filter conditions. **Roll Up** requires $d$ Joins and $2d + 1$ filter conditions. The complexity of inserting additional classifications (**Add Dimension**) does not depend on $d$.

For the denormalized relational solution, only the number of filter conditions for **Dice** and **Cube Join** depends on $d$ and only **Cube Join** requires a join.

## 5.2     MapReduce

The MapReduce solution for **Roll Up** in pseudocode is :

```
map(fact):
    if fact.cube == 'Test':
        key_classes = []
        for classification in fact.classifications:
            if classification['dimension'] in [0,1]:
                key_classes.append(classification)
        emit({classifications : key_classes, cube: fact.cube},fact)

reduce(key, facts):
    result = {cube : key.cube, value: 0,
                classifications: key.classifications};
    for fact in facts:
        result.value += fact.value;
    return result;
```

For **Roll Up**, we generate the key in the map function to group the facts from the first classifications with dimension $\leq \max(\lfloor d/2 \rfloor, 1)$. Then, the reduce function calculates the sum of the facts. We use the map function to filter the facts for **Dice** and to add a classification to each fact for **Add Dimension**. For **Cube Join**, we use all classifications of a fact as key in the map function. Therefore, the reduce function gets two fitting facts – one from each cube – and produces the joined fact. This is known as the Standard Repartition Join [2]. The complexity of the MapReduce queries does not depend on $d$ except for the number of filter conditions for **Dice** and the key construction of **Roll Up**.

### 5.3   MongoDB Query Language

Besides MapReduce, MongoDB offers a custom query language[6]. However, this language does not support joins and advanced aggregation features. Therefore, **Cube Join** and **Roll Up** can not be mapped to the MongoDB query language. However, the Aggregation Framework[7] extends the MongoDB query language. This enables us to perform the **Cube Join** and **Roll Up** queries.

**Dice** is mapped using the filter functionality of the MongoDB query language:

```
{'$and': [{'cube': 'Test'},
        {'classifications': {'$elemMatch':
            {'dimension': 0, 'value': {'$lte': 5}}}},
        {'classifications': {'$elemMatch':
            {'dimension': 1, 'value': {'$lte': 5}}}}
    ]}
```

The `$elemMatch` operator enables conditions for subdocuments. The number of conditions in this query is $d + 1$.

**Roll Up** is mapped using the Aggregation Framework, which uses pipelining of operators:

```
{'$match' : {'cube' : 'Test'}},
{'$unwind' : '$classifications'},
{'$match' : { 'classifications.dimension' :
    { '$in' : [0,1]}}},
{'$group' : { '_id' : '$_id', 'classifications' :
    {'$push' : '$classifications'},
        'value' : { '$first' : '$value'}}},
{'$group' : { '_id' : "$classifications",
    'value' : { '$sum' : '$value'}}},
{'$project' : {'value' : 1, 'classifications' : '$_id'}}
```

Here, we use `$match` to find the facts of the desired cube. Then, we split up the array containing the classifying subdocuments using `$unwind`. This produces copies of the fact document for each element of the array. These documents

---

[6] http://www.mongodb.org/display/DOCS/Querying
[7] http://docs.mongodb.org/manual/applications/aggregation/

contain only one classification instead of the classification array. Then, `$match` finds the classifications to group by. After that, `$group` reverses the `$unwind` operation. Now, each document contains only the classifications we want to group by. The final `$group` is the actual aggregation and `$project` produces the right output format. The number of classifications to group by is $\max(\lfloor d/2 \rfloor, 1)$.

**Add Dimension** filters the facts and adds a classification with `$push`. The complexity of **Add Dimension** does not depend on $d$.

For **Cube Join**, we utilize the Aggregation Framework again:

```
{'$match' : {'$or' : [{"cube" : 'Test'},
                      {"cube" : 'Test2'}]}},
{'$project' : {'classifications' : 1,
    'leftvalue' :
      {'$cond' : [{'$eq' : ['$cube', 'Test']}, '$value', 0]},
    'rightvalue' :
      {'$cond' : [{'$eq' : ['$cube', 'Test2']}, '$value', 0]}}},
{'$group' : {'_id' : "$classifications",
    'leftvalue' : {'$sum' : '$leftvalue'},
    'rightvalue' : {'$sum' : '$rightvalue'}}},
{'$project' : {'leftvalue' : 1, 'rightvalue' : 1,
    'classifications' : '$_id'}}
```

Conditional values (`$cond`) split up the value attributes of facts and store them in new attributes `leftvalue` and `rightvalue`. Then, we group by all classifications to match facts from one cube to the corresponding facts from the other cube. Therefore, each group contains two documents. Then, we can sum up the `leftvalue` and `rightvalue` attributes, because they either contain the desired value or 0. Again, the complexity does not depend on $d$.

## 5.4 RDF

Since version 1.1, SPARQL has supported aggregation and updates. Therefore, we are able to map all queries to SPARQL.

**Dice** needs $d$ `FILTER` operators to find the desired facts. The number of triples in this query is $3d + 2$:

```
SELECT ?fact ?value ?dimvalue0 ?dimvalue1 WHERE {
?fact cube:value ?value ; cube:cube "Test" .
?fact cube:classification [:dimension 0; :value ?dimvalue0] .
FILTER( ?dimvalue0 <= 5 )
?fact cube:classification [:dimension 1; :value ?dimvalue1] .
FILTER( ?dimvalue1 <= 5 )
}
```

**Roll Up** uses `GROUP BY` to aggregate values and needs $3d + 2$ triples for the query. For **Add Dimension**, the `INSERT` statement generates and stores 4 additional triples for each fact in the respective cube. Finally, **Cube Join** requires $6d + 4$ triples in the query.

## 6   Evaluation

In this section, we present the results of our evaluation for PostgreSQL, SQLite, MongoDB and Jena. For MongoDB, we evaluated it's query language as well as MapReduce.

A Python implementation of our benchmarking framework is available for download on our Homepage[8]. Data sets and queries are generated automatically for a desired problem size. As this framework is based on a simple data model and well defined queries, it can be extended to evaluate other solutions. Additionally, unit tests guarantee that all evaluated solutions adhere to the defined semantics.

### 6.1   Hardware and Software Configuration

We used a quad-core computer[9] with a 2.5 inch hard drive[10] and two 4 GB DDR3 RAM modules with 1333 MHz for the evaluation. The operating system was Ubuntu 12.04 64 bit with the OpenJDK 64 bit server VM (Java version 1.6.0). We evaluated MongoDB version 2.2.0, PostgreSQL version 9.1.5, SQLite version 3.7.9 and Jena Fuseki version 0.2.3 with the TDB back end. We used 64 bit versions of the systems if they were available and did not modify the standard configuration. As journaling is activated in the standard configuration of MongoDB, all systems guarantee the durability of stored data.

### 6.2   Evaluation Results

We evaluated data cubes with sizes $n^d$ ranging from $10^3$ to $10^6$. These cube sizes cover most of the data sets in ProHTA. For **Insert**, we measured the time to store a cube of the desired size in an empty database. We executed all other queries on a database containing two cubes of the desired size. SQLite does not start the query execution until the results are fetched. Therefore, we measured the time for executing the query and returning the results.

We ran each test 25 times (if possible) and compared the results to rule out caching effects. Tests with a long run time were executed only three times. We compared the results and used the first one for the evaluation if there were no significant differences.

With PostgreSQL, subsequent test runs showed significantly different results. This can be attributed to optimizations that were employed after a certain number of queries. We eliminated this behavior by running `ANALYZE` before each query. That way, PostgreSQL employed the optimizations for every query. We did not include the time for running `ANALYZE` in the results. This is valid, as we assume that statistical data for optimizations exists in our data management system.

Table 1 and Fig. 1 show the results of our evaluation. For each test run, we present in Table 1 the average time (in seconds) and the standard deviation if

---

[8] http://www6.cs.fau.de/pb/
[9] Intel(R) Core(TM) i5-2540M CPU @ 2.60GHz
[10] Seagate Momentus / max.: 7.200 rpm / buffer: 16 MB / bus: S-ATA II (S-ATA 300)

**Table 1.** Time (s) and standard deviation for the queries

| Query | Dimensions | MongoDB QL | MapReduce | PostgreSQL | SQLite | Jena | PostgreSQL 1T | SQLite 1T |
|---|---|---|---|---|---|---|---|---|
| Insert | 3 | **0.011** (0.003) | See MongoDB QL | 1.830 (0.024) | 0.291 (0.016) | 1.818 (0.126) | 0.540 (0.010) | 0.165 (0.022) |
|  | 4 | **0.290** (0.067) |  | 22.71 (0.255) | 1.213 (0.105) | 18.86 (0.910) | 5.323 (0.075) | 0.247 (0.051) |
|  | 5 | 11.86 (3.164) |  | 276.3 (1.887) | **5.515** (1.118) | 264.8 | 53.88 (0.503) | 1.103 (0.020) |
|  | 6 | 215.3 |  | 2631 | **61.80** | 8193 | 510.1 | 8.676 |
| Dice | 3 | **0.004** (2.9e-4) | 0.044 (0.008) | 0.012 (0.004) | 0.010 (0.003) | 0.068 (0.016) | 0.003 (0.001) | 8.6e-4 (4.4e-4) |
|  | 4 | **0.048** (0.006) | 0.370 (0.051) | 0.072 (0.003) | 0.057 (0.003) | 0.528 (0.009) | 0.016 (0.004) | 0.007 (9.8e-5) |
|  | 5 | **0.403** (0.219) | 3.652 (0.050) | 0.897 (0.008) | 0.699 (0.005) | 5.809 (0.071) | 0.074 (0.004) | 0.075 (3.2e-4) |
|  | 6 | **3.202** (0.370) | 37.90 (0.341) | 11.86 (0.133) | 8.668 (0.074) | 77.01 (7.103) | 0.494 (0.022) | 0.803 (0.004) |
| Roll Up | 3 | 0.015 (0.003) | 0.060 (0.008) | 0.004 (0.002) | **0.003** (3.0e-5) | 0.036 (0.011) | 0.002 (8.7e-4) | 0.001 (6.8e-4) |
|  | 4 | 0.188 (0.005) | 0.594 (0.004) | **0.041** (0.004) | 0.064 (3.8e-4) | 0.433 (0.010) | 0.009 (0.004) | 0.015 (2.8e-4) |
|  | 5 | 0.998 (0.665) | 6.154 (0.096) | **0.408** (0.003) | 0.860 (0.003) | 4.844 (0.079) | 0.050 (0.004) | 0.163 (5.9e-4) |
|  | 6 | 29.86 (8.629) | 62.60 (0.244) | **4.227** (0.283) | 17.77 (0.257) | 81.80 (4.824) | 0.351 (0.003) | 4.749 (0.027) |
| Cube Join | 3 | **0.045** (0.006) | 0.199 (0.024) | **0.044** (0.003) | 0.220 (0.007) | 15.82 (0.027) | 0.016 (0.005) | 0.003 (5.3e-5) |
|  | 4 | **0.425** (0.007) | 2.597 (0.194) | 0.502 (0.008) | 24.68 (0.301) | — | 0.105 (0.006) | 0.039 (0.003) |
|  | 5 | **4.570** (0.071) | 13.26 (0.727) | — | — | — | 0.940 (0.040) | 0.441 (8.5e-4) |
|  | 6 | — (Doc. Sz.) | **374.6** | — | — | — | 10.77 (0.044) | 5.247 (0.031) |
| Add Dim. | 3 | 0.180 (0.066) | 0.133 (0.021) | **0.043** (0.009) | 0.309 (0.026) | 0.391 (0.015) | 0.240 (0.010) | 0.160 (0.008) |
|  | 4 | 2.317 (0.392) | 1.365 (0.247) | **0.319** (0.050) | 0.990 (0.288) | 0.910 (0.019) | 0.405 (0.018) | 0.178 (0.027) |
|  | 5 | 27.96 (6.667) | 13.27 (3.263) | 4.436 (0.211) | **3.819** (0.344) | 7.023 (0.734) | 1.751 (0.351) | 0.190 (0.026) |
|  | 6 | 280.7 | 90.42 | 46.48 | **36.04** | 270.7 | 16.98 (0.619) | 0.166 (0.008) |

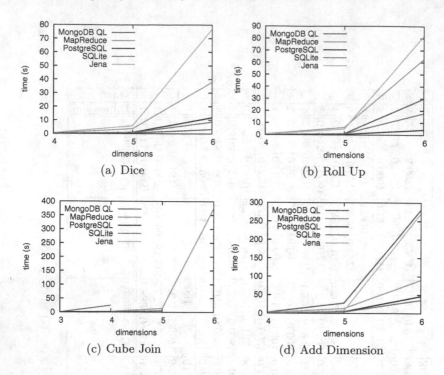

**Fig. 1.** Evaluation of the queries

we were able to run the test 25 times. We aborted each query except **Insert** aft 600 seconds ("–" in the table). "PostgreSQL 1T" and "SQLite 1T" are the denormalized relational solutions. We did include these solutions in our evaluation despite being not flexible enough to manage heterogeneous data. The fastest solution for each problem size and query is printed in boldface (excluding the denormalized solutions). As an overview, we depicted the results in Fig. 1. Because of space limitations, we omitted **Insert** and for clarity reasons, we did not plot the results for the denormalized relational solutions.

PostgreSQL performs the **Cube Join** very fast by using hash joins. However, for larger cubes ($d > 4$) the available memory was not sufficient. Therefore, PostgreSQL was not able to perform the **Cube Join** in less than 600 seconds. MongoDB was not able to perform the **Cube Join** for $d > 5$. This was because the Aggregation Framework stores all results in one single document and the result exceeded the hard coded maximum document size ("– Doc. Sz." in the table). To be able to perform the cube join for $d = 5$, we increased the maximum document size from 16MB to 256MB in the MongoDB code. However, we were not able to increase this limit further because of integer overflows. For Jena, we extended the Java heap space to 4GB to be able to evaluate **Add Dimension** for $d = 6$.

In conclusion, MongoDB seems to be most suitable for managing simulation data in ProHTA. The MongoDB query language is fast and MapReduce is the only solution that allows for large **Cube Joins**. PostgreSQL is very fast for **Roll Up** queries and for adding dimensions but is slow for **Insert**. Jena is too slow and can not perform **Cube Joins** for cubes with more than $10^3$ facts.

## 6.3   Evaluating a Prefilled Database

In this section, we evaluate the dependency between the performance and the amount of data in the data management system. To this end, we evaluated **Dice** for cubes with $10^4$ facts for an empty database and for a database prefilled with 100 cubes containing $10^4$ facts each. This amount of data is realistic for a large healthcare simulation project.

**Table 2.** Time (s) and standard deviation of the dice query

| Load | 0 | $10^4 \cdot 100$ |
|------|------|------|
| MongoDB QL | 0.048 (0.006) | 0.040 (0.007) |
| MapReduce | 0.370 (0.051) | 5.882 (0.054) |
| PostgreSQL | 0.072 (0.003) | 0.064 (0.005) |
| PostgreSQL 1T | 0.016 (0.004) | 0.016 (0.006) |
| SQLite | 0.057 (0.003) | 0.081 (0.004) |
| SQLite 1T | 0.007 (9.8e-5) | 0.018 (0.005) |
| Jena | 0.528 (0.009) | 0.595 (0.052) |

Table 2 presents the results. The difference for the MongoDB query language is within the limits of the error of measurement. However, MapReduce depends heavily on the amount of data in the database. This is because MapReduce has to process each document in the database. In MongoDB, MapReduce and the filter from the MongoDB query language can be combined. That way, the amount of data in the database does not influence the performance of MapReduce queries. PostgreSQL showed no significant difference between the two experiments. Jena and SQLite showed a small dependency on the amount of data in the database. Therefore, the amount of data in the ProHTA simulation project does not influence whether or not the solutions are suitable.

## 7   Conclusions and Future Work

This paper proposed a benchmark to evaluate solutions for storing heterogeneous multidimensional statistical data. The benchmark is based on the data management of a large scale healthcare simulation project. The data model in this project is based on the EAV approach. As EAV is a widespread solution to create generic data models [10], our benchmark is valid for a large number of applications besides statistical data.

We simplified the data model to be able to map it easily to a large number of different data management solutions. However, the simplified data model is still close enough to the real model to be able to use the evaluated solutions in the ProHTA project.

We created a set of well-defined queries, which can be mapped to various query languages. These queries are based on realistic and typical queries from health-care simulations. The mapping of queries enables evaluating the expressiveness of the target query languages as well as their performance.

No clear winner has emerged from our evaluation. As expected, all solutions were far slower than the denormalized relational approach. However, as a generic data model is required, we can not rely on this solution. SQLite was even faster than MongoDB for inserting the facts; PostgreSQL and Jena were very slow. For **Dice**, the MongoDB query language was faster than the relational solutions and more than 20 times faster than Jena. For **Roll Up**, PostgreSQL performed fastest. Despite being slow, MapReduce was the only solution that was able to perform the **Cube Join** for $d > 4$. For MapReduce, there was no need to create indexes, therefore it performed **AddDimension** faster than the MongoDB query language. Jena showed the slowest performance for each query and was not able to perform the **Cube Join** for $d > 3$.

In future work, we need to evaluate other systems to be able to decide which solution suits the ProHTA data management problem best. The dependency between performance and amount of data in the database should be evaluated more thoroughly. In addition to that, we need to evaluate how the queries can be performed in parallel. This is because we did not utilize this strength of MapReduce and MongoDB in our benchmark yet. Also, sparse data cubes with a high number of dimensions need to be evaluated.

Using this benchmark, we can define a catalog of criteria to find the best system for a wide range of data management problems with certain characteristics. Additionally, we can evaluate optimization strategies for each solution.

**Acknowledgements.** This project is supported by the German Federal Ministry of Education and Research (BMBF), project grant No. 13EX1013B.

# References

1. Baumgärtel, P., Le██, R.: Towards data and data quality management for large scale healthcare simulations. In: Conchon, E., Correia, C., Fred, A., Gamboa, H. (eds.) Proceedings of the International Conference on Health Informatics, pp. 275–280. SciTePress - Science and Technology Publications (2012) iSBN: 978-989-8425-88-1
2. Blanas, S., Patel, J.M., Ercegovac, V., Rao, J., Shekita, E.J., Tian, Y.: A comparison of join algorithms for log processing in mapreduce. In: Proceedings of the 2010 ACM SIGMOD International Conference on Management of Data, SIGMOD 2010, pp. 975–986. ACM, New York (2010)
3. Cooper, B.F., Silberstein, A., Tam, E., Ramakrishnan, R., Sears, R.: Benchmarking cloud serving systems with ycsb. In: Proceedings of the 1st ACM Symposium on Cloud Computing, SoCC 2010, pp. 143–154. ACM, New York (2010)

4. Cudre-Mauroux, P., Kimura, H., Lim, K.T., Rogers, J., Madden, S., Stonebraker, M., Zdonik, S.B., Brown, P.G.: Ss-db: A standard science dbms benchmark (2012) (submitted for publication)
5. Darmont, J., Boussaïd, O., Bentayeb, F.: DWEB: A data warehouse engineering benchmark. In: Tjoa, A.M., Trujillo, J. (eds.) DaWaK 2005. LNCS, vol. 3589, pp. 85–94. Springer, Heidelberg (2005)
6. Dean, J., Ghemawat, S.: Mapreduce: simplified data processing on large clusters. Commun. ACM 51(1), 107–113 (2008)
7. Djanatliev, A., Kolominsky-Rabas, P., Hofmann, B.M., Aisenbrey, A., German, R.: Hybrid simulation approach for prospective assessment of mobile stroke units. In: SIMULTECH 2012 - Proceedings of the 2nd International Conference on Simulation and Modeling Methodologies, Technologies and Applications, pp. 357–366 (2012)
8. Floratou, A., Teletia, N., Dewitt, D.J., Patel, J.M., Zhang, D.: Can the elephants handle the nosql onslaught? In: Proceedings of the VLDB Endowment, vol. 5 (2012)
9. Jain, R.: The art of computer systems performance analysis. John Wiley & Sons, Inc. (1991)
10. Lenz, R., Elstner, T., Siegele, H., Kuhn, K.A.: A practical approach to process support in health information systems. Journal of the American Medical Informatics Association 9(6), 571–585 (2002)
11. Nadkarni, P.M., Marenco, L., Chen, R., Skoufos, E., Shepherd, G., Miller, P.: Organization of heterogeneous scientific data using the eav/cr representation. Journal of the American Medical Informatics Association 6(6), 478–493 (1999)
12. Pavlo, A., Paulson, E., Rasin, A., Abadi, D.J., DeWitt, D.J., Madden, S., Stonebraker, M.: A comparison of approaches to large-scale data analysis. In: Proceedings of the 2009 ACM SIGMOD International Conference on Management of Data, SIGMOD 2009, pp. 165–178. ACM, New York (2009)
13. Stonebraker, M., Bear, C., Çetintemel, U., Cherniack, M., Ge, T., Hachem, N., Harizopoulos, S., Lifter, J., Rogers, J., Zdonik, S.: One size fits all? - part 2: benchmarking results. In: Proceedings of the 3rd Conference on Innovative Data Systems Research, CIDR (2007)
14. Tudorica, B., Bucur, C.: A comparison between several nosql databases with comments and notes. In: 2011 10th Roedunet International Conference, RoEduNet, pp. 1–5 (June 2011)

# Extracting Deltas from Column Oriented NoSQL Databases for Different Incremental Applications and Diverse Data Targets

Yong Hu and Stefan Dessloch

University of Kaiserslautern, Germany
{hu,dessloch}@informatik.uni-kl.de

**Abstract.** This paper describes the Change Data Capture (CDC) problems in the context of column-oriented NoSQL databases (CoNoSQLDBs). Based on analyzing the impacts and constraints caused by the core features of CoNoSQLDBs, we propose a logical change data (delta) model and the corresponding delta representation which could work with different incremental applications and diverse data targets. Moreover, we present five feasible CDC approaches, i.e. Timestamp-based approach, Audit-column approach, Log-based approach, Trigger-based approach and Snapshot differential approach and indicate the performance winners under different circumstances.

**Keywords:** CoNoSQLDB, delta model, CDC approaches.

## 1 Introduction

"Big Data" currently receives significant attention in research and industry [1], [2]. The characteristics of Big Data are generally summarized as "3V" [2], namely, volume, velocity and variety. To handle the challenges and problems caused by Big Data, new types of data storage systems called "NoSQL databases" have emerged. These systems claim to manage tremendous data volume in off-the-shelf commodity hardware with support for elastic scalability, high availability and robust fault tolerance.

A "Column-oriented NoSQL Database" (CoNoSQLDB) organizes the data in a structured way and stores the data which belong to the same "column" continuously on disk. Such systems are conceived as key-value stores with support for new data models, new IRUD (Insert, Retrieve, Update and Delete) operational semantics, single-row transactions and data expiration constraints. Well known examples are "BigTable" [7] which was proposed by Google in 2004 and "Cassandra" [10] which was inspired by Amazon's "Dynamo" [6].

In contrast to traditional RDBMS, data analysis tasks of CoNoSQLDB are usually implemented by programmers at the application layer as CoNoSQLDBs lack a powerful data processing engine. MapReduce (MR) [5] is a parallel computing framework which is mostly utilized when in combination with CoNoSQLDBs. The desired data

B. Catania, G. Guerrini, and J. Pokorný (Eds.): ADBIS 2013, LNCS 8133, pp. 372–387, 2013.

analysis processing is embodied in user-defined *Map* and *Reduce* functions and the framework takes charge of data partitioning, parallel task scheduling and execution.

Frequently, MapReduce jobs are repeated and perform a complete recomputation of the analysis task to deal with changes in data sources (also known as "starting from scratch"). However, this approach is inefficient when the size of change data is much smaller than the size of the whole base data. The most cost-effective way is "incremental recomputation", which only propagates data changes into the data target or analysis result. In the relational context, this strategy is widely used by incremental maintenance for materialized views [8] and Data warehousing/ETL [3], [4]. In the MapReduce context, our previous work [11] has proven the feasibility of incremental MapReduce jobs and investigated an approach for incrementally maintaining aggregate self-maintainable MapReduce jobs for a wide range of data transformation and analysis tasks.

Nevertheless, before incremental computation can be applied, deltas need to be discovered and extracted to indicate the modifications of data sources. This data processing is termed *Change Data Capture* (CDC) [3] which generally includes the following steps: 1). Detecting deltas; 2). Classifying delta objects into appropriate categories based on the modification events; 3). Customizing delta outputs based on application logic; 4). Delivering deltas. As we will see later, each of these steps is influenced and constrained by the peculiarities of the data source, the characteristics of the incremental applications and the properties of the data target. For example, to incrementally calculate the aggregation function such as *SUM*, both data source and data target can be relational tables, and the CDC should produce the before and after state of updated tuples. Figure 1 describes the relationships between data source, CDC, incremental application and data target.

**Fig. 1.** CDC data flow

In this paper, we study the CDC problem in the CoNoSQLDBs context. The incremental applications can be multiple types, such as view maintenance techniques, procedures to incrementally recompute the aggregation functions and self-maintainable MapReduce jobs. We do not restrict the data target to a specific data storage system, it could be either a CoNoSQLDB, a RDBMS or a file system. Our goal is to produce a set of deltas which can be used by different incremental applications and can be applied to diverse data targets. However, to achieve that, several challenges arise:

- Different from RDBMSs, CoNoSQLDBs support new data models and new IRUD operational semantics, e.g. CoNoSQLDBs do not distinguish update from insertion. In consequence, the original deltas derived from CoNoSQLDBs are incomplete (e.g. the operation types of deltas are uncertain). However, some incremental applications, e.g. materialized view maintenance techniques, require a complete delta set [8]. Hence, a suitable delta model needs to be built to refine the original deltas into a series of operation-specific categories.

- Data values in CoNoSQLDBs may be time-sensitive and have expiration properties. Hence, it is possible that CDC cannot capture or only partially obtain the deleted data changes. We will discuss this problem in Section 3 and Section 4.
- The data target can be heterogeneous to the data source (CoNoSQLDBs). For example, CoNoSQLDBs support *schema flexibility* which means the schema of a source table can be varied on-the-fly. However, if the data target is RDBMS, the incremental application is incapable to propagate the deltas without first varying the schema. Hence, deltas need to contain sufficient metadata besides values.

The remainder of this paper will address these problems and is structured as follows: In Section 2, we describe the core properties of CoNoSQLDBs. We introduce the logical delta model and the corresponding delta representation in Section 3. In Section 4, five CDC approaches, namely, Timestamp-based approach, Audit-column approach, Log-based approach, Trigger-based approach and Snapshot differential approach are described. We give the performance for each CDC approach in Section 5. The related work is discussed in Section 6 and Section 7 makes the conclusion.

## 2    Characteristics of CoNoSQLDBs

As CoNoSQLDBs play the role of data sources which are accessed by CDC to detect and extract deltas, in this section, we will indicate the new features of CoNoSQLDBs

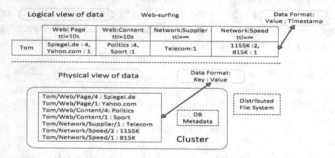

**Fig. 2.** Example of CoNoSQLDB

compared to RDBMSs. Although CoNoSQLDBs support a lot of new properties, we only summarize the characteristics of CoNoSQLDBs on two aspects which could heavily affect the design of our delta model and the implementation of CDC approaches:

- **Data model**. In addition to the concepts of table, row and column, CoNoSQLDB introduces a new concept called "*column-family*". In analogy with RDBMS, a column-family can be considered as a "table" and columns which belong to the same column-family will be stored continuously on disk. Each table in a CoNoSQLDB is partitioned based on the row keys and each data object is stored as a key-value pair in a distributed file system. For a given tuple, one column may store multiple data versions which are sorted by the corresponding timestamp (TS). Moreover, users can indicate the *ttl* (time-to-live) property for each column-family to denote the life time of a data object. When data items expire, CoNoSQLDBs will make them invisible. The actual data deletion occurs at data compaction time.

- **Data operational model**. Different from RDBMS, a CoNoSQLDB does not distinguish update between insert operations. A new data value for a specific column will be generated by the "Put" command[1]. When issuing a "Put" command, the user needs to denote the parameters such as row key, column-family name, column name and value[2]. Following the data model of CoNoSQLDB, data deletion can be classified into various granularities, namely, *data version*, *column* and *column-family*[3]. A delete operation will not delete data right way but insert a marker called "*tombstone*" to mask the data values whose TSs are equal to or less than the TS of tombstone. The actual data deletions take place at data compaction time.

Figure 2 shows an example defined in "HBase" to illustrate the aforementioned characteristics. The "Web-Surfing" table records the information when a user browses the internet. It contains two column families and each column family includes 2 columns. For row "Tom", the "Web:Page" column contains two data versions with the corresponding timestamps. The ttl for the "Web:Page" column is 10 seconds, which indicates "Yahoo.com:1" will be invisible at 11 seconds.

## 3    Delta Model

Generally, the design and implementation of CDC are influenced by three aspects, namely, the peculiarities of the data sources, the characteristics of incremental applications and the properties of the data targets. To maintain the materialized views in the traditional RDBMSs context [8], both data source and data target are relational tables. Due to the data manipulation language (DML) supported by RDBMS, change data is usually classified into insert deltas and delete deltas (update deltas are represented as a set of delete deltas and insert deltas). To incrementally load and recompute the transformation results in the DW/ETL context, the strategies for maintaining materialized views are then utilized in distributed manners. Normally, the captured deltas are often "partial deltas" [4]. Partial deltas may lack attribute values, e.g. the previous state of an updated tuple is unknown, or the type of partial deltas may be uncertain, e.g. inserted tuples cannot be distinguished from updated ones. Although partial delta lacks some kinds of information, it is very useful to maintain a key sort of tables, i.e. dimension tables in DW [4].

In our scenario, the original deltas derived from CoNoSQLDBs, namely, new data values generated by "Put" commands and the tombstones produced by "Delete" commands are partial deltas. Different from the DW scenario, the partial deltas are produced because of the characteristics of CoNoSQLDBs rather than a specific CDC

---

[1]  "Put" is used by HBase [9] which is an open-sourced implementation of "BigTable" and Cassandra uses "Set", in the rest of paper we use "Put" to represent both.

[2]  Timestamp (TS) can also be a parameter of "Put" command. However, in this paper, we only consider the situation in which TS is automatically generated by CoNoSQLDBs.

[3]  A *row deletion command* is supported by Cassandra but not by HBase. However, a row deletion can be translated into a set of column-families deletions.

approach. Hence, to work with different incremental applications, the original delta sets have to be further refined and a suitable delta model needs to be built. As the data target can be heterogeneous to the data source (CoNoSQLDBs), different metadata information has to be included in the CDC output. In this section, we propose our logical delta model and describe the corresponding delta representation.

### 3.1     Logical Delta Model

To generate the delta sets which can be used by the different incremental applications and can be applied to the diverse data targets, each delta object derived from CoNoSQLDBs is modeled as a quadruplet: (*identifier, operation, time-dimension, value*).

- **Identifier** includes the information such as row key, column-family name and column name which are derived from the source data.
- **Operation** indicates the type of data modification which was performed. As CoNoSQLDBs do not distinguish between insertion and update and supports various deletion granularities, the original (basic) operation types supported by CoNoSQLDBs are "Put", "DeleteVersion", "DeleteColumn" and "DeleteFamily". However, these operation types reflect an incomplete delta set, e.g. no indications of deleted deltas caused by ttl expiration. Furthermore, these operation types lack the ability to work with diverse incremental applications and various data targets, e.g. RDBMSs do not understand the operational semantics of "Put". Hence, we further refine and extend the basic operation types into *Insert, $Update_{old}$, $Update_{new}$, $Update_{partial}$, Upsert, $Delete_{pcomm}$, $Delete_{comm}$* and *$Delete_{ttl}$* where *Insert* indicates deltas are generated by insertion; *$Update_{old}$* and *$Update_{new}$* denote the delta is generated via update and represents the state before and after update, respectively; *$Update_{partial}$* describes the delta is generated via update but only with the state after update; *Upsert* denotes the delta is produced via insertion or update; both *$Delete_{pcomm}$* and *$Delete_{comm}$* denotes the delta is generated via delete commands. The difference is that *$Delete_{pcomm}$* indicates the delta records the tombstone information and *$Delete_{comm}$* denotes the delta contains the actual deleted value; *$Delete_{ttl}$* depicts the deltas are generated through ttl expiration.
- **Time-dimension** is composed of two sorts of information, i.e. timestamp (TS) which denotes when data modification occurs, and time-to-live (ttl) which indicates how long a data value in a column can exist.
- **Value** represents the delta value.

From the previous description, we can notice that our logical delta model represents deltas at various completeness levels. We refer the deltas which are incomplete as *partial deltas*. One delta object is considered as a partial delta due to 1.) The operation type is ambiguous; 2.) The delta value only contains metadata; 3.) The delta value is incomplete. We give a formal definition for partial deltas derived from CoNoSQLDB as follows:

**Definition 1 (Partial Deltas).** Let $R(rk, cfs:cols)$ be a CoNoSQLDB table with row-key $rk$ and a set of $cfs:cols$ (column-families:columns). For one $rk$, each column may contain multiple data versions. Let $R_{old}$ denote the state of $R$ at time point $t_1$ and $R_{new}$

the state of $R$ at time point $t_2$, where $t_1 < t_2$. During $t_1$ and $t_2$, several data modifications take place. Partial deltas of $R$ during time span $t_1$ and $t_2$ are a seven-tuple of sets $(R_{ins}, R_{upn/upo}, R_{pup}, R_{ups}, R_{comdel}, R_{pcomdel}, R_{ttldel})$ where:

- $R_{ins} \subseteq R_{new}$ indicates a set of data values inserted into $R$; (**insertion**)
- $R_{upn/upo}$ with $R_{upn} \subseteq R_{new}$ and $R_{upo} \subseteq R_{old}$ denotes a set of tuples updated in $R$; (**update**)
- $R_{pup} \subseteq R_{new}$ denotes a set of data values updated in $R$ only with the current state; (**partial update**)
- $R_{ups} \subseteq R_{new}$ indicates a set of data values either inserted or updated in $R$; (**upsert**)
- $R_{comdel} \subseteq R_{old}$ indicates a set of data values deleted from $R$. $R_{comdel}$ embodies the actual deleted data values; (**command deletion**)
- $R_{pcomdel}$ indicates delta values which contain only metadata derived from tombstones; (**partial command deletion**)
- $R_{ttldel} \subseteq R_{old}$ denotes a set of data values which are eliminated due to ttl expiration. (**ttl deletion**)

## 3.2   Delta Representation

The delta representation is the actual representation of the logical delta model which reflects the delta structures and delta values. Before we go into details regarding the "Delta" table architecture, we will first discuss how we represent delete information for ttl expiration. Different from deletion due to the delete commands, CoNoSQLDBs process the ttl expiration implicitly. The value of ttl is stored as metadata for each column-family. When a user tries to read the contents of a table, CoNoSQLDBs will mask the data items which expire the ttl and only return the valid data. The actual data deletion occurs at data compaction time. Moreover, the data which are deleted due to ttl expiration will not be recorded in the log file. Hence, it's very difficult to detect the ttl expiration simultaneously when it happens.

We considered two alternatives which can be used to represent the $R_{ttldel}$. One is to implicitly represent $R_{ttldel}$, i.e. each delta object embodies the ttl information. For example, if the delta output is stored in a structured format, each delta tuple contains an additional column to indicate the ttl information. In this way, the incremental applications have to decide whether the deltas are invalid. The second way is to explicitly represent the $R_{ttldel}$, i.e. to calculate the $R_{ttldel}$ when generating $R_{ins}$, $R_{upn/upo}$, $R_{pup}$ and $R_{ups}$. Although the data deletions due to ttl expiration have not happened yet, this strategy can be used to forecast the future data modifications. Someone can argue the "future-deleted data" will be seen by an application when it obtains the delta objects. However, this situation will only happen when the application reads the deltas without any time constraints. We will choose the second strategy, as the applications can obtain the $R_{ttldel}$ with less data processing. The generic structure of a "Delta" table is shown in Figure 3:

- The **Operation** column indicates the operation-dimension of delta objects. Values from 1 to 8 will be assigned to indicate the corresponding delta sets. The meaning of each number is illustrated in table 1;
- The **Identifier** column represents the row key extracted from the source table;
- The **Metadata** column contains the metadata of the source table such as column-family name and column name;
- The **Op$_{TS}$** column denotes when a specific data modification took place;
- The **Val$_{cur}$** column contains the current data value after each modification and the **Cur$_{TS}$** column denotes when that data value is created. To store the information of $R_{pdelcom}$, the $Val_{cur}$ column will be assigned by the TS of the responding tombstone and the $Cur_{TS}$ column is empty.
- The **Val$_{upo}$** and **Upo$_{TS}$** columns represent the previous state of the updated data. The $R_{pup}$ is treated as a special case of $R_{upn/upo}$ where the $Val_{upo}$ and $Upo_{TS}$ columns are empty.

The deltas can be stored in either a distributed file system, a RDBMS or a CoNoSQLDB. We choose to persist delta values with the associated metadata as a table in a CoNoSQLDB because of its "update-in-place" property, the range scan facility and a better scalability. Each delta object is hence represented by a row-key with a set of columns adhering to one or more column families. The row-key consists of the information of *Operation, Identifier, Metadata* and *Op$_{TS}$* with a specific order, e.g. a row-key can have either *Op$_{TS}$* or *Metadata* as a prefix. The reason for this design is that each row-key of CoNoSQLDBs must be unique and to facilitate the range scan when extracting desired delta values, as CoNoSQLDBs usually store the rows lexicographically. To handle the situations when a ttl deletion and a new data generation happen at the same time, the operation type must be included in the row key. We create a column-family "Value" with two columns, i.e. "*Current*" and "*Up$_{old}$*" to record the values of $Val_{cur}$, $Cur_{TS}$, $Val_{upo}$ and $Upo_{TS}$.

| Operation | Identifier | Metadata | Op$_{TS}$ | Val$_{cur}$ | Cur$_{TS}$ | Val$_{upo}$ | Upo$_{TS}$ |
|---|---|---|---|---|---|---|---|

**Fig. 3.** Delta representation

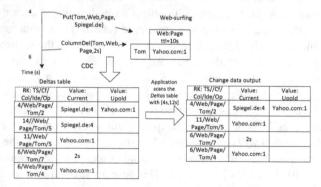

**Fig. 4.** CDC example

**Table 1.** Operation types

| OpType | Corresponding representations |
|--------|-------------------------------|
| 1 | Insertion |
| 2 | Update |
| 3 | Upsert |
| 4 | Deletion due to delete command (contain deleted value) |
| 5 | Deletion due to ttl expiration |
| 6 | Tombstone information for version deletion |
| 7 | Tombstone information for column deletion |
| 8 | Tombstone information for column-family deletion |

The "Deltas" table in Figure 4 illustrates the delta representation. We use the combination of $Op_{TS}/Metadata/Identifier/Operation$ as the row-key for "Deltas" table. The source table is called "Web-surfing" which contains one column "$Web:Page$". In the example, there are one "Put" operation which generates a new data version for row key $Tom$ and one "Column-Delete" operation which deletes all the data versions under "$Web:Page$" column for $Tom$ whose TSs are less than or equal to 2 seconds. The first "Put" operation is treated as an update as the row-key $Tom$ and the corresponding column have already existed. The "Deltas" table records $R_{pcomdel}$ and $R_{comdel}$ for representing column deletion. For $R_{pcomdel}$, one tuple which depicts the tombstone is created (6/Web/Page/Tom/7). And for $R_{comdel}$, a new tuple is inserted to indicate the actual deleted data (6/Web/Page/Tom/4). $R_{ttldel}$ (e.g. 14/Web/Page/Tom/5) is calculated when generating $R_{ins}$, $R_{upn/upo}$, $R_{pup}$ and $R_{ups}$. As we have integrated $Op_{TS}$ which indicates when a specific data modification took place into the row key, it's very easy for an application to obtain a desired delta set by checking if the TSs of row keys are included in a specific time interval. The corresponding example is shown in Figure 4.

From the example, we can notice that the CDC produces a historical delta set rather than the latest data changes. Using this strategy, the incremental application and the data target could always obtain a complete set of changes and are capable to choose the desired parts to work with.

## 4    CDC Approaches

In this section, we describe five feasible CDC approaches, namely, Timestamp-based approach, Audit-column approach, Log-based approach, Trigger-based approach and Snapshot differential approach, which are exploited to detect and extract deltas from CoNoSQLDBs. With the exception of the Timestamp-based approach, these are well-known CDC techniques in RDBMS and DW environment, which we have adopted to our change data capture approaches. Due to their inherent characteristics, each CDC approach can produce deltas only at a certain complete delta set, e.g. Timestamp-based approach cannot generate $R_{ins}$. Moreover, the limitations and pitfalls for each CDC approach will be discussed.

## 4.1    Timestamp-Based Approach

Besides data values, each data object in CoNoSQLDBs also contains a TS which denotes when that data object (version) is generated by the "Put" command. Hence, it is feasible to scan a table with a certain time interval as a selection criterion to discover the delta candidates. Figure 5 describes this approach. A MapReduce job is applied to scan the table with timing constraint [2s, 7s], which indicates to return the data values whose timestamps are larger than or equal to 2 seconds and less than or equal to 7 seconds. The corresponding output is shown at the right hand side of the figure. Although the Timestamp-based approach is easy and straightforward, there are two drawbacks. First, CDC cannot distinguish between update and insertion (e.g. 4/Web/Page/Tom/3). Second, $R_{pcomdel}$ and $R_{comdel}$ cannot be detected. Moreover, capturing the current $R_{ttldel}$ depends on whether the previous CDC results are persistent.

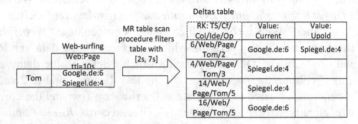

**Fig. 5.** Timestamp-based approach

## 4.2    Audit-Column Approach

The Audit-column approach can be used to address some drawbacks of the Timestamp-based approach. Each source table is appended with auxiliary columns to record the data insertion time. Figure 6 depicts this approach. Now, "Spiegel.de:4" is categorized into $R_{pup}$ rather than $R_{ups}$ (in Figure 5). Although this approach can differentiate insertion from update, $R_{pcomdel}$ and $R_{comdel}$ are still missing and capturing the current $R_{ttldel}$ still depends on the previous CDC results. Moreover, unless such a column is already part of the database design for auditing process, the application logic needs to be added or changed to maintain the audit column. More implementation details will be introduced in Section 5.

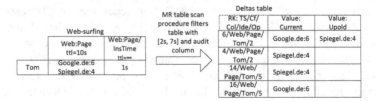

**Fig. 6.** Audit-column

## 4.3    Log-Based Approach

For recovery and durability reasons, each work node in CoNoSQLDBs maintains a WAL (Write Ahead Log). The content of WAL can be used to extract deltas. The left hand side of Figure 7 shows a WAL segment of the "Web-surfing" table. Each log entry is organized as a key-value pair. The key part is composed of log sequence number, node ID, log writing time and table name. The value part contains the information of the data modification. Different from RDBMS, a WAL in CoNoSQLDBs doesn't distinguish between update and insertion and only records the "tombstones". Moreover, data objects (versions) which are deleted because of ttl expiration are not recorded in the WAL.

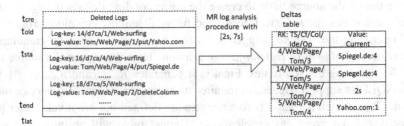

**Fig. 7.** Log-based approach

Log entries will be automatically deleted by CoNoSQLDB when the data modifications have been made persistent on disk. Hence, without a specific configuration, CDC may only get a partial WAL for analysis. In Figure 7, we use $t_{cre}$, $t_{old}$ and $t_{lat}$ to denote the creation time of log, the oldest and the latest timestamps among the existing log entries, respectively. $[t_{sta}, t_{end}]$ indicates the time interval in which CDC wishes to capture deltas. A MapReduce log analysis procedure is applied to scan the partial WAL of the "Web-surfing" table with $[t_{sta}=2s, t_{end}=7s]$. We can notice the log analysis procedure cannot decide whether "Spiegel.de:4" is generated via insertion or update, as it is the oldest log entry for row "*Tom*" under "*Web:Page*" column. Moreover, the log analysis procedure cannot capture the deleted data which are generated during $[t_{sta}, t_{end}]$ but the log entries resides in $[t_{cur}, t_{sta}]$ (e.g. 5/Web/Page/Tom/4). The intuitive solution is to extend the scanning time interval from $[t_{sta}, t_{end}]$ to $[t_{cur}, t_{end}]$. Nevertheless, as the WAL remains partial, it is still impossible to detect the inserted deltas, and the deleted deltas are still incomplete. Hence, a full WAL needs to be saved if the complete delta sets are required.

## 4.4    Trigger-Based Approach

A database trigger allows to automatically execute DML (data manipulation language) or other operations in response to certain events on a particular table. Hence, detecting and storing deltas can be easily implemented by triggers, if they are supported by CoNOSQLDBs. Figure 8 shows an example of the Trigger-based approach.

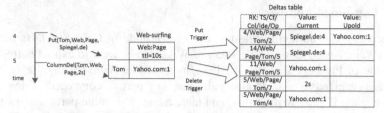

**Fig. 8.** Trigger-based approach

Two triggers are defined to monitor "Put" and "Delete" events on the "Web-surfing" table. When a "Put" command is issued, the "Put" trigger will first check the related data values on the source table to create the corresponding delta sets ($R_{upo/upn}$ or $R_{ins}$). Then it calculates the $R_{ttldel}$ to forecast the future deletions. For monitoring the "Delete" operation, the "Delete" trigger needs to first detect the deletion granularity, i.e. *data version-*, *column-* or *column-family deletion*. Then it accesses the source table to find the matching deleted tuples. Finally, it generates $R_{comdel}$ and $R_{pcomdel}$. As yet, no CoNoSQLDBs support triggers to monitor ttl expiration. Hence, triggers cannot capture the $R_{ttldel}$ which are valid before triggers are defined but are invalid after triggers are activated. To avoid this problem, the "Deltas" table will be initialized with the state of source table before triggers are activated (e.g. 11/Web/Page/Tom/5).

### 4.5    Snapshot Differential Approach

The Snapshot differential approach is typically utilized to capture deltas from legacy system or unsophisticated system. However, this approach can also be seen as a possible CDC candidate to detect and capture deltas from CoNoSQLDBs. Snapshot differential processing generally contains two phases: 1) taking snapshots from data source at different timestamps; 2) comparing snapshots to extract change data. For backup purposes, most CoNoSQLDBs support commands to dump valid contents into the distributed file system, e.g. the "export" command of HBase. The "valid content" means the data values which are not masked by tombstones or have expired.

Figure 9 describes this approach. First, two snapshots of the "Web-surfing" table are taken at $t_1$ and $t_2$, respectively. Then, two snapshot images are sent into the comparison procedure. In the snapshot comparison phase, all tuples which appear in $S_2$ but not in $S_1$ are considered as insertion deltas. In contrast, tuples which exist in $S_1$ but not in $S_2$ are treated as deletion deltas. Tuples which exist in both $S_1$ and $S_2$ but differ from each other are considered as update deltas. However, as the snapshot content doesn't contain any tombstone information, CDC cannot decide whether a data value is deleted by "Delete" command or ttl expiration if $t_2$ is greater than the time of ttl expiration. When this situation occurs, we always consider the deleted data changes are generated via ttl expiration.

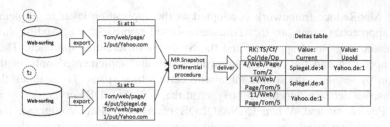

**Fig. 9.** Snapshot differential

## 4.6   Summary

From the previous description, we can notice that the Timestamp-based approach is the most convenient approach by only scanning the source table with the denoted time spans. However, this approach cannot distinguish between inserted and updated data changes. The Audit-column approach is an improvement of the Timestamp-based approach but auxiliary columns need to be created to record the data insertion time. The pitfalls of these two approaches are: First, both of these approaches lack the ability to extract deleted deltas. Second, a full table scan will be very inefficient if only a small portion of source table is modified. In the Log-based approach, the completeness of deltas depends on the completeness of WAL. However, storing and analyzing the whole WAL is space-consuming and time-consuming. When a CoNoSQLDB provides the active capabilities such as triggers, the Trigger-based approach can be adopted as an alternative of the Log-based approach. The Snapshot differential approach is usually seen as "a last resort" for capturing deltas. As the volume of source data grows, more and more data has to be extracted and larger and larger comparisons have to be performed.

## 5   Implementation and Performance

In this section, we describe the implementation details and show the performance charts to indicate the processing speed of each CDC approach. The experiments are performed on a six-node cluster (Xeon Quadcore CPU at 2.53GHz, 4GB RAM, 1TB SATA-II disk, Gigabit Ethernet) running Hadoop (version 2.0.2-appending) [13] and HBase (version 0.92) [9], respectively. Test data sets are generated through the *workloada* of YCSB (Yahoo-Cloud-Service-Benchmark) [14]. At the *"Loading"* phase, one million tuples are inserted into a predefined "usertable" in HBase. The "usertable" contains one column family named "cf1" with 10 columns. The ttl property for "cf1" is set to 30 minutes. At the *"Transaction"* phase, we vary the proportion of data modifications applied to "usertable" through the properties of *"updateproportion"* and *"insertproportion"*. As we can only indicate data modification proportion of update and insertion inside *workloada*, we write our own program to generate the delete operations.

The MapReduce framework is adopted as the application layer to implement the CDC approaches, such as the Timestamp-based approach, the Audit-column approach, the Log-based approach and the Snapshot differential approach. The Time-based approach and the Audit-column approach are implemented only in the Map function by scanning the source table with certain time spans. To maintain the audit columns, we define a *"preCheck"* trigger at the CoNoSQLDB server side to monitor each "Put" command. When CoNoSQLDB receives a "Put" command, the *"pre-Check"* will first check the corresponding columns. If no values exist, the insertion time for such a column is created. For the Log-based approach, the whole WAL is maintained. To achieve that, we set the property *"hbase.master.logcleaner.ttl"* (used to define the life cycle of log) to a long time and incrementally copy the HBase log into a specific archive. For implementing the Trigger-based approach, two triggers i.e. *"prePut"* and *"preDelete"* are defined. The functionalities of these two triggers have been described in Section 4.

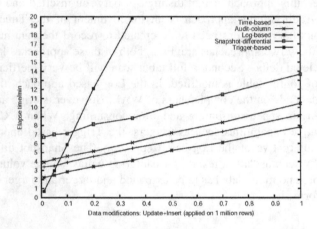

**Fig. 10.** Update+Insertion

We utilize two separate charts to depict the CDC performance for data generation and data deletion in Figure 10 and Figure 11, respectively, as the Timestamp-based approach and the Audit-column approach are not capable to extract $R_{pcomdel}$ and $R_{comdel}$. The range of execution time of five CDC approaches is between 2 to 20 minutes which consists of initializing MapReduce jobs, scanning NoSQLDB tables, CDC data processing and loading CDC results. Normally, we need 30s to 40s to initialize MapReduce jobs in the cluster. The low performance of scanning HBase table via MapReduce may be caused by the sequential scan mechanism of HBase [15]. As HBase is memory-intensive, when Garbage Collection is executed by JVM (Java Virtual Machine) during loading CDC results, any client requests are stalled and the server will pause for a significant amount of time [15].

In both charts, the Trigger-based approach shows the best performance if the fraction of change data is tiny. The possible reason is that the triggers do not cache any table locations. Hence, triggers have to find the correlated working regions for every invocation. In Figure 10, the Audit-column approach is slower than the Timestamp-based approach as it needs to process the audit columns. For both figures, it is at first supervising to see that the Log-based approach shows a better performance than the other approaches for the middle and high percentages of data modifications, even if it needs to process more volume of data (in our experiment, WAL contains the log information for a "usertable" table and a "Deltas" table). The reason is that reading data from HDFS is approximately 5 times [12] faster than reading the same amount of data from HBase when utilizing the MapReduce framework. Moreover, the header of each WAL maintains the metadata information, such as table name and log creation time. Hence, the log analysis procedure can filter unnecessary log entries very rapidly. The Snapshot approach shows a bad performance compared to the Log-based approach, as it needs to scan the whole table twice to dump the contents and one MapReduce snapshot comparison needs to be executed.

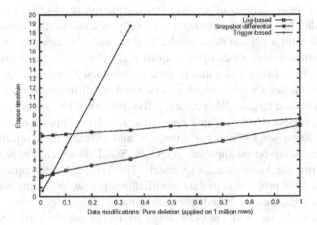

**Fig. 11.** Deletion

# 6    Related Work

In the context of data warehouses, there are some popular CDC approaches, such as Audit Columns, Database Log Scraping, Trigger-based approach and Snapshot Differential which are utilized by the most ETL tools [3]. The "Audit columns" are additional columns which are appended to a relational table and are refreshed when a table gets modified. "Database Log Scraping" analyzes transactional redo-log to obtain the change data. The "Trigger-based" approach utilizes triggers to capture deltas. "Snapshot differential" estimates snapshots taken at different time points to extract the data changes. However, all of these approaches are designed under the restrictions and circumstances of RDBMSs. Hence, some of these approaches cannot be directly utilized in the context of CoNoSQLDBs. First, DML of two systems differs heavily, i.e. CoNoSQLDBs do not distinguish insertion from update and support multiple deletion granularities. Second, the functionalities of DML and DDL (data definition language) of CoNoSQLDBs have overlaps, e.g. the schema of a table can be changed when generating a new tuple. Third, CoNoSQLDBs support versioned data and each data object can have time expiration properties.

In the context of CoNoSQLDBs, state of the art of CDC is ambiguous. Few publications mention the technologies about how to detect and extract deltas from CoNoSQLDBs. Our previous work [12] can be seen as the first study of the feasible CDC approaches to capture the deltas from CoNoSQLDBs. However, in our previous work, we assumed both data source and data target are CoNoSQLDBs and restricted the operational semantics of the incremental application in the scope of "relational algebra".

## 7     Conclusion

Change data capture (CDC) is a prerequisite data processing task of incremental maintenance of data warehouses or data transformation results. To our knowledge, our work is the first intensive study of CDC problems in the context of CoNoSQLDBs. We built a logical delta model to deal with the aspects, such as schema flexibility, multiple granularities of operations e.g. deletions, versioned/time-stamped data and expiration of data items due to time constraints. Moreover, our logical delta model is a general delta model which can be used by different incremental applications and diverse data targets. We presented five feasible CDC approaches, i.e. *Timestamp-based approach*, *Audit-column approach*, *Log-based approach*, *Trigger-based approach* and *Snapshot Differential approach* and indicate their pitfalls and drawbacks. The Timestamp-based approach is a CoNoSQLDBs-specific approach which is independent from the Audit-column approach. The Trigger-based approach is always the best choice if the proportion of data modification is low. When the fraction of data modification is high, the Log-based approach is more preferable.

In further work, we will consider ways to incrementally load and shrink the change data table. Moreover, we will investigate if our logical delta model is also reasonable for other NoSQLDBs, e.g. document-oriented NoSQLDBs.

**Acknowledgements.** This work is a sub-project of "Virga" project which is supported by Google Inc. with a Google Research Award.

## References

1. Jacobs, A.: The pathologies of big data. Commun. ACM 52(8), 3644 (2009)
2. Eaton, C., et al.: Understanding Big Data. McGraw-Hill Companies (April 2012)
3. Kimbal, R., Kastera, J.: The Data Warehouse ETL Toolkit: Practical Techniques for Extracting, Cleaning, Conforming, and Delivering Data. Published by Wiley Publishing, Inc. (2004)
4. Joerg, T., Dessloch, S.: View Maintenance using Partial Deltas. In: BTW, pp. 287–306 (2011)
5. Dean, J., Ghemawat, S.: MapReduce: Simplified Data Processing on Large Clusters. In: OSDI, pp. 137–150 (2004)

6.  DeCandia, G., et al.: Dynamo: Amazons highly Available Key-value Store. In: SOSP 2007, Stevenson, Washington, USA (2007)
7.  Change, F., et al.: Bigtable: A Distributed Storage System for Structured Data. In: OSDI, pp. 205–218 (2006)
8.  Gupta, A., Mumick, I.: Maintenance of Materialized Views: Problems, Techniques, and Applications. IEEE Data Eng. Bull. 18(2), 318 (1995)
9.  Apache HBase, http://hbase.apache.org/
10.  Apache Cassandra, http://cassandra.apache.org/
11.  Joerg, T., et al.: Incremental Recompilations in MapReduce. In: CloudDB 2011, Glasgow, Scotland, U.K. (2011)
12.  Hu, Y., Qu, W.: Efficiently Extracting Change Data from Column Oriented NoSQL Databases. In: ICS 2012, Hualian, Taiwan (2012)
13.  Apache Hadoop, http://hadoop.apache.org
14.  https://github.com/brianfrankcooper/YCSB/
15.  http://mail-archives.apache.org/mod_mbox/hbase-user/

# Very Large Graph Partitioning
# by Means of Parallel DBMS

Constantin S. Pan and Mikhail L. Zymbler

South Ural State University, Chelyabinsk, Russia

**Abstract.** The paper introduces an approach to partitioning of very
large graphs by means of parallel relational database management system (DBMS) named PargreSQL. Very large graph and its intermediate
data that does not fit into main memory are represented as relational
tables and processed by parallel DBMS. Multilevel partitioning is used.
Parallel DBMS carries out coarsening to reduce graph size. Then an initial partitioning is performed by some third-party main-memory tool.
After that parallel DBMS is used again to provide uncoarsening. The
PargreSQL's architecture is described in brief. The PargreSQL is developed by authors by means of embedding parallelism into PostgreSQL
open-source DBMS. Experimental results are presented and show that
our approach works with a very good time and speedup at an acceptable
quality loss.

## 1 Introduction

Nowadays graph mining plays an important role in modeling complicated structures, such as chemical compounds, protein structures, biological and social networks, the Web and XML documents, circuits, etc. Being the one of the topical
problems of graph mining, graph partitioning is defined as follows [9]. Given a
graph $G = (N, E)$, where $N$ is a set of weighted nodes and $E$ is a set of weighted
edges, and a positive integer $p$, find $p$ subsets $N_1, N_2, \ldots, N_p$ of $N$ such that

- $\cup_{i=1}^{p} N_i = N$ and $N_i \cap N_j = \emptyset$ for $i \neq j$,
- $W(i) \approx W/p, i = 1, 2, \ldots, p$, where $W(i)$ and $W$ are the sums of the node
  weights in $N_i$ and $N$, respectively;
- the *cut size*, i.e. the sum of weights of edges crossing between subsets is
  minimized.

The usual way to do partitioning is through several recursive steps of bisection.
So our approach was aimed at this particular kind of partitioning problem.

A very large graph, comprising of billions vertices and/or edges, is the most
challenging case of partitioning because the graph being partitioned and all the
intermediate data does not fit into main memory.

In this paper we introduce an approach to partitioning of very large graphs
by means of parallel relational database management system (DBMS). The rest

B. Catania, G. Guerrini, and J. Pokorný (Eds.): ADBIS 2013, LNCS 8133, pp. 388–399, 2013.

of the paper is organized as follows. Section 2 briefly discusses the related work. Section 3 provides a short intro to PargreSQL parrallel DBMS that we use for the graph partitioning. Section 4 presents our approach. The results of experiments are shown in section 5. Section 6 contains concluding remarks and directions for future work.

## 2   Related Work

A significant amount of work has been done in the area of graph-based data mining, including graph partitioning problem [1].

The classical algorithm based on a neighborhood-search technique for determining the optimal graph partitioning is proposed in [16]. Multilevel approach to graph partitioning is suggested in [14]. There are many sophisticated serial and parallel graph partitioning algorithms have been developed [9]. One of the first parallel graph partitioning algorithms based upon multilevel approach was proposed in [15]. An approach to graph partitioning based upon genetic alorighms investigated in [17]. In [7] authors suggested graph partitioning algorithm without multilevel approach.

A parallel graph partitioner for shared-memory multicore architectures is discussed in [28]. Parallel disk-based algorithm for graph partitioning is presented in [29]. In [26] distributed graph partitioning algorithm is suggested. Cloud computing-based graph partitioning is described in [6].

There are software packages implementing various graph partitioning algorithms, e.g. Chaco [12], METIS and ParMETIS [13], KaFFPa [25], etc.

In [3] ParallelGDB, a system to handle massive graphs in a shared-nothing parallel system is described. Pregel [20] is a Google's distributed programming framework, focused on providing users with a natural API for programming graph algorithms.

Existing graph data mining algorithms typically face difficulties with respect to scalability (storing a graph and its intermediate data in main memory). Mining by means of relational DBMS and SQL allows to overcome main memory restrictions. This approach supposes bringing mining algorithms to data stored in relational database instead of moving stored data to algorithms of third-party tools. Using relational DBMS for mining we have got all its services "for free": effective buffer management, indexing, retrieval, etc. However it demands uneasy "mapping" of graph mining algorithms onto SQL with cumbersome source code.

In recent years a number of graph mining algorithms have been implemented using relational DBMS. In [22] a scalable, SQL-based approach to graph mining (specifically, interesting substructure discovery) is presented. A framework for quasi clique detection based upon relational DBMS is described in [27]. Frequent SQL-based subgraph mining algorithms proposed in [5,10]. Database approach to handle cycles and overlaps in a graph is investigated in [2,4]. We also have to mention non-DBMS oriented but disk-based system for computing efficiently on very large graphs described in [18].

Despite the benefits of using a DBMS, it still could be incapable of effectively processing *very large* graphs. Our contribution is applying a *parallel* DBMS to graph partitioning (what was done for the first time, to the best of our knowledge).

## 3    PargreSQL Parallel DBMS

Currently, open-source PostgreSQL DBMS [24] is a reliable alternative for commercial DBMSes. There are many research projects devoted to extension and improvement of PostgreSQL. PargreSQL DBMS is a version of PostgreSQL adapted for parallel processing. PargreSQL utilizes the idea of *fragmented parallelism*[1] [8] in cluster systems (see fig. 1). This form of parallelism supposes horizontal fragmentation of the relational tables and their distribution among the disks of the cluster.

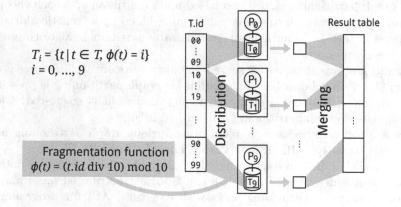

$$T_i = \{t \mid t \in T,\ \phi(t) = i\}$$
$$i = 0, ..., 9$$

Fragmentation function
$$\phi(t) = (t.id \text{ div } 10) \bmod 10$$

**Fig. 1.** Fragmented parallelism

The way of the table fragmentation is defined by a *fragmentation function*, which for each record of the table returns the ID of the processor node where this record should be placed.

A query is executed in parallel on all processor nodes as a set of parallel *agents*. Each agent processes its own fragment and generates a partial query result. The partial results are merged into the resulting table.

The PargreSQL architecture (see fig. 2) implies modifications to the source code of existing PostgreSQL's subsystems as well as implementation of new subsystems to provide parallel processing.

Modified PostgreSQL's engine instance is installed on every node of a cluster system. The *par_Storage* subsystem provides fragmentation and replication of tables among the disks of the cluster. The *par_Balancer* subsystem is responsible

---

[1] Also known as *partitioned parallelism*, but that term is not used in the paper to avoid possible confusion with *graph partitioning*.

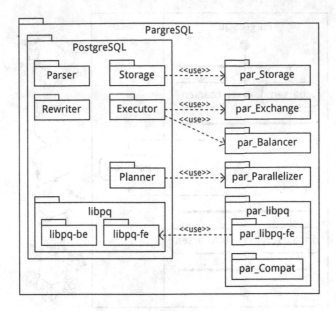

**Fig. 2.** PargreSQL architecture

for load balancing of the engine instances. The *par_Exchange* subsystem provides the EXCHANGE operator [11,19], which is in charge of data transmission among PostgreSQL's engine instances during the query execution and encapsulates all the parallel work. The *par_Parallelizer* subsystem inserts EXCHANGEs into appropriate places of the query execution plan made by PostgreSQL's engine instance. The *par_libpq* provides an API that is transparent to PostgreSQL applications.

## 4   Applying PargreSQL to Graph Partitioning

This section describes an approach to bisecting of very large graphs with PostgreSQL parallel DBMS. *Bisection* of a graph is a basic case of graph partitioning problem ($p = 2$, partitioning into two sets). Multi-way partitioning is performed by recursively applying bisection.

Graph partitioning with PargreSQL is depicted in fig. 3. We use the multilevel partitioning scheme [14], which includes three steps.

On the first step, *coarsening*, we reduce the input graph to the size that existing graph partitioning tools could deal with. A relational table (list of edges) that represent the graph is split into horizontal fragments and PargreSQL executes SQL-queries which reduce its size. During the coarsening step we collapse the heaviest edges of the graph, reducing excessive vertices and edges and aggregating their weight. Coarsening can be repeated multiple times to get smaller graphs. This step is implemented in PargreSQL by means of representing the input graph as a relational table and quering this table using SQL.

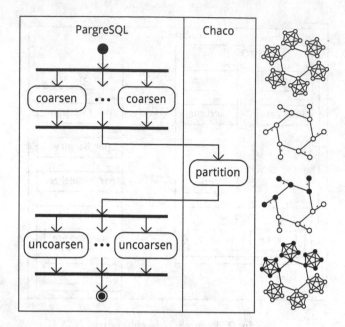

**Fig. 3.** Graph partitioning scheme

On the second step, *initial partitioning*, we export the graph from the database and feed it to a third-party graph partitioning utility (Chaco [12]) which performs the initial partitioning using one of the well-known algorithms (Kernighan-Lin, inertial, spectral, etc.). The result of the initial partitioning is imported into the database as a relational table containing the list of the graph's vertices with their partitions assigned.

**Table 1.** Data structure for graph partitioning

| Relational table | Description |
|---|---|
| GRAPH(A, B, W) | The original fine graph<br>A, B: edge's ends, W: weight |
| MATCH(A, B) | The matching of the fine graph<br>A, B: edge's ends |
| COARSE_GRAPH(A, B, W) | The coarse graph<br>A, B: edge's ends, W: weight |
| COARSE_PARTITIONS(A, P) | The partitions of the coarse graph<br>A: vertex, P: partition |
| PARTITIONS(A, P, G) | The partitions of the fine graph<br>A: vertex, P: partition, G: gain |

On the final step, *uncoarsening*, we refine the coarse results using another sequence of SQL queries. This step is repeated as many times as the coarsening was performed. During this step we first map coarse results on the larger graph, and then apply a local optimization to improve the results. PargreSQL calculates a table of two columns: vertex and its partition.

Since we apply a relational DBMS, our algorithms suppose that the data is stored in relational tables (see tab. 1). These tables are uniformly distributed among cluster nodes by means of $\psi(t) = \lfloor \frac{t.A \times n}{|E|} \rfloor$ fragmentation function, where $n$ is the number of nodes in the cluster and $t.A$ is used as the fragmentation attribute. Every instance of the PargreSQL's engine processes its own set of fragments of these tables.

Since our approach does not implement the initial partitioning itself, we will discuss further only the coarsening and uncoarsening steps.

### 4.1   Coarsening

We divided the coarsening problem into two subproblems: **find a matching** and **collapse the matching** (see fig. 4).

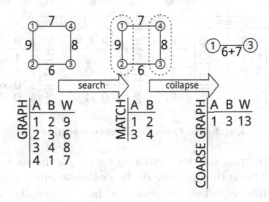

**Fig. 4.** An example of **coarsening**

During the **find** substep we search for the heaviest matching (independent edge set) in the graph. Since we use a simple greedy algorithm for that, the matching could turn out not to be actually the heaviest. This matching is stored into a database table as a list of edges for later use.

The **collapse** substep implements the removal of all the edges found on the previous substep. The edges in question get collapsed into vertices (see fig. 5).

### 4.2   Uncoarsening

The uncoarsening step is also implemented inside PargreSQL as a series of relational operations. The implementation consists of three substeps: **propagate the partitions, calculate the gains** and **refine the gains** (see fig. 6).

```
select least(newA, newB) as A, greatest(newA, newB) as B, sum(W) as W
from (
  select
    coalesce(match2.A, GRAPH.A) as newA,
    coalesce(MATCH.A, GRAPH.B) as newB,
    GRAPH.W
  from
    GRAPH, left join MATCH on GRAPH.B=MATCH.B
    left join MATCH as match2 on GRAPH.A=match2.B)
where newA<>newB group by A, B
```

**Fig. 5.** The **collapse** substep implementation

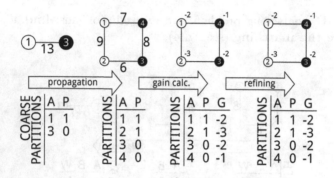

**Fig. 6.** An example of **uncoarsening**

The idea of the **propagate** step is to map coarse partitions onto the finer version of the graph. This is the opposite of the **collapse** substep of the coarsening process. The matching saved earlier gets used here to uncollapse the edges and mark their ends as belonging to the corresponding partitions (see fig. 7).

```
select a, p from COARSE_PARTS
union
select match.b, part.p
from MATCH as match, COARSE_PARTS as part
where match.a = part.a
```

**Fig. 7.** The **propagate** substep implementation

Right after the partition propagation we could end up with a nonoptimal solution. So we need to know, which vertices should go to the opposite partition. This is calculated for each vertex as the *gain*, which is the difference between the total weight of the edges connecting the vertex with ones from the other partition and the total weight of the edges connecting the vertex with ones from the same partition.

The gain is calculated as

$$\text{gain}(v) = \text{ext}(v) - \text{int}(v),$$

where

$$\text{ext}(v) = \sum_{(v,u)\in E, P(v)\neq P(u)} w(v,u),$$

$$\text{int}(v) = \sum_{(v,u)\in E, P(v)=P(u)} w(v,u).$$

Basically, the gain means how much better the overall solution would be if we excluded this vertex $v$ from its current partition $P(v)$. In case $\text{gain}(v) > 0$ we would want to move $v$ to the opposite partition. The gain calculation implemented in PL/pgSQL as shown in fig. 8.

```
select PARTITIONS.A, PARTITIONS.P, sum(subgains.Gain) as Gain
from
  PARTITIONS left join (
    select GRAPH.A, GRAPH.B,
    case when ap.P = bp.P then -GRAPH.W
      else GRAPH.W end as Gain
    from
    GRAPH left join PARTITIONS as ap on GRAPH.a = ap.A
    left join PARTITIONS as bp on GRAPH.b = bp.A
  ) as subgains
  on PARTITIONS.A = subgains.A or PARTITIONS.A = subgains.B
group by PARTITIONS.A, PARTITIONS.P
```

**Fig. 8.** The gain calculation

```
select * from PARTITIONS
where
  P = current and
  G = (
    select max(G)
    from PARTITIONS
    where P = current)
limit 1
into V
```

```
update PARTITIONS
  set G = G + W * (case when P = V.P then 2
    else -2 end)
from (
  select case when A = V.A then B else A end,
  W from GRAPH
  where B = V.A or A = V.A) as neighbors
where neighbors.A = PARTITIONS.A;

update PARTITIONS
  set G = -G, P = 1 - P
where A = V.A;
```

(a) Vertex picking                    (b) Vertex moving

**Fig. 9.** Refining

During the **refine** substep we switch back and forth between the partitions, each time picking a vertex which has the largest positive gain and moving it to the opposite partition. This repeats until we cannot find such vertices any more. This method is based on the heuristic proposed in [16]. Its implementation in PL/pgSQL is shown in fig. 9.

## 5   Experiments

We have conducted a series of experiments using SKIF-Aurora supercomputer of South Ural State University [21]. We tried to partition a street network graph[2] of $10^5$ vertices and we got an embarassingly parallel implementation where every fragment of the graph's database represenation gets processed independently of the other fragments. Since the problem has the complexity of $O(n^3)$, this kind of "trick" gives us the superlinear speedup that you can see on fig. 10, but at a price of reduced accuracy of the results.

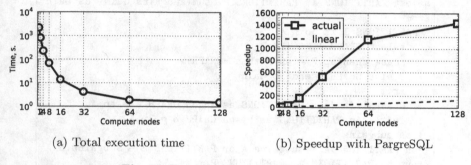

(a) Total execution time               (b) Speedup with PargreSQL

**Fig. 10.** Execution time with PargreSQL

Relative execution time of coarsening and uncoarsening did not show any dependency on configuration (see fig. 11).

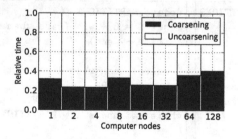

**Fig. 11.** Relative execution time

---

[2] Luxemburg street map from
http://www.cc.gatech.edu/dimacs10/archive/streets.shtml

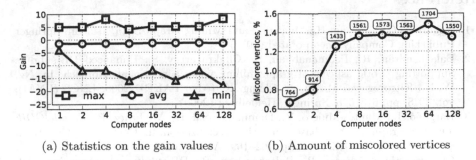

(a) Statistics on the gain values          (b) Amount of miscolored vertices

**Fig. 12.** The quality of the partitioning

We have investigated exactly how much worse the results would get as we increased the number of computer nodes in the system. The statistics on the quality of partitioning is shown in fig. 12.

## 6  Conclusions

The paper introduces an approach to partitioning of very large graphs, comprising of billions of vertices and/or edges. Most of the existing serial and parallel algorithms suppose that the graph being partitioned and all the intermediate data fit into main memory, so they cannot be applied directly for very large graphs.

Our approach assumes using PargreSQL parallel relational DBMS. A very large graph is represented as a relational table (list of edges). PargreSQL DBMS carries out the coarsening of the graph. The coarsened graph and the intermediate data generated during the process fit into main memory, so its initial partitioning could be performed by some third-party tool (e.g. Chaco). In the end PargreSQL performs the uncoarsening of the coarse partitions.

PargreSQL is implemented on the basis of PostgreSQL open-source DBMS. PargreSQL utilizes the idea of fragmented parallelism and implies modifications in the source code of the existing PostgreSQL's subsystems as well as implementation of new subsystems to provide parallel processing.

Because of using a DBMS our approach will work even in cases when traditional tools may fail due to memory limits. Parallel query processing provides us with a very good time and speedup of graph partitioning at an acceptable quality loss.

In the future we would like to explore applying PargreSQL to other problems connected with mining very large graphs from different subject domains and try some more sophisticated partitioning schemes, not only the bisection.

**Acknowledgment.** The reported study was partially supported by the Russian Foundation for Basic Research, research projects No. 12-07-31217 and No. 12-07-00443.

# References

1. Aggarwal, C.C., Wang, H.: Managing and Mining Graph Data, 1st edn. Springer Publishing Company, Incorporated (2010)
2. Balachandran, R., Padmanabhan, S., Chakravarthy, S.: Enhanced DB-subdue: Supporting subtle aspects of graph mining using a relational approach. In: Ng, W.-K., Kitsuregawa, M., Li, J., Chang, K. (eds.) PAKDD 2006. LNCS (LNAI), vol. 3918, pp. 673–678. Springer, Heidelberg (2006)
3. Barguñó, L., Muntés-Mulero, V., Dominguez-Sal, D., Valduriez, P.: *ParallelGDB*: a parallel graph database based on cache specialization. In: Desai, B.C., Cruz, I.F., Bernardino, J. (eds.) IDEAS, pp. 162–169. ACM (2011)
4. Chakravarthy, S., Beera, R., Balachandran, R.: DB-subdue: Database approach to graph mining. In: Dai, H., Srikant, R., Zhang, C. (eds.) PAKDD 2004. LNCS (LNAI), vol. 3056, pp. 341–350. Springer, Heidelberg (2004)
5. Chakravarthy, S., Pradhan, S.: DB-FSG: An SQL-based approach for frequent subgraph mining. In: Bhowmick, S.S., Küng, J., Wagner, R. (eds.) DEXA 2008. LNCS, vol. 5181, pp. 684–692. Springer, Heidelberg (2008)
6. Chen, R., Yang, M., Weng, X., Choi, B., He, B., Li, X.: Improving large graph processing on partitioned graphs in the cloud. In: Proceedings of the Third ACM Symposium on Cloud Computing, SoCC 2012, pp. 3:1–3:13. ACM, New York (2012)
7. Delling, D., Goldberg, A.V., Razenshteyn, I., Werneck, R.F.F.: Graph partitioning with natural cuts. In: IPDPS, pp. 1135–1146. IEEE (2011)
8. DeWitt, D.J., Gray, J.: Parallel Database Systems: The Future of High Performance Database Systems. Commun. ACM 35(6), 85–98 (1992)
9. Fjallstrom, P.: Algorithms for graph partitioning: A survey (1998)
10. Garcia, W., Ordonez, C., Zhao, K., Chen, P.: Efficient algorithms based on relational queries to mine frequent graphs. In: Nica, A., Varde, A.S. (eds.) PIKM, pp. 17–24. ACM (2010)
11. Graefe, G.: Encapsulation of parallelism in the volcano query processing system. In: Garcia-Molina, H., Jagadish, H.V. (eds.) SIGMOD Conference, pp. 102–111. ACM Press (1990)
12. Hendrickson, B.: Chaco. In: Padua (ed.) [23], pp. 248–249
13. Karypis, G.: Metis and parmetis. In: Padua (ed.) [23], pp. 1117–1124
14. Karypis, G., Kumar, V.: Multilevel graph partitioning schemes. In: ICPP (3), pp. 113–122 (1995)
15. Karypis, G., Kumar, V.: A parallel algorithm for multilevel graph partitioning and sparse matrix ordering. J. Parallel Distrib. Comput. 48(1), 71–95 (1998)
16. Kernighan, B.W., Lin, S.: An efficient heuristic procedure for partitioning graphs. The Bell System Technical Journal 49(1), 291–307 (1970)
17. Kim, J., Hwang, I., Kim, Y.-H., Moon, B.R.: Genetic approaches for graph partitioning: a survey. In: Krasnogor, N., Lanzi, P.L. (eds.) GECCO, pp. 473–480. ACM (2011)
18. Kyrola, A., Blelloch, G., Guestrin, C.: Graphchi: Large-scale graph computation on just a pc. In: Proceedings of the 10th USENIX Symposium on Operating Systems Design and Implementation, OSDI 2012, Hollywood (October 2012)
19. Lepikhov, A.V., Sokolinsky, L.B.: Query processing in a dbms for cluster systems. Programming and Computer Software 36(4), 205–215 (2010)
20. Malewicz, G., Austern, M.H., Bik, A.J.C., Dehnert, J.C., Horn, I., Leiser, N., Czajkowski, G.: Pregel: a system for large-scale graph processing. In: Elmagarmid, A.K., Agrawal, D. (eds.) SIGMOD Conference, pp. 135–146. ACM (2010)

21. Moskovsky, A.A., Perminov, M.P., Sokolinsky, L.B., Cherepennikov, V.V., Shamakina, A.V.: Research Performance Family Supercomputers 'SKIF Aurora' on Industrial Problems. Bulletin of South Ural State University. Mathematical Modelling and Programming Series 35(211), 66–78 (2010)
22. Padmanabhan, S., Chakravarthy, S.: HDB-subdue: A scalable approach to graph mining. In: Pedersen, T.B., Mohania, M.K., Tjoa, A.M. (eds.) DaWaK 2009. LNCS, vol. 5691, pp. 325–338. Springer, Heidelberg (2009)
23. Padua, D.A. (ed.): Encyclopedia of Parallel Computing. Springer (2011)
24. Pan, C.: Development of a parallel dbms on the basis of postgresql. In: Turdakov, D., Simanovsky, A. (eds.) SYRCoDIS. CEUR Workshop Proceedings, vol. 735, pp. 57–61. CEUR-WS.org (2011)
25. Sanders, P., Schulz, C.: Engineering multilevel graph partitioning algorithms. In: Demetrescu, C., Halldórsson, M.M. (eds.) ESA 2011. LNCS, vol. 6942, pp. 469–480. Springer, Heidelberg (2011)
26. Sanders, P., Schulz, C.: Distributed evolutionary graph partitioning. In: Bader, D.A., Mutzel, P. (eds.) ALENEX, pp. 16–29. SIAM/Omnipress (2012)
27. Srihari, S., Chandrashekar, S., Parthasarathy, S.: A framework for SQL-based mining of large graphs on relational databases. In: Zaki, M.J., Yu, J.X., Ravindran, B., Pudi, V. (eds.) PAKDD 2010, Part II. LNCS, vol. 6119, pp. 160–167. Springer, Heidelberg (2010)
28. Sui, X., Nguyen, D., Burtscher, M., Pingali, K.: Parallel graph partitioning on multicore architectures. In: Cooper, K., Mellor-Crummey, J., Sarkar, V. (eds.) LCPC 2010. LNCS, vol. 6548, pp. 246–260. Springer, Heidelberg (2011)
29. Trifunovic, A., Knottenbelt, W.J.: Towards a parallel disk-based algorithm for multilevel k-way hypergraph partitioning. In: IPDPS. IEEE Computer Society (2004)

# Author Index